Science and Technology in Korea

The M.I.T. East Asian Science Series
Nathan Sivin, general editor

1. Ulrich Libbrecht, *Chinese Mathematics in the Thirteenth Century: The Shu-shu chiu-chang of Ch'in Chiu-shao*
2. Shigeru Nakayama and Nathan Sivin (eds.), *Chinese Science: Explorations of an Ancient Tradition*
3. Manfred Porkert, *The Theoretical Foundations of Chinese Medicine: Systems of Correspondence*
4. Sang-woon Jeon, *Science and Technology in Korea: Traditional Instruments and Techniques*

Science and Technology in Korea

Traditional Instruments and Techniques

Sang-woon Jeon

The MIT Press
Cambridge, Massachusetts, and London, England

This book was set in Monotype Baskerville,
printed on Fernwood Opague,
and bound in G.S.B. S/535/34
by Halliday Lithograph Corp.
in the United States of America.

Library of Congress Cataloging in Publication Data

Chōn, Sang-un
Science and technology in Korea.

(The M.I.T. East Asian science series, v. 4)
Based largely on the author's Han'guk kwahak kisulsa (A history of science and technology in Korea), 1966.
Bibliography: p.
1. Science—History—Korea. 2. Technology—History—Korea. I. Title.
II. Series: Massachusetts Institute of Technology. M.I.T. East Asian science series, v. 4.
Q127.K8C49 509'.519 73-18480
ISBN 0-262-10014-2

ISBN: 0-262-60052-8 (Paperback)

To my family

Contents

The M.I.T. East Asian Science Series

One of the most interesting developments in historical scholarship over the past two decades has been a growing realization of the strength and importance of science and technology in ancient Asian culture. Joseph Needham's monumental exploratory survey, *Science and Civilisation in China,* has brought the Chinese tradition to the attention of educated people throughout the Occident. The level of our understanding is steadily deepening as new investigations are carried out in East Asia, Europe, and the United States.

The publication of general books and monographs in this field, because of its interdisciplinary character, presents special difficulties with which not every publisher is fully prepared to deal. The aim of the M.I.T. East Asian Science Series, under the general editorship of Nathan Sivin, is to identify and make available books which are based on original research in the Oriental sources, and which combine the high methodological standards of Asian studies with those of technical history. This series will also bring special editorial and production skills to bear on the problems which arise when scientific equations and Chinese characters must appear in close proximity, and when ideas from both worlds of discourse are interwoven. Most books in the Series will deal with science and technology before modern times in China and related Far Eastern cultures, but manuscripts concerned with contemporary scientific developments or with the survival and adaptation of traditional techniques in China, Japan, and their neighbors today will also be welcomed.

Foreword

Perhaps we have thought about science and technology in Korea over the past decade or so most often in connection with debates on such fashionable topics as "technology transfer" and "the imperatives of modernization." This book is appearing at a time when we can no longer be unaware of how poorly founded our assumptions about modernization are. It is no longer possible to argue convincingly that the model of technological and social development idealized out of the Industrial Revolution in England, the United States, and certain parts of Western Europe is the sole means by which the traditional societies of Asia, Latin America, and Africa can hope to survive. Technocratic bureaucratization, suicidal urbanization, and identity defined by consumption may indeed be the wave of the future, but there is much doubt that such a future will be capable of supporting human life, and little ground for optimism that freedom and the dignity of the individual will survive it.

But what are the alternatives? Must they be invented from scratch? Can we learn only from our own trial and error—or China's? It is not so hopeless as that. Modernization (or, in its latest avatar, technology transfer) is one of the oldest and most continuous processes in human history. Realignments of society to adjust to the impact of new technologies are as ancient as the emergence of the first specialist potters or metalworkers. Conversely, the adoption of iron in one civilization after another must have been perceived as imperative for survival (we can study such perceptions among Bushmen today). The deforestation of large portions of North China by about A.D. 1050 provides a much simpler and in many ways more comprehensible instance of the failure of social restraints than we are likely to find in our recent experience. It is easy to demonstrate, although this is not the place to do it, that China's contemporary originality and independence are in no small part due to the strength of its own long engineering tradition (still proving very useful in the large low-capital, labor-intensive sectors of its economy) and to its thousands of years' experience at mobilizing very large numbers of people for public engineering projects. These examples from the past may not be as flashy as the latest computer simulation, but people who pay attention to them are much less likely to lose sight of the enormous role human values and unpredictable choices play in change.

Korea's science and technology are worth knowing and thinking about in connection with technology transfer for special reasons. Unlike China, Korea's styles in thinking systematically and objectively about nature and in developing instruments and techniques of material culture were always defined in the shadow of a large and sophisticated nearby civilization. The Korean experience differs from Japan's in that its influences from China flowed in more freely and directly, across a shared land border or a short stretch of sea. It was from Korea, in fact, that new sciences and arts were carried into Japan during the early centuries until regular contact between Japan and China became possible. As recent Korean and Japanese scholarship begins to cohere, it is becoming plain that we have not yet adequately recognized what a great part immigrant Koreans played in the formative phases of Japanese civilization as men of learning, craftsmen, and indeed nobles. Korea thus presents for our reflection the case of a country seeking to maintain its identity against pressures too imminent to be shut out. Granted, Peking five hundred years ago was immensely further from Seoul than Washington is today, if we measure the cost of communication in wealth, time, and effort; but it was close enough to serve the linked functions of source, market, and threat. If we have hardly begun to define the parameters of survival and social health in an unpredictably changing world, Korea has always been faced with the practical necessity of making these parameters actual, of building them into society. There are bound to be universal implications in its patterns of accepting, rejecting, copying, adapting, and creating. But before introducing this book as a starting point for evaluation of such patterns, let me pose a more general question still.

It is natural to ask what application insights derived from the study of traditional science and engineering can have, not merely to the transfer of knowledge from one society to another, but to the most general processes of technological change. I would argue that there is nothing unique about "westernization." The transition from traditional hierarchic social modes and attitudes across the threshholds of the Scientific, Industrial, and Information Revolutions to the secular and rationalizing perspectives of "modernity" is wrenching not only for members of traditional societies. The great watershed is the one on the foreside of the Scientific Revolution, and it must be traversed in the consciousness of every individual human being, whether through education, passive experience, or personal discovery. Since the Scientific Revolution, the power of number and measure applied to the multiplicity of phenomena have generated an ability to squeeze nature into new shapes, to make, alter, and destroy, that was unthinkable earlier. In traditional Europe before the Renaissance, in India, Islam, the Far East, scientific knowledge was meant on the whole for understanding of the natural order and of humanity's place in it (or for spiritual enlightenment, which is not entirely a different thing), and was not expected to have practical applications except in astrology and medicine. Technology was crafts. In these respects most ancient traditions have more in common with each other than any one (even Europe's) has with modern science and its application. Many traditions persist even in the rich countries long after custom has been pronounced dead and we are assured that society is postindustrial. Traditional ways of looking at the world and of making things will play a large part in our lives so long as natural materials are fabricated in part by the direct application of human hands, and so long as people need (or yearn for) a greater unity of mind than they can personally find in the relations of modern science to contemporary humanities and of both of these to everyday life.In the poor countries the need is even more basic. To present the notion of the tractor as an agricultural panacea to an African people who have not yet moved from the hoe to the plow is perverse; but it is not at all uncommon south of the Sahara for consultants on modernization and the men of affairs they advise to forget about the hoe and do just that. The process of adaptation and change for survival can never be adequately represented by a model in which new ideas or techniques are imposed on a passive and neutral civilization. It is always a matter of interaction with competing ideas and techniques which are already established within the recipient society and which indeed play a part in its definition of itself. If we are to understand the adaptation or its lack, we must remain aware of the competition. Cases from the past make accessible a fine texture of day-to-day decision and gradual reorientation of value that helps us learn what to look for, how to see past our prejudices and preconceptions, in working through today's problems.

Jeon Sang-woon's exploration of his own nation's scientific and technological traditions does not carry us to the end of this quest; but, starting very near the beginning, he has demonstrated the promise of his subject matter. This book is not a history of Korean science and technology, if by that we mean a reconstruction of the interplay of forces which accounts for Korea's own balance between constancy and change or between fluctuating social circumstances and the unfolding inner potentialities of ideas and techniques. That labor has hardly begun, and we will be fortunate if we taste its fruit in our lifetimes. *Science and Technology in Korea* aims, modestly and sensibly, to hasten the day by proving the importance of the work and by setting the preliminaries in order. The author surveys and evaluates the primary records, aided by a large descriptive literature in Korean and Japanese. His experience in museum work has prompted particular stress on the implements and instruments of science and the useful arts.

The purpose of this book is to review the technical accomplishments of Korean civilization as measured by the yardstick of the present day. This is the yardstick with which historians of science always begin their reconnaissances of unknown territory, whatever misgivings may prompt them to set it aside once they have learned the lay of the land. Many of the accomplishments pointed out in this book need to be compared with what was being done elsewhere at the same time, to avoid either missing or inflating the significance of subtle variations. Jeon Sang-woon's knowledge of Chinese and Japanese science make him a reliable guide in this respect. He is a Korean, and his pride in certain inventions and techniques is perceptibly greater than if he were a foreigner writing about Korean science. He knows that he is addressing a world-wide readership most of whom did not dream before they picked up his book that Korea is entitled to exert any claim upon the universal history of science. He knows that many educated people in Europe and the United States are just recovering from the shock of learning Joseph Needham's lesson, that the Chinese tradition is as indispensable as that of the early West in determining the potentialities of science. This book opens up still another range of awareness by demonstrating that peripheral societies must be examined with equal seriousness if we are not to overlook real originality. The author also knows that this implication will be equally surprising to most of his fellow Koreans. In Korea today the power to exploit nature is seen as an importation, as foreign in its essence. Few people are aware that, say, Korea in 1400 may very well have had the most advanced astronomical observatories in the world. Is it possible that science is not fundamentally Caucasian and Judeo-Christian (and all sorts of other things Koreans are not) after all?

This is only one of a wide range of very general questions which this book suggests, and which we ignore at peril of straitened understanding. It also leaves us with some rather specific questions which point to issues in the relations of science and its background about which we do not know enough. By way of illustration, let me dwell upon a typical cluster of these questions for a moment.

It is well known that printing with movable cast-metal type was perfected in Korea. Movable metal type was first tried in Korea in the fourteenth century or perhaps a bit earlier but was not at first practical for printing many copies. The crucial step in development was firmly imbedding the types so that their surfaces remained level after many impressions. That this technique should have been worked out in the Far East, where production of books by woodblock printing was routine from ca. 1000, seems much more obvious a state of affairs than it actually is. Printing with metal type is obviously economical with the alphabetic languages of the West, in which one can make repetitive use of very large numbers of types cast from a small number of common molds. Its economy diminishes as the size of the font increases. In traditional China the cost of labor for cutting woodblocks was not higher, and was often lower, than that of casting fonts of several thousand different types, setting, and sorting them. The capital investment required was orders of magnitude lower. The ability to print small editions with woodblocks to meet immediate demand, store the blocks, and pull a few more copies from time to time was also valued. Extended storage would have negated the great advantages of movable type, which depended upon rapid turnover. Movable type began to replace blocks once and for all only late in the nineteenth century. By that time a large demand for cheap books had developed, and ability to print the smallest possible characters within the limits of legibility became an advantage in price competition. The publishers who supplied this market were capitalized to an extent that was new, means were first beginning to develop for truly national distribution, and the cost of mechanical printing with modern presses was falling below the cost of hand labor. There is hardly room in the modern world for the large woodblock character with its flavor of calligraphic aesthetics, set 150 to 250 on each uncluttered page (see p. 177 for an example).

Jeon Sang-woon's account suggests that metal type was perfected in the royal court rather than elsewhere in Korea because a protracted program of development could not otherwise have been funded. In the early Far East the normal pattern of private manufacture ruled out reinvestment of profits, either in expansion or in what we would call research. If not consumed, they usually flowed into land instead. It was normal, especially in China and Korea, that elaborate technical innovations which needed protracted effort could originate only in the imperial workshops, or to a much smaller extent in the ateliers of very wealthy amateurs. Movable type is an excellent example of the sort of invention which could not have reached perfection without the sustained enthusiasm of powerful monarchs. It took two of them, in fact—T'aejong and Sejong—to carry the art past the intractable problem of imbedding the types. Once quality fully comparable with that of woodblocks had been attained, movable-type printing proved especially suited to Korean needs. Unlike China, in Korea the very small reading public (and a royal disinclination toward increasing literacy) meant that only small editions were needed, and hardwood for printing blocks was scarce and expensive in the vicinity of the capital.

I am suggesting that Korea's contributions to science and technology were strongly conditioned by what Gregory Henderson has described as "enormous centralization of politics, administration, values, and even emotion." This centralization is perhaps the most constant theme of Jeon's book, and to my mind the most noteworthy. One discovery after another was ordered, supported, and promulgated (often with little effect) by royal decree. Technical projects interlocked in ways which would have been most unusual in China. For instance, the improvement of metal type was assigned to a task force of people who had distinguished themselves in the making of astronomical instruments (see pp. 46 and 180). In 1592 the manufacture of firearms was undertaken by a team whose members included a military training officer, a craftsman, a sailor, and two slaves, commanded by Korea's greatest naval hero, Yi Sun-sin, who at the same time was involved in improving the famous wood-canopied "turtle ships" (pp. 205, 216). Admiral Yi's projects depended on local initiative but still had the knowledge and concurrence of the central government. The continuity of centralization as a habit of mind in Korea today is reflected in the author's tendency to praise and blame ancient kings for their willingness or unwillingness to patronize innovation, as though the research grants of today should also have been the norm half a millennium ago. I am sure I will not be the only reader to come away from this book curious as to whether any important technique was invented in Korea by a private individual working at his own behest on his own premises. Certainly a great deal of Korea's everyday technology, like everyone else's, was anonymous, but how about more elaborate devices and instruments of the sort Jeon describes? It may be that the picture is distorted because more documents concerning the central government have survived than those which reflect private initiative. This question we must leave for the Korean scholarship of the future, noting only that the many private literary collections Jeon cites do not paint a very different picture than the government documents. Even with room allowed for eventual reinterpretation, Jeon has shown that in the Korean record we can find a wealth of data for the study of centripetal tendencies in science and technology.

To sum up, not only the subject matter but the author's world view make this book part of a revolution now under way in the history of science and technology. It used to be the rule when people were trying to think out the whole sweep of the development of modern science that the subject resolved itself along and off a main line to which everything else was subsidiary. This main line was the European tradition. It was thought of as quintessentially Occidental, although connected at its origins with the archaic Near East. Islam was seen as an obliging caretaker for European scientific learning while the West was sorting itself out in the interval between the Roman Empire and the Renaissance. There was some difference of opinion as to whether Indians and Chinese were capable of doing science at all, or whether there were actually scientists of real stature blossoming unseen in the general atmosphere of mysticism and mandarinism. It was not at all unusual for books modestly entitled "A History of" some science or of the science of some period to ignore altogether the natural philosophy and computational techniques of any non-European culture. Nor is it unheard of today, although better awareness and a desire to avoid embarrassment conspire to encourage the insertion of "Western" or some similar word in titles of such parochial treatises.

The work of the past decades reveals a different picture, in which a single main line no longer appears. We are much less ignorant of the constant links between civilizations over the course of recorded history. We are much more aware that the European tradition of science was only one of several (including old Middle Eastern, Chinese, and Indian) which converged and blended in Islam, so that crucial insights and skills learned from the East in the late Middle Ages were truly unprecedented in the Occident. We now have a long list of probable Chinese, Islamic, and Central Asian inventions which, when they finally reached Europe, changed the face of the world and the nature of man's work in it. What we begin to make out is a complex picture with many streams repeatedly converging and diverging, continually changing their original characters, combining to form swelling tributaries which flow one by one, at their own rates, into what may or may not one day become the single river of universal science. It would be a mistake to imagine that we can forecast the shape of a science which is no longer distorted by the weight of historical accidents which happened in particular societies. Perhaps we can do no more than recognize that it will be very different from modern science, whose bias toward mechanism and whose irksome dichotomies of mind and body, particle and wave, can be traced to historical necessities of an earlier time in Europe but can be overcome only at the cost of major perturbations.

In the course of thinking about the great non-European scientific traditions not as exotic flowers but rather as related parts of the modern world's prehistory, it is natural enough that we begin to cast our nets wider still. The few major civilizations which have been studied so far were all centers for the convergence of ideas. We already know that well-documented Korean innovations had their effects in China. The evidence is gradually mounting that still other discoveries considered Chinese in origin because they were first recorded there actually were made in Southeast or Central Asia. Our understanding of the range of possibilities of scientific thought and of the relations of science and society is bound to be broadened every time we study another culture. No matter what the form of a culture's attempt to comprehend nature, it can throw light on central questions. This concern should hardly be abandoned to exotic tastes; it ought to be considered the meat and potatoes of the history of science. Jeon Sang-woon has worked long, hard, and successfully to establish the claim of Korean science and technology to serious and thoughtful attention. I believe that in doing so he has helped measurably to widen our mental horizons on science as a world phenomenon.

N. Sivin
Technology Studies Program
M. I. T.

Preface

This book is intended to be a systematic approach to certain aspects of traditional Korean scientific and technological activity that I consider important. At present a comprehensive historical approach is not to be expected. There is too little basic research in this field, and too few books on Korean history and culture based on reliable source materials have been written in any Western language. I have made an effort to sketch the role of Korean science and technology as the bridge between the Chinese and Japanese sciences and civilizations as well as Korea's unique contributions to the history of science.

This volume is based largely on my book *Han'guk kwahak kisulsa* (A history of science and technology in Korea, 1966), but reflects much additional research, thinking, and writing.

In this treatise, as a general rule, Korean, Japanese, and Chinese personal names are given with family name first, as is customary. In the romanization of Far Eastern languages, the McCune-Reischauer system for Korean, the Hepburn system for Japanese, and the Wade-Giles transcription for Chinese have been strictly followed, except for certain simplifications of the latter which are widely used by Sinologists today. When available and accurate, I cite the authors' own translations of the titles of Korean, Japanese, and Chinese secondary sources. Translations from primary sources are my own. Most English terms for Korean government offices and official posts are based on translations by Professor Edward W. Wagner of Harvard University.

I am deeply indebted to many scholars and friends for their help. My sincere thanks go to Professor Wagner, to the late Professor Martin Levey of the State University of New York at Albany, and to Dr. Joseph Needham, F.R.S., F.B.A., of Gonville and Caius College, Cambridge University, for their constant encouragement and invaluable advice during the preparation of this book, and for many improvements in the clarity of the draft. I wish to express my special appreciation to Professor Nathan Sivin of the Massachusetts Institute of Technology, who gave me great assistance and encouraged me to publish this book in the M.I.T. East Asian Science Series. I am grateful also to Professors Hong I-sŏp of Seoul and Yabuuti Kiyosi of Kyoto, and to my friends Drs. Yagi Eri and Nakayama Shigeru of Tokyo and Yu Yŏng-ik (Young Ick Lew) of Cambridge, Massachusetts, who advised and assisted me in scholarly matters. With regard to English style, I wish to acknowledge my obligation to Mr. Hwang Ton (astronomy and meteorology), Mr. Song Sang-yong (all chapters), Mr. Yi Pyŏng-hun (astronomy), Mr. Kim Yong-guk (geography), and Mr. Sim Chae-hun (physics) of Seoul, and to Mr. William R. Shaw (introduction and geography) and Mr. James K. Ash (meteorology and physics) of Cambridge, Massachusetts. I was fortunate to have access to the excellent facilities of the Seoul National University Library and the Harvard-Yenching Library. I am grateful to Assistant Librarian Kim Sung-ha of the latter institution, who assisted me with reference works. Financial support from the Asia Foundation Seoul Office, Ford Foundation funds of the Korean program of the East Asian Research Center, Harvard University, Harvard-Yenching Institute, and the Research Foundation of the State University of New York at Albany is gratefully acknowledged; without it, this book would have been impossible. I am also thankful for the self-sacrificing services of my wife Jeon Chŏng-yŏn (Pak Ok-sŏn) who assisted me in gathering materials and typing the first draft.

Finally my acknowledgment goes to the authors and publishers who kindly permitted me to quote from their works and to reproduce their illustrations. Chapter 2, "Meteorology," appeared in an earlier form in the journal *Yon'gu nonmunjip* (Theses collection) of the Sungshin Women's Teachers College, and is included here by permission of the editors. And my thanks are extended to the staff of the M.I.T. Press for their painstaking services.

Introduction

Generally speaking, the history of science in Korea has been only a tributary of the mainstream of scientific developments in traditional China. The progress of science in Korea therefore depended upon the fluctuating interplay between external stimulations and demands on one hand and internal necessities on the other. As with many other aspects of Korean culture, Chinese science and technology were not accepted without modifications. Attempts were always made to fit foreign inventions to local needs and conditions. More often than not, these attempts resulted in new inventions and discoveries.

The first obvious and decisive influence from China followed the establishment by the Chinese Han Dynasty in 108 B.C. of the colony of Lo-lang (Nangnang 樂浪) in northern Korea. This Chinese colony introduced the Han's superior metal technology into Korean civilization, stimulating an entirely new series of technological developments in the utilization of iron and bronze and leading to the emergence of a native civilization in the lower reaches of the Naktong River farther south. The basis of this native civilization was the production of iron. Another important product of this new native culture, commonly referred to as "Kimhae earthenware," was derived from the "gray ceramics" of China. Chinese technology introduced through the medium of the Lo-lang colony left its influence in the tombs of Koguryŏ 高句麗 and the solid gold artifacts of Silla 新羅.[1] However, the tombs of Koguryŏ are in the main distinctly native in style, as are the artifacts of Silla.

The influence of the native cultural development in the Three Kingdoms era is shown in the construction of the Ch'ŏmsŏngdae 瞻星臺 observatory and the Sŏkkuram 石窟庵 cave temple of Silla, as well as in the beautiful metalwork produced in that kingdom. The Ch'ŏmsŏngdae observatory closely follows the traditional

[1] For instance, the Ch'ilchi 七枝 sword, now kept at the Isonokami-jingū 石上神宮 Shrine in Japan, is an iron sword forged in 369 A.D. in Paekche 百濟 as a gift to the reigning monarch in Japan. The artistic standard of the fine gold inlay of sixty-one Chinese characters, found on both sides of the iron blade, is definitely superior to that of Lo-lang artistry, after which it is no doubt patterned.

Chinese graduated scale used in astronomical observations.[2] The amazing talent of the Silla people finds its expression in the artful assimilation of external influences and acquired knowledge. For instance, the square and linear outlines of the Chinese observatory are transformed into the simple but beautiful curved lines native to Korea. Of course, the plastic beauty of the Ch'ŏmsŏngdae observatory is not the only merit of this remarkable achievement of Silla culture. While it is true that the astronomers of Silla studied the astronomy and calendrical science of China, they did not satisfy themselves with learning and imitation alone. The Ch'ŏmsŏngdae observatory is a product of their effort to create and develop a new field of learning. The scientists of Silla made fruitful use of the open-dome observatory for observation of the movements of celestial bodies and measurements of the solstitial and equinoctial points. Such an advanced astronomical observatory could not be found even in contemporary China. This is without doubt the highest scientific achievement of Korea to that time.

The second relic of the high cultural attainments of the Silla kingdom, the artificial cave temple of Sŏkkuram, represents in its curious design, beautiful sculptures, and intricate architectural plan the very essence of the arts and sciences of ancient Korea. It is said that Sŏkkuram is a copy of a Buddhist cave temple in China. But whereas the Chinese temple made use of a natural grotto, the Sŏkkuram temple is an entirely artificial creation, built around a dome and comprising stone structures of circular, triangular, hexagonal, and octagonal shape. The metalworking technique of the creative artisans of Silla is demonstrated by a Buddhist bell of incomparable beauty. The bell combines the functions of the traditional *chung* 鐘 and *to* 鐸 bells of old China in an original manner, in addition to providing evidence of the metallurgical skill of the Silla artisans.

The technology of Koryŏ 高麗 was largely based on the tradition and heritage of Silla but also received constant and strong stimuli from China of the Sung era. The outstanding achieve-

[2]It is also believed to show the influences of the astronomical observatory of Paekche.

ments of the Koryŏ era in Korea are its celadon ceramics and wood-block printing. The development of wood-block printing in Koryŏ was motivated by the wish to repel Khitan and Mongol invaders with the spiritual aid of Buddhist scriptures. The resulting work, the famous Koryŏ printing block of the Tripitaka, is the oldest and greatest extant achievement of this art.

Koryŏ celadons, on the other hand, were meant entirely for the aesthetic enjoyment of the aristocracy of that dynasty. They are characterized by a unique and artful inlay technique that was greatly admired by contemporary Chinese scholars. Koryŏ celadon, at least in its first stage, was no doubt a copy of the ceramic art of Sung China. In its final form, however, the Koryŏ porcelain surpassed in beauty and execution anything produced from the ceramic kilns of the Sung Dynasty. The use of inlays for decoration of ceramic works was one of the original inventions of the Koryŏ artisans.

Signs of original and creative endeavors are found in various areas of national life in the early part of the succeeding Yi Dynasty. In 1403, King T'aejong 太宗 ordered the making of the famous bronze printing type of Kemi 癸未 in the face of strong opposition from his courtiers. In fact, however, the extant copies of books printed with the Kemi bronze type are in no way superior to the xylographs of the Koryŏ era. In addition, the casting of bronze types must have been much more time-consuming than the carving of wood blocks. The additional expense of time, money, and labor clearly was not worth the result achieved. But this project served as a matrix for the development of the renowned Yi Dynasty printing of the next era, that of King Sejong 世宗, the greatest patron of culture among the Yi Dynasty kings.[3]

Mention should also be made of the invention of rain gauges and water level marks in the reign of King Sejong of the Yi

[3]It has become established among Western scholars that the first recorded invention of movable metal type took place in Korea of the Koryŏ era (1230s) and was refined upon in the Yi Dynasty, but it is still too early to answer the question of how much lasting influence the invention and development of movable metal type in the early Yi era had on the external world.

Dynasty, during the years 1441–1442. These inventions were naturally accompanied by the perfection of methods for the quantitative measurement of precipitation. Such developments were a result of the efforts of Yi Dynasty scientists and scholars to attain a statistical understanding of the natural environment of Korea with its seasonally uneven distribution of rainfall. The scientists and officials of the Bureau of Astronomy (Sŏun'gwan 書雲觀) engaged in the work of compiling meteorological data gathered from outposts in the provinces, counties, and towns of the realm for more than 400 years, until the downfall of the dynasty at the beginning of the twentieth century.

The scientific achievements of the Sejong era were not confined to these. The signs of advances in science, arts, and learning are evident in all fields of natural science and humanities, including astronomy, geography, ceramics, architecture, medicine, and agriculture, in spite of the Yi Dynasty's political subordination to China.

The peak of scientific achievement in the Yi Dynasty occurred in the fifteenth century and was followed by a gradual decline. A crushing blow came at this stage of scientific development in the form of the invasion of Hideyoshi from Japan, a national disaster of the first magnitude. After this traumatic experience, scientific development in Korea lost its independent and creative base. This was soon followed by the introduction of Chinese and, for the first time, Western, learning.

We can discover the origin of Korean science in the technical tradition of the craftsmen who passed down their practical experiences and skill privately from generation to generation. They devoted themselves only to the search for phenomena and neglected theoretical explanations. The result of attaching importance to empirical research instead of theory was that technique was developed apart from any systematic experimental method.

It was very difficult for those craftsmen, who were officials supported by their government, to have any opportunity to follow their own bent because they had to engage in the practical research and manufacture dictated by governmental policy. In addition craftsmen were placed in the society's strict class system and were given little freedom of thought or movement.[4] What ingenious craftsmen there were often failed to record and preserve their professional experiences or to transmit what they had acquired to those outside their group.

Their work was made more mechanical and less creative than it might have been because of the lack of societal pressures and rewards for innovation and exploration. Their search was to recover the knowledge of the past, the formulae and secret prescriptions and methods handed down to them. There was thus a slow attrition of knowledge as the meaning of old texts and prescriptions became less intelligible to later generations and as war and destruction took their toll of manuscripts.

In the seventeenth and eighteenth centuries, many Korean scholars attempted to elicit systematic explanations for their traditional techniques. They advocated practical learning (*sirhak* 實學) under the slogan *silsa kusi* 實事求是 (verification of truth on the basis of factual studies), and accepted some of the little European science that came their way. They thus began what might have become a scientific reformation because they were under the influence of modern science and technology in Europe through Ch'ing China, where the same kind of movement had been going on for some time, partly as a reaction to mystical tendencies in late Confucian philosophy.

The *sirhak* scholars, however, were unable to systematize their technical inheritance in spite of their efforts, because they

[4] *Yangban* 兩班 was the highest social class in Yi society, superior to the three lower grades; *chungin* 中人 (middle people), *sangmin* 常民 (commoners), and *ch'ŏnmin* 賤民 (lowest class). The *yangban* supplied all the high-ranking government officials for the state and army. All land other than that owned by the royal family, the government, and the Buddhist temples was in the hands of the *yangban*. The *chungin* were skilled men, such as accountants, meteorological and astronomical observers, court painters, interpreters, and secretaries. Farmers, craftsmen, and merchants belonged to the commoner class, and slaves, actors, *kisaeng* 妓生 (Japanese, *geisha*), and Buddhist monks belonged to the lowest class. See Edwin O. Reischauer and John K. Fairbank, *East Asia: The Great Tradition* (Boston, 1958), pp. 427–428.

could not learn and master what were after all manual traditions. Even if such knowledge could have been learned from books, the ancient books were lost to them. As a result, they relied for the most part on Chinese texts. Their goal was not antiquarianism but efficient techniques. Their efforts resulted from a movement of institutional reformation in policy. Moreover, they were so absorbed in the civilization, culture, and institutions of Ch'ing China and Europe that they stuck to idealistic notions of reformation that were not suited to actual circumstances. This was a natural consequence of their social role and position.

Therefore, many works written by scholars belonging to the school of practical learning were not of as much practical use as less programmatic works compiled afresh by scholars on the basis of broad personal observation and reading.[5] The efforts of the *sirhak* movement led in the main to unrealistic proposals, and came to an end in nothing more than the compilation and enumeration of records. Their achievements had very little scientific value, but to the historian of science they are priceless. Without the efforts of the *sirhak* scholars, the records of the history of science and technology in Korea would not have been collected and systematized. There are few studies of Korean science and technology of the later Yi Dynasty except those produced by the school of practical learning.

Research into the history of Korean science presents many problems. Lack of reference materials is of course one of the most serious. To say the least, absence of historical records has made it difficult to determine the scientific value of this or that invention. It is understood that the Unified Silla period between the eighth and tenth centuries saw the development of science reach a new peak. Two accounts of the time remain: *Samguk sagi* 三國史記 (History of the Three Kingdoms) and *Samguk yusa* 三國遺事 (Reminiscences of the Three Kingdoms). Mainly dealing with political developments within the royal palace,

[5]Such as *T'ien kung k'ai wu* 天工開物 (The exploitation of the works of nature) in seventeenth-century China and Chŏng Ch'o's 鄭招 *Nongsa chiksŏl* 農事直說 (Theories and practice of farming) in fifteenth-century Korea.

both fail to mention even a single name of a representative scientist or craftsman of the time.

The only surviving record of the Koryŏ Dynasty (918–1392), the *Koryŏsa* 高麗史 (History of the Koryŏ Dynasty) is similarly frustrating. Much is said and written about Koryŏ celadon, but neither the process of manufacture nor the names of its makers finds a place in this history.

The only part of Korean history on which conventional sources shed some light is the Yi Dynasty (1393–1910). Eagerness for nation-building and preparations to move the capital to Seoul led the way for development of new techniques. Politically, the new kingdom that replaced Koryŏ inclined to China, but political subservience could not affect independent minds. Indeed, scientists of the Yi Dynasty displayed strong originality of thought and intellectual integrity. This trend was so marked in the fifteenth century that science as such flowered on an unprecedented scale. The man chiefly responsible for this was King Sejong, renowned for his patronage of science. The thirty-two years of his rule until his death in 1450 were so marked by new inventions, discoveries, and priorities that *Yijo sillok* 李朝實錄 (Veritable records of the Yi Dynasty) alone lists nearly one hundred important and minor scientists and technicians. The official records of the last Korean dynasty make it clear that fifteenth-century Korea saw a whole gamut of scientific endeavors flourish. The degree of achievement was such that few further improvements were made until the advances of the eighteenth century. These materials have been little studied in either East Asia or the West.

The scientific culture of Korea was first given a comparative evaluation in Yu Kil-chun's 兪吉濬 *Sŏyu kyŏnmun* 西遊見聞 (Travels to the Western world), written in 1896. Academic research was, for the first time, extended to this aspect of Korean culture in 1910 by Wada Yūji 和田雄治, a Japanese meteorologist who was staying in Korea at the time. Wada wrote a series of papers on astronomy and meteorology in *Kankoku kansokusho gakujutsu hōbun* 韓國觀測所學術報文 (Journal of the Korean meteorological observatory). His papers, written in Japanese and summarized in several European languages, drew more attention in the West and Korea than they did in Japan.

From 1900, when Western missionaries working in Korea first organized the Korean branch of the Royal Asiatic Society, the scientific culture of Korea began to gain more attention from Europeans. Their articles in the *Transactions* of the Society dealt with several fields of scientific culture: N. H. Bowman, "The History of Korean Medicine" (1915); W. C. Rufus, "Astronomy in Korea" (1936); H. H. Underwood, "Korean Boats and Ships" (1934); T. L. Boots, "Korean Weapons and Armor" (1934) deserve particular mention.

Up to 1944, research of this kind was done solely by foreigners. From that year, the first attempt by a Korean scholar bore fruit. In 1944, Hong I-sŏp 洪以燮 published *Chōsen kagakushi* 朝鮮科學史 (A history of Korean science), which represented a fresh attempt in Japanese to present to the Korean reading public a comprehensive review of the historical development of science in Korea.

Japan's defeat in the Second World War offered Korean scholars a chance to renew their efforts in exploring their traditional culture. Despite growing enthusiasm, however, practically no serious studies were made on broader subjects of scientific history until the 1960s.[6]

The historical study of Korean science gradually became more active in the 1960s.[7] Examining the outpouring of works in history of science, Hong in 1960 published the survey article "Korean Studies of Natural Science," written in English. He

[6]That is, with the exception of Hong's revised Korean edition of the *Chōsen kagakushi* (*Chosŏn kwahaksa*, 1946) and Kim Tu-jong's 金斗鍾 *Han'guk ŭihaksa* 韓國醫學史 (A history of Korean medicine) of 1948.

[7]Yi Tŏk-pong's 李德鳳 thesis, "Han'guk saengmulhakŭi sachŏk koch'al" 韓國生物學의 史的考察 (The dawn of biological knowledge in Korea), written in 1959, paved the way for systematic studies on this subject in the 1960s. From an economic historian's standpoint, Ko Sŭng-je 高承濟 analyzed problems in premodern ore and salt-mining industries. This book, first published in 1959, is entitled *Kŭnse han'guk sanŏpsa yŏn'gu* 近世韓國產業史硏究 (A study of the industrial history of the Hermit Kingdom of Korea).

was soon followed by Kim Tu-jong's *Han'guk ŭihak palchŏne taehan kumi mit sŏnambang ŭihakŭi yŏnghyang* 韓國醫學發展에 대한 歐美 및 西南方醫學의 影響 (The influence of Western and Central Asian medicine on the development of Korean medicine). There was also a series of papers dealing with such separate subjects as the development of metallic types and the art of printing, ceramic arts, and architectural art, but these were discussed mainly for their aesthetic value. In 1961, Yi Kwang-nin 李光麟 published *Yijo surisa yŏn'gu* 李朝水利史研究 (A study of irrigation in the Yi Dynasty); and in 1964, Yi Ch'un-yŏng 李春寧 published *Yijo nongŏp kisulsa* 李朝農業技術史 (A history of agricultural technology in the Yi Dynasty).[8] These diversified and yet ultimately interrelated researches established a foundation on which a broader and more comprehensive study of Korea's history of science could progress.[9]

Study of the history of Korean science has grown with increasing momentum during the later sixties. In 1966, Jeon Sang-woon published *Han'guk kwahak kisulsa* 韓國科學技術史 (A history of science and technology in Korea), a synthesis of research to date. Another development in 1966 was the publication of Kim Tu-jong's *Han'guk ŭihaksa* 韓國醫學史 (A history of Korean medicine) and *Han'guk ŭihak munhwa taeyŏnp'yo* 韓國醫學文化大年表 (Chronology of Korean medical science). They constitute a summing up of Kim's lifelong study of this subject. This work and Miki Sakae's 三木榮 *Chōsen igakushi oyobi sitsubyōshi* 朝鮮醫學史及疾病史 (A history of Korean medicine and disease) published in 1955 are masterpieces in

[8]This is a revised and expanded version of his earlier *Chosŏn nongŏp kisul sosa* 朝鮮農業技術小史 (Short history of agricultural technology in Korea) of 1950.

[9]Meanwhile, increasing interest was also being shown for early geography and cartography, which up to this time had been the exclusive domain of Japanese scholars. In 1965–1966, Hŏ Sŏn-do 許善道 published "Yŏmal-sŏnch'o hwagiŭi chŏllaewa paltal" 麗末鮮初 火器의 傳來와 發達 (Introduction and development of firearms in Korea, [1356–1474]), and "Yijo chunggi hwagiŭi paltal" (The development of firearms in Korea [1474–1591]). Hŏ's contribution in this field provided a turning point in studies of military technology as a whole, which up to that time had revolved around discussions of Admiral Yi Sun-sin and his naval maneuvers.

this field. Joseph Needham has included Korean developments in his renowned history of Chinese science and technology.

In 1969, *Kwahak kisulsa* 科學技術史 (History of science and technology) was published as a volume of *Han'guk munhwasa taege* 韓國文化史大系 (Korean cultural history series) by the Korean Classic Research Institute, Korea University. It is expected to build a bridge between the history of special sciences and that of general science, and to delineate important problem areas for future study.

1

Astronomy

Characteristics of Astronomy in Korea

Efforts to regulate agricultural production, which was the staff of national life, combined with the influence of the ancient Chinese astrological concept that the phenomena of the heavens were related to affairs on earth, early led to the development of *ch'ŏnmun* 天文 (heaven study) in Korea. However, astronomy in ancient Korea remained essentially subsidiary to the Chinese mainstream, in spite of occasional original observations and independent researches. This was perhaps inevitable, since astronomy in Korea had its origin in China and developed, like much else in ancient Korea, within the framework of Chinese culture and civilization. Thus just as China's astronomical studies in the Han and later periods chiefly dealt with the science of the calendar, so the astronomers of Korea through the successive dynastics mainly concerned themselves with the introduction and assimilation of the fruits of ancient Chinese calendrical studies. However, Korea's distinct geographical, climatological, and politicosocial environment compelled departure in many aspects from the basic tradition of Chinese astronomy. We can note peculiarly Korean traits in the methods of astronomical observations as well as in the manner of treatment of results.

Several lofty peaks were attained in the history of astronomy in Korea, alternating with periods of stagnation. Thus the remarkable achievements in many fields of natural science, including astronomy, during the Three Kingdoms and Unified Silla eras were followed by the "medieval" slump of astronomical science in the Koryŏ era, which in turn preceded a period of renaissance in the early Yi Dynasty period. Before the renaissance could reach full bloom, unfortunately, a succession of foreign incursions intervened. The subsequent endeavors at revival of the science in the reigns of Yŏngjo 英祖 and Chŏngjo 正祖 had not reached full scope when the development of Chinese-style astronomy in Korea was terminated in the late ninteenth century.

Western astronomy, which was introduced through China in the late Ming and early Ch'ing periods, did not take root in Korea amidst the prevailing political confusion and economic disruption that finally led to the extinction of the Yi Dynasty in the early twentieth century.

Astronomical Concepts of the Universe

COSMOGRAPHY OF ANCIENT CHINA

The celestial planisphere called *Ch'ŏnsang yŏlch'a punyajido* 天象列次分野之圖 (Chart of the constellations and the regions they govern), which was carved on a stone slab in the fourth year of the founder of the Yi Dynasty (1396), carries an interesting account by Kwŏn Kŭn 權近 concerning cosmological ideas of ancient China.[1] Kwŏn Kŭn made mention of six masters and their theories concerning this subject.

The first of them is the *hun t'ien* 渾天 (celestial sphere) theory of Chang Heng 張衡; the second is the *kai t'ien* 蓋天 (canopy heaven) theory, or the system of the *Chou pi* 周髀 book; and the third is the theory of *hsüan yeh* 宣夜, whose exponent, as well as its general principles, remains unknown. The fourth is the *an t'ien* 安天 theory, advanced by Yü Hsi 虞喜 in his book, *An t'ien lun* 安天論 (Discussions on the conformation of the heaven), and the fifth the *hsin t'ien* 昕天 (diurnal revolution) theory of Yao Hsin 姚信. The last is the *ch'iung t'ien* 穹天 (vast heaven) theory advocated by Yü Sung 虞聳.

Kwŏn Kŭn, however, dismisses all but the first of these as "childish and heterodox," indicating that Chang Heng's *hun t'ien* theory, based on armillary sphere practice, was the only

[1]See p. 26 and Joseph Needham, *Science and Civilisation in China*, vol. 3 (Cambridge, England, 1959), p. 279 (references to this work will hereafter be abbreviated *SCC*, followed by the appropriate volume and page numbers). A full translation of the inscriptions will be found in W. C. Rufus, "The Celestial Planisphere of King Yi Tai-Jo," *Journal of the Royal Asiatic Society, Korea Branch*, 1913, *4*.3: 23–72.

cosmological proposition that received support among scholars of the Yi Dynasty.

The first references to these astronomical ideas are found in the *Li chi yüeh ling* 禮記月令 and *Shun tien* 舜典, and more detailed accounts by Ts'ai Yung 蔡邕 in the reign of Emperor Ling 靈帝 of the Later Han appear in the Standard History of the Chin 晉書.[2] Kwŏn Kŭn's cosmological theories are derived from the last-mentioned book.

Of the six theories, only two, the *hun t'ien* and *kai t'ien* theories, survived until the Later Han, and these two, as representing the cosmological notions of ancient China, seem to have greatly influenced the astronomical concepts of the Silla and Koryŏ dynasties in Korea and the early period of the Yi Dynasty. They are further dealt with in Yi Sun-ji's 李純之 *Chega yŏksangjip* 諸家曆象集 (Collected discourses on astronomy and meteorology of the Chinese masters).[3]

KAI T'IEN AND HUN T'IEN THEORIES

The *kai t'ien* theory has two explanations, which probably represent earlier and later stages in its formation. The first explanation views the round sky and the square earth as flat and parallel, with the heavens moving around the earth on the hub of the North Pole while the sun revolves about the pole on circular orbits of varying radii according to the different seasons. The second explanation regards the heavens as a bamboo hat enclosing the earth like an inverted basin. The surfaces of both the heavens and the earth are curved to form parallel vaults, with the highest points beneath the North Pole. The *kai t'ien* theory was of long standing, dating back to the age of Chou.[4]

The *hun t'ien* theory, on the other hand, is much more

[2]See, for instance, Needham, *SCC*, vol. 3, pp. 210 ff., and Shigeru Nakayama, *A History of Japanese Astronomy, Chinese Background and Western Impact* (Cambridge, Mass., 1969), pp. 35–36.

[3]*Chega yŏksangjip* (Collected discourses on astronomy and calendrical science of the Chinese masters), vol. 1, *Ch'ŏnmun* (Astrology).

[4]Nōda Chūryō, *Shūhi sankei no kenkyū* (An inquiry concerning the *Chou pi suan ching;* Kyoto, 1933), pp. 60 ff.; Needham, *SCC*, vol. 3, pp. 210–216; Nakayama, *Japanese Astronomy*, pp. 26–35.

advanced astronomically than the *kai t'ien* theory and first appears in Chang Heng's *Hun t'ien i* 渾天儀. According to this theory, the cosmos is composed of the heavens, which surround the earth much as the white of an egg embraces the yolk. This cosmos rotates on an axis running through the North and South Poles, with all the celestial bodies laid out around it. Chang Heng likened the celestial motion to that of a wheel that spins so fast that its entire appearance is "blurred" (*hun* 渾).[5] He went on to say that the heavens had a circumference of 365¼ degrees (*tu* 度, *do* in Korean), half of it above the earth and the other half below. He established in this spherical cosmos a North and a South Pole, an equator and an ecliptic, the angle of intersection between the latter two being 24 *tu*. Although the contemporary skeptic Wang Ch'ung 王充 criticized an implication that the sun, which moved above the earth during the daytime, was compelled to proceed underwater during the night, the *hun t'ien* theory received support from a majority of astronomers of Han and subsequent ages.[6] On the other hand, the *kai t'ien* theory failed to keep a wide following after the Han era.[7]

These two main cosmological theories were introduced to Korea together with Han civilization and are presumed to have entrenched themselves in the astronomical world-concept of the Korean people in the fourth or fifth century. Their influence is to be found in the remains first of Koguryŏ, then of Paekche and Silla, as shown in the paintings of various celestial bodies found in the tombs of Koguryŏ, and in the domed ceiling of the Sŏkkuram grotto of Silla, which seem to embody the round-heaven 天圓, square-earth 地方 concept of the primary *kai t'ien* theory.[8]

[5]See, for instance, Needham, *SCC*, vol. 3, pp. 217–219; and Nakayama, *Japanese Astronomy*, pp. 35–39.
[6]See Needham, *SCC*, vol. 3, pp. 210 ff.
[7]Nōda, *Shūhi sankei no kenkyū*, pp. 60 ff.; and Nakayama, *Japanese Astronomy*, pp. 26–35.
[8]Nakamura K., "Kokuri jidai no kofun ni tsuite—sono seishō hekiga no kōsatsu o chūshin to shite" (On the paintings of various celestial bodies in ancient tombs of the Koguryŏ period), *Kōkogaku ronsō* (Theses on Arche-

The *p'alch'ŏkp'yo* 八尺表 (eight-foot gnomon), which is based on the *Chou pi*, occupies an important position as an observational instrument of ancient astronomy. Observatories based on the system of the *Chou pi*, which heavily stressed gnomon shadow observations, were first built in Paekche, were then copied by Silla, and became the the model of the Kyŏngju Ch'ŏmsŏngdae 瞻星臺, the only extant observatory of the Silla era in Korea. The construction of the Ch'ŏmsŏngdae seems to be based on the concept of a round heaven and a square earth.[9]

The *hun t'ien* theory of Chang Heng seems to have been inherited by succeeding ages without change or modification, in Korea as well as in China. Kwŏn Kŭn gives an account of Chang Heng's *hun t'ien* theory as it appeared in the treatise on harmonics and calendrical astronomy of the Chin dynastic history (*Chin shu* 晉書) and also of similar theories advanced by Ho Ch'eng-t'ien 何承天 of the Liu Sung era and Tsu Keng-chih 祖暅之 of the Liang period. All in all, it appears that the *hun t'ien* theory held sway until the early Yi Dynasty era in Korea as the orthodoxy of ancient astronomy.[10]

Very similar to the *hun t'ien* theory are doctrines advanced by two Koreans, Kim Mun-p'yo 金文豹 and Yi Kyŏng-ch'ang 李慶昌. Their *sado* 柶圖 and *chuch'ŏn* 周天 theories both appear in the *Chunggyŏngji* 中京志 (Gazetteer of the middle capital).[11] According to the *sado* theory, the cosmos is in the shape of a *sa* 柶, or ladle, the external rim of which represents the heavens while the earth occupies a square area in the middle. The earth is completely surounded by the sky. In the midst of all the celestial bodies is the Polestar, surrounded by the twenty-eight lunar mansion constellations. The winter solstice occurs when the sun travels through water most of the day, making the day short. The vernal equinox is caused by the

---

ology), 1937, *4*: 375 ff.; Yoneda Miyoji, *Chōsen jōdai kenchiku no kenkyū* (Studies in ancient Korean architectural remains: Tokyo, 1944), pp. 131 ff.

[9]Jeon Sang-woon, "Samguk mit T'ongil Sillaŭi ch'ŏnmun ŭigi" (On astronomical instruments in the Three Kingdoms and Unified Silla periods), *Komunhwa* (Korean antiquity), 1964, *3*: 18–22.

[10]*Chega yŏksangjip*, vol. 1, *Chŏnmun*.

[11]*Chunggyŏngji* (Gazetteer of the middle capital), ch. 10, supplement.

sun's emerging out of the water after half a day. Other seasons are explained similarly. Kim Mun-p'yo's *sado* theory, in the final analysis, may be regarded as a blend of traditional theories of yin and yang and the Five Elements and the round-heaven, square-earth theory described above.

Yi Kyŏng-ch'ang's *chuch'ŏn* theory is similar to, but generally more advanced than, the traditional *hun t'ien* theory. It pictures the sky as spherical, with the usual circumference of $365\frac{1}{4}$ degrees (*do*). The North Pole is 36 *do* from the zenith; it and the South Pole serve as the axis of rotation of the sky. (The zenith is 55 *do* from the North Pole). The distance between the summer solstitial points on the equator and ecliptic is 24 *do*. The south point is 31 *do* away from the winter solstice. The stars together with the sun and the moon rotate leftward (westward) around the earth at one revolution per day. The sun moves a little more slowly, falling behind by one degree each day, and the moon, moving more slowly, lags 13 *do* daily. Setting the length of one year at $365\frac{1}{4}$ days the sun gains $5\frac{235}{940}$ days per year while the moon loses $5\frac{499}{940}$ days each year. These figures are based on the Metonic cycle of 7 evenly spaced intercalated months every 19 years, or an average of one each 33 months.[12]

Yi Kyŏng-ch'ang, who was a philosopher-astronomer of the middle Yi Dynasty era (1554–1617), stated this theory in his *Chuch'ŏn tosŏl* 周天圖說 as part of an explanation for his astronomical clock, which he called the *sŏn'gi okhyŏng* 璇璣玉衡. Yi is said to have invented this device from studies of the ancient theories in the hope of making it conform perfectly with the astronomical phenomena then known to exist.[13] His *chuch'ŏn* theory is similar in many respects to the principles embodied in the armillary sphere manufactured in the reign of King Sejong 世宗. Such expressions as "gaining one degree per day" and "lagging so many degrees daily" are often repeated in

[12]W. C. Rufus, "Astronomy in Korea," *Transactions of the Royal Asiatic Society, Korea Branch*, 1936, *26*: 27–28; Nathan Sivin, "Cosmos and Compution in Early Chinese Mathematical Astronomy," *T'oung Pao*, 1969, *55*: 1–73.

[13]*Chunggyŏngji*, ch. 10, supplement.

those portions of *Sejong sillok* 世宗實錄 (Veritable records of the King Sejong era) that deal with the armillary sphere.

Rotating-Earth Theories of Yi Dynasty Scholars

NEW MORPHOLOGICAL CONCEPTS OF THE UNIVERSE

The transmission of Western astronomy to Korea began in July of 1631 (ninth year of King Injo 仁祖) when Chŏng Tu-wŏn 鄭斗源 and his party brought back from Ming China the *T'ien wen lüeh* 天問略 (Explication of the celestial sphere) by Emmanuel Diaz.[14] In 1645 (twenty-third year of Injo), Prince Sohyŏn brought back from the Ch'ing court Chinese editions of astronomical and calendrical works by Adam Schall von Bell, the German Jesuit in Peking, and a similar service was rendered the following year by Kim Yuk 金堉. In this way, the astronomers of the Yi Dynasty acquired knowledge of contemporary works on astronomy in the West.[15]

Between 1631 and 1634, the Ming court commissioned a host of scholars, including Hsü Kuang-ch'i 徐光啓 and Li Chih-tsao 李之藻 as well as the Jesuits Johann Schreck, James Rho (Lo Ya-ku), and Adam Schall von Bell, to summarize in 135 chapters of Chinese text the entire Western system of astronomy. This huge work was called *Ch'ung-chen li shu* 崇禎曆書 (Treatises on astronomy and calendrical science compiled in the Ch'ung-chen reign period). It was printed during the Ch'ing Dynasty as *Hsi-yang hsin-fa li shu* 西洋新法曆書 (Treatises on calendrical science according to new Western methods), in 100 volumes. Much of it was based on the geoheliocentric cosmology of Tycho Brahe.[16] The calendrical works brought back by

14*Munhŏn pigo* (Comprehensive study of civilization), ch. 1, pp. 4b–5a; Hong I-sŏp, *Chosŏn kwahaksa* (A history of Korean science; Seoul, 1946), pp. 236, 242–243.

15Hong, *Chosŏn kwahaksa*, pp. 243–245; Yamaguchi Seiji, "Shōken Seshi to Tōjakubō" (On Prince Sohŏn and Adam Schall von Bell), *Seikyū gakusō*, 1931, *5*: 112–117, and "Shinchō ni okeru zai-Shi Ōjin to Chōsen shisin" (Jesuits and Korean envoys in Ch'ing China), *Shigaku zasshi*, 1933, *44*, no. 7, pp. 13–15.

16Yabuuchi Kiyoshi, "Kinsei Chūgoku ni tsutaerareta seiyō tenmongaku"

Kim Yuk from his trip to Ch'ing China are presumed to have been the *Hsi-yang hsin-fa li shu*.[17]

The Jesuits' Tychonic cosmological system was propagated in Korea during the reign of King Hyojong 孝宗 in the Sihŏn 時憲 (Shih-hsien) Calendar (1653). This superseded the conventional *hun t'ien* theory in Korea, establishing itself by the early eighteenth century as the orthodox cosmological doctrine of the Yi Dynasty. The even more obsolete Aristotelian theory of 12 concentric celestial spheres (*shih-erh chung t'ien* 十二重天) advanced by Matteo Ricci, on the other hand, finds acceptance in the *Munhŏn pigo* "Sangwigo" 文獻備考 象緯考 (Study of astronomy and meteorology, Comprehensive study of civilization).[18] Essentially, the theory of the twelve heavenly spheres was nothing but the typical Aristotelian-Ptolemaic system as widely accepted in Islam and the late medieval West.

ARMILLARY SPHERE OF YI MIN-CH'ŎL AND SONG I-YŎNG AND THE THEORY OF KIM SŎNG-MUN

In the Yi Dynasty there existed together with the recently imported Western system of cosmology two systems based on traditional astronomical concepts, the *samdaehwan kongbu* 三大丸空浮 (Three spherical bodies floating in space) theory of Kim Sŏng-mun 金錫文 and the rotating-earth theory of Hong Tae-yong 洪大容 地轉說, of which an account is given in Pak Chi-wŏn's 朴趾源 *Yŏrha ilgi* 熱河日記 (Diary of a mission to Jehol, 1783).[19] According to Kim Sŏng-mun, the sun, the earth, and the moon were all spherical and floating in space. The work by Pak Chi-wŏn, however, fails to give details on the origin or the content of the theory advanced by Kim Sŏng-mun, who is said to have flourished in the seventeenth century. There is, however, sufficient evidence that the concept of a spherical

(The introduction of western astronomy into China), *Kagakushi kenkyū*, 1955, *32*: 15–16.

[17]See *Munhŏn pigo*, ch. 1, pp. 5b–6a.

[18]Ibid., ch. 1, pp. 13a–14a; *Sŏngho sasŏl yusŏn* (Sŏngho's detailed discourses, classified), ch. 1, "Heaven and Earth."

[19]*Yŏnamjip* (Collected writings of Yŏnam = Pak Chi-wŏn), ch. 14, *Yŏrha ilgi* (Diary of a mission to Jehol).

earth had an established place in the astronomical scholarship of the Yi Dynasty.

Moreover, there is physical proof of the existence of this school of thought in the middle Yi period. This is the terrestrial globe inserted in an armillary sphere in an astronomical clock, the *sŏn'gi okhyŏng*, constructed in 1669 by Song I-yŏng 宋以穎. The terrestrial sphere, attached to a clock mechanism, made precisely one rotation each day. According to available records,[20] two astronomical clocks were manufactured between 1664 and 1669 (fifth and tenth years of King Hyŏnjong 顯宗) by Yi Min-ch'ŏl 李敏哲 and Song I-yŏng, respectively. In both these devices, it is recorded, a terrestrial globe was connected to the polar axis of an armillary sphere so as to rotate one full turn in exactly one day. This may indicate that the contemporary astronomers of Korea had independently arrived at the discovery of the diurnal rotation of a spherical earth. The Jesuits had discussed this rotation, but only to refute it.

While there is no way of judging today what influence was exerted by the terrestrial spheres of Yi Min-ch'ŏl and Song I-yŏng upon the thought of Kim Sŏng-mun, it is certain at least that the terrestrial spheres must have been based upon theoretical considerations. Yi and Song were both leaders of astronomy in the Yi Dynasty, and their astronomical clocks were apparently adjusted to conform to the results of their observations and theoretical researches. These astronomical clocks were presumably further improved upon by the removal of the *kyuhyŏng* 窺衡, or sighting tube, which was obviously unnecessary indoors, and the mounting of a terrestrial sphere connected to the rotating polar axis so as to establish a link between terrestrial movements and those of other celestial bodies.

We have no way of knowing today how such a strong conviction about the rotation of the earth was arrived at and would never have been acquainted with the three spherical bodies theory of Kim Sŏng-mun had it not been for the *Yŏrha ilgi* of

[20]Jeon Sang-woon, "Sŏn'gi okhyŏnge taehayŏ" (On armillary spheres with clockwork, in the Yi Dynasty), *Komunhwa*, 1963, *2*: 3–10. Same article summarized in Japanese, *Kagakushi kenkyū*, 1962, no. 63, pp. 137–141.

Pak Chi-wŏn. There is a strong possibility, therefore, that their theory of a rotating earth was never made public or recorded in book form first hand. Like the three spherical bodies theory, it has survived only through the work of Pak Chi-wŏn.

ROTATING-EARTH THEORY OF HONG TAE-YONG AND PAK CHI-WŎN

The rotating-earth theory developed by Hong Tae-yong and Pak Chi-wŏn was as follows in its outline:[21]

> Everything in nature, such as the sun, the moon, the stars, even the teardrop, is round, and there is no creation of God that is square. It is accordingly inferred that the earth, one of the creations of nature, is also round or spherical. As a whole, form is round while virtue 德 (nature) is square, merit 事功 is in motion while emotion is still. If the earth were suspended stationary in space, silent and without motion, it would soon decay and perish. . . .

They considered, somewhat along the lines of the three spherical bodies theory, that the sun, the moon, and the earth belonged to the same category of matter, and that inasmuch as there was nothing in the universe that was not round, so the earth too must be round. They further supposed that the earth must be turning and in motion, for otherwise there would be nothing alive or in motion on earth, such as the trees and the running streams and rivers. The essential reasoning is not wholly dissimilar to the auxiliary argument of Copernicus and his predecessors that the earth is round and that a motion peculiar to a round body is rotation.

These Yi Dynasty astronomers are also stated to have discovered that "since the earth in its daily rotation covered a wide area of 90,000 *ri* 里 in the space of 12 double hours 時 the speed of terrestrial motion was faster than that of the thunderbolt or the cannonball."[22] They thus computed that terrestrial rotation proceeds at a great speed.

Of the annual revolution of the earth, however, they make no mention, and confine their remarks to the following definitions:

[21] *Yŏnamjip*, ch. 114: *Yŏrha ilgi.*

[22] *Tamhŏnsŏ* (Writings of Tamhŏn = Hong Tae-yong), *Naejip* (part one), ch. 4.

A rotation of the earth makes a day, *il* 日 : a revolution of the moon around the earth makes a month, *sak* 朔: a revolution of the sun around the earth makes a year, *se* 歲: a revolution of Jupiter around the earth makes a dozen years, *ki* 紀: and a concordance cycle for eclipses takes 513 years, a *hoe* 會."[23]

Evidently, the Yi Dynasty astronomers had no notion of the annual revolution of the earth, which in their thinking formed the center of the universe, around which all other celestial bodies revolved. They attempted to describe cosmological matters in highly metaphysical terms, as in the expression, "shape is round while virtue is square, merit is in motion while emotion is still. . . ." And Pak Chi-wŏn said: "The Westerners merely pointed out the spherical shape of the earth but failed to note its rotation, because they failed to recognize that every spherical body has a rotating motion."[24]

Some contend from these facts that the rotating-earth theory of Hong Tae-yong and Pak Chi-wŏn is original.[25] However, both Hong and Pak probably knew late medieval European arguments for the diurnal rotation of the earth through the Jesuits in China, with whom Hong had personal contact.[26] Pak Chi-wŏn was reacting to the fact that James Rho, who transmitted these arguments, proceeded to refute them. The fact that Yi Min-ch'ŏl and Song I-yŏng had installed a revolving globe in the armillary clock of 1669 refutes the view that Hong

[23] *Yŏnamjip*, ch. 4. On this cycle, see N. Sivin, "Cosmos and Computation," p. 14.
[24] Ibid.
[25] For instance, Tamura Sennosuke, *Tōyōjin no kagaku to gijutsu* (Science and technology of the East Asian peoples; Tokyo, 1958), pp. 145–149.
[26] See Ch'ŏn Kwan-u, "Hong Tae-yongŭi chijŏnsŏlŭi chae-gŏmt'o" (Reexamination of the rotating-earth theory of Hong Tae-yong), in *Hyosŏng Cho Myŏng-gi paksa hwagap kinyŏm pulgyo sahak nonch'ong* (Commemorative studies on Buddhism and history for Dr. Cho Myŏng-gi's sixtieth birthday, Seoul, 1962), p. 649; and Yabuuchi Kiyoshi, "Richō gakusha no chikyū kaitensetsu" (On the rotating-earth theory of the scholars in the Yi Dynasty), *Chōsen gakuhō* 1968, *49*: 432–433. Both these articles suggest knowledge of Copernicanism, but that is both unlikely and unnecessary. The diurnal rotation of the earth was discussed in a widely disseminated treatise of James Rho on the planetary motions (1634). See Nathan Sivin, "Copernicus in China," in *Colloquia Copernicana II* (Warsaw et al., 1973).

and Pak were the initiators of the rotating-earth theory more than a century later.

Astronomical Charts

STAR MAPS IN THE THREE KINGDOMS PERIOD

Astronomical charts were symbols of royal authority, and under the ancient dynasties the results of astronomical observations were recorded in the form of star maps. Koguryŏ was probably the first Korean state to prepare a star map. The celestial planisphere mentioned earlier, prepared by Kwŏn Kŭn in the fourth year of the first monarch of the Yi Dynasty (1395), is believed to derive its main content from a star chart of Koguryŏ, carved on a stone slab, which was thrown into the Taedong River and lost when Koguryŏ fell to the combined forces of T'ang and Silla.[27] A paper print or rubbing seems to have survived until the Yi Dynasty. The chart shows 1,464 stars,[28] which agrees with the number of stars (283 constellations and 1,465 stars) appearing in a chart of the period of the Three Kingdoms in China (222–238), that is, the star map of the three schools of astronomers,[29] which seems to have been brought over to Koguryŏ in the latter half of the fourth century.[30] The map includes almost all the stars visible to the naked eye.

Koguryŏ's obviously extensive knowledge about stars and constellations also survives in astral maps painted on the inner walls of ancient tombs,[31] of which one example is the constella-

[27]A full translation of this inscription will be found in Rufus, "The Celestial Planisphere of King Yi Tai-Jo," p. 31, and in Joseph Needham and Gwei-djen Lu, "A Korean Astronomical Screen of the Mid-Eighteenth Century from the Royal Palace of the Yi Dynasty (Chosŏn Kingdom, 1392 to 1910)," *Physis*, 1966, *3*, 2: 141–143.

[28]W.C. Rufus and Celia Chao, "A Korean Star-Map," *Isis*, 1944, *35*: 316–326.

[29]*Sui shu*, ch. 19, pp. 2a–2b; Needham, *SCC*, vol. 3, p. 264.

[30]But Ch'en Cho first made a map of the stars according to the three schools of astronomers in A.D. 310. See *Chega yŏksangjip*, vol. 1, *Ch'ŏnmun*.

[31]See Nakamura, "Kokuri jidai no kofun ni tsuite," pp. 375 ff.; Ikeuchi Hiroshi, *T'ung-kou* (The ancient site of Kao-Kou-Li [Koguryŏ] in Chi-An District, Northeastern China; Tokyo and Hsin-Ching, 1938), vol. 2,

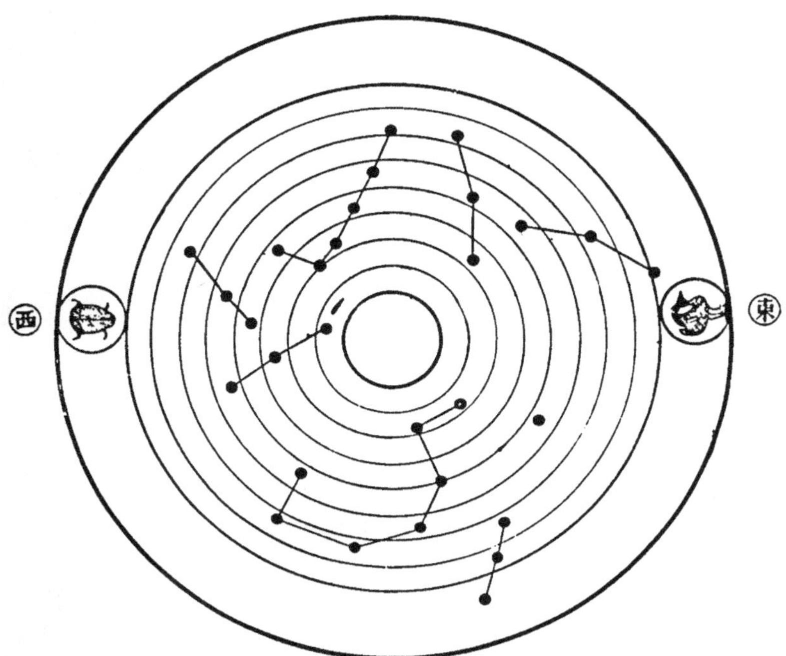

Fig. 1.1. Constellation map on the ceiling of the main chamber of the Tomb of the Dancers, example of a descriptive diagram from the Koguryŏ tombs. From Nakamura Kiyoe, "Kokuri jidai no kofun ni tsuite."

tion map on the ceiling of the main chamber of the Muyong tomb 舞踊塚 (see Fig. 1.1). Seven of the 28 lunar mansion constellations are depicted there with fair precision, with lines connecting adjacent stars in the constellations. A similar example survives on the ceiling of the main chamber of the Kakchŏ tomb 角抵塚.[32]

The star charts in the Koguryŏ tombs represented the constellations, as well as the sun and the moon, on circles, placing

pp. 11, 16; and Hori Junji, "Kokuri kofun no seishinzu ni tsuite" (On the painting of the celestial bodies in the ancient tombs of Koguryŏ), *Shiseki*, 1954, *42*: 370–391.

[32]Nakamura, "Kokuri jidai no kofun ni tsuite," pp. 382–383.

the sun to the east and the moon to the west, and depicting the four "celestial palaces" centered on the cardinal points: the Blue Dragon, Black Tortoise and Snake, White Tiger, and Red Phoenix.

In the Silla era (692), a Buddhist monk named Tojŭng 道證 brought a star chart home from a visit to T'ang China.[33] While the chart is presumed to have been prepared in conformity with the *kai t'ien* theory of Li Shun-fêng 李純風,[34] there is actually little difference from the star chart of Ch'en Cho 陳卓 that had come earlier to Koguryŏ.

Although the knowledge of Paekche astronomers is presumed to have been comparable to that of Koguryŏ, we have no surviving records to attest to this. In 554, Paekche sent a professor of calendrical science,[35] Kodŏk Wangson 固德王孫, to Japan, and later in 602 (third year of King Mu 武王), Kwallŭk 觀勒 visited Japan, bearing with him calendars and astronomical literature.[36] From this it appears that Paekche scholars made independent observations on the basis of records brought over from China and used them in the preparation of their star charts. It is on record that Paekche astronomers took an active part in the construction of astronomical observatories in Japan.[37]

THE FOURTEENTH-CENTURY CELESTIAL PLANISPHERE OF SEOUL

In *Koryŏsa* 高麗史 (History of the Koryŏ Dynasty), there is mention of star charts made by O Yun-bu 伍允孚, but no description of them.[38] In any event, copies of the stone star

[33] *Samguk sagi* (History of the Three Kingdoms), ch. 8, p. 6b.

[34] Rufus, "Astronomy in Korea," p. 14.

[35] *Paksa* (Japanese *hakase,* Chinese *po-shih*). This term is equivalent to the academic title of "doctor" in contemporary Korean and Japanese usage, but in ancient Far Eastern countries it indicated among other things the highest teaching position in technical subjects.

[36] *Nihon shoki* (Chronicle of Japan), ch. 19 and 22; Nakayama, *Japanese Astronomy*, p. 9.

[37] Mikami Yoshio, "Nihon kagaku no tokushitsu: Tenmon" (Characteristics of Japanese science; astronomy), in *Tōyō shichō no tenkai* (The development of Oriental thought; Tokyo, 1936), p. 9.

[38] *Koryŏsa* (History of the Koryŏ Dynasty), ch. 122, p. 25a.

Fig. 1.2. A Koryŏ bronze mirror (ca. eleventh century) represents the universe. The central figure is the sun, with the four cardinal points: the Blue Dragon, Black Tortoise, White Tiger, and Red Phoenix. The second inner circle represents the eight cardinal points and symbolic inscriptions of the eight trigrams. The third circle represents the twelve double hours; the fourth, the twenty-eight constellations; and the outermost circle, the twenty-four fortnightly periods. Diameter 17 cm. Author's collection.

charts of Koguryŏ were handed down to Koryŏ,[39] as can be seen from Fig. 1.2.

Yi Sŏng-gye 李成桂, who established the Yi Dynasty by toppling the Koryŏ Dynasty, desired to acquire new star charts as symbols of the royal authority of the new dynasty.

39 *Yangch'onjip* (Collected works of Yangch'on = Kwŏn Kŭn), ch. 22, Poem on the astronomical screen.

His wish was fulfilled through several years' efforts of eleven royal astronomers, namely, Kwŏn Kŭn 權近, Yu Pang-t'aek 柳方澤, Kwŏn Chung-hwa 權仲和, Ch'oe Yung 崔融, No Ŭl-chun 盧乙俊, Yun In-yong 尹仁龍, Chi Sin-wŏn 池臣源, Kim T'oe 金堆, Chŏn Yun-gwŏn 田潤權, Kim Cha-su 金自綏, and Kim Hu 金候. This is the Celestial Planisphere in stone named *Ch'ŏnsang yŏlch'a punyajido* 天象列次分野之圖 (Chart of the constellations and the regions they govern). Yi Sŏng-gye, the Royal Progenitor (T'aejo 太祖), chanced to acquire one of the rare copies of the earlier star chart of Koguryŏ and ordered copies to be made of it. The Sŏun'gwan 書雲觀, the Bureau of Astronomy, incorporated secular changes on the basis of new observations. They then compiled a new *Chungsŏng-gi* 中星記 (Record of meridian transits), including the results of new observations and determinations, and it was on the basis of this that the celestial planisphere was carved upon a stone slab in December 1395.[40]

The extant stele of the planisphere (see Fig.1.3) is now preserved at the Ch'anggyŏngwŏn 昌慶苑 Royal Garden in Seoul. Made of black marble, it measures 2.01 m high and 1.23 m wide. At the center of a large circle is the North Pole, which is surrounded by a number of smaller circles, including the equator and the ecliptic. Along the circumference of the larger circle are recorded the names of the 28 constellations. By means of the radiating lines connecting the determinative stars of each constellation with the North Pole, the relative position of each star can be determined easily.

In an account of observations are recorded the polar distance in degrees to the 28 constellations, descriptions concerning the stars passing the meridian at dusk and dawn of each of the 24 fortnightly periods, the 12 regions governed by the stars, the solar and lunar mansions, discourses on the heavens, and the process by which the chart was prepared, with the official positions and personal names of those who took part in the

[40]Ibid., ch.22, and the stone slab of the *Ch'ŏnsang yŏlch'a punyajido* (Chart of the constellations and the regions they govern), preserved in Ch'anggyŏng Palace, Seoul.

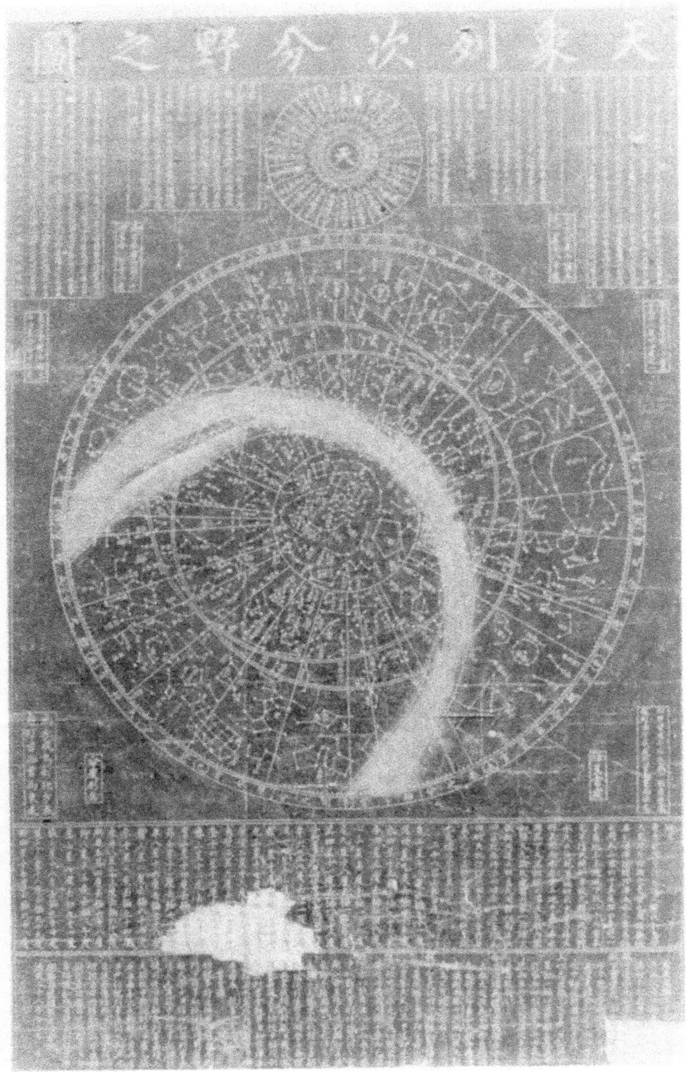

Fig. 1.3. A printed copy of the Seoul planisphere. Full translation of the inscriptions can be found in W. C. Rufus, "The Celestial Planisphere of King Yi Tai-Jo" and Needham, "A Korean Astronomical Screen of the Mid-Eighteenth Century from the Royal Palace of the Yi Dynasty."

project.[41] There was no attempt to make fresh observations of the constellations because the Koguryŏ star chart gave a fairly accurate and comprehensive picture, which was easily improved by incorporating secular deviations and changes.

There is also a record of a celestial planisphere said to have been carved in stone in 1433, but no details are available.[42] Perhaps it contained the same constellations that were represented on the celestial globe completed in April of 1437 (nineteenth year of Sejong).[43] The circumference of this globe measured 10 *ch'ŏk,* 8 *ch'on,* 6 *p'un* (234.5 cm), and it contained the constellations and the stars north and south of the equator.[44]

In the reign of King Sukchong 肅宗, a copy was made of the fourteenth-century planisphere, in order to preserve the original inscription, much of which had been defaced in the course of repeated foreign wars.[45] The reproduced stele, measuring 211 cm × 108.5 cm, was identical to the original in all respects except that the title of the stele was moved to the top. It is entirely possible that this stone reproduction was copied from one of 120 rubbings made in 1571.[46] These two stone planispheres were housed together in the newly built Hŭmgyŏnggak 欽敬閣 (Pavilion of respectful veneration) of the Bureau of Astronomy in 1770. In form, they much resemble the famous Chinese Soochow planisphere of 1247.[47]

TRANSMISSION OF WESTERN ASTRONOMICAL CHARTS INTO KOREA

In 1708 the Bureau of Astronomy prepared a copy of the *Ch'ih tao nan pei tsung hsing t'u* 赤道南北總星圖 of Adam Schall von Bell and presented it to the King.[48] It contained 1,812

[41]See Rufus and Chao, "A Korean Star-Map," pp. 316–326.
[42]*Munhŏn pigo,* ch. 2, p. 31b.
[43]*Sejong sillok* (Veritable records of the King Sejong era), ch. 77, p. 9b. This and other "Veritable Records" of various reigns are part of the *Yijo sillok* cited in the Bibliography.
[44]*Munhŏn pigo,* ch. 2, p. 31b.
[45]Ibid., ch. 3, p. 8a.
[46]*Sŏnjo sillok,* ch. 5, pp. 6a, 7a.
[47]*Munhŏn pigo,* ch. 3, pp. 7b–8a; W. C. Rufus and Tien Hsing-chih, *The Soochow Astronomical Chart* (Ann Arbor, 1945).
[48]Ibid., ch. 3, pp. 3b-6a.

observable stars—16 stars of the first magnitude, 67 stars of the second magnitude, 216 stars of the third magnitude, 522 stars of the fourth magnitude, and 572 stars of the fifth magnitude. No copy of this chart, known by the name of *Kŏnsangdo* 乾象圖 (*Ch'ien hsiang t'u*), is extant. It was probably the first duplicate in the Yi Dynasty of a star chart prepared under Western influences.

The *Honch'ŏn chŏndo* 渾天全圖 (Complete map of the celestial sphere), a block-printed astronomical chart of the latter half of the Yi Dynasty (late eighteenth century) shown in Fig.1.4, was also presumably derived from Western influence. While it is traditional in its general appearance, resembling the astral charts of the T'aejo era, the accompanying accounts concerning the number of stars say that there are 1,449 stars in 336 constellations, and that the perpetually invisible stars around the South Pole number 121 in 33 constellations. This chart includes drawings and explanations of the solar system, solar and lunar eclipses, the hours of sunrise and sunset, and syzygies, as well as the waxing and waning of the moon and the relative magnitude of each planet and its distance from the earth. In the diagram of the solar system are represented telescopic views of the sun, moon, and Five Planets. Saturn is accompanied by five satellites and Jupiter by four. The principles of solar and lunar eclipses are explained with diagrams. The so-called old planetary system 七政古圖 describes the cosmic system of Ptolemy while the new planetary system 七政新圖 represents that of Tycho Brahe.

There are other astral charts prepared under the influence of Western astronomy, such as a copy of the star map of Ignatius Kögler of 1723, comprising 300 constellations and 3,083 stars, executed by Kim T'ae-sŏ 金兌瑞 and An Kukpin 安國賓 in 1742 and now preserved in the Pŏpchu Temple 法住寺;[49] *Hwangdo nambuk yang-ch'ong-sŏngdo* 黃道南北兩總星圖 (General map of the stars in both northern and southern

[49]See Yi Yong-pŏm, "Pŏpchusa sojangŭi sinpŏp ch'ŏnmun tosŏle taehayŏ" (On the astronomical map of 1743 preserved in Pŏpchu temple, Korea), *Yŏksa hakpo*, 1966, *31*: 1–66; *32*: 59–119.

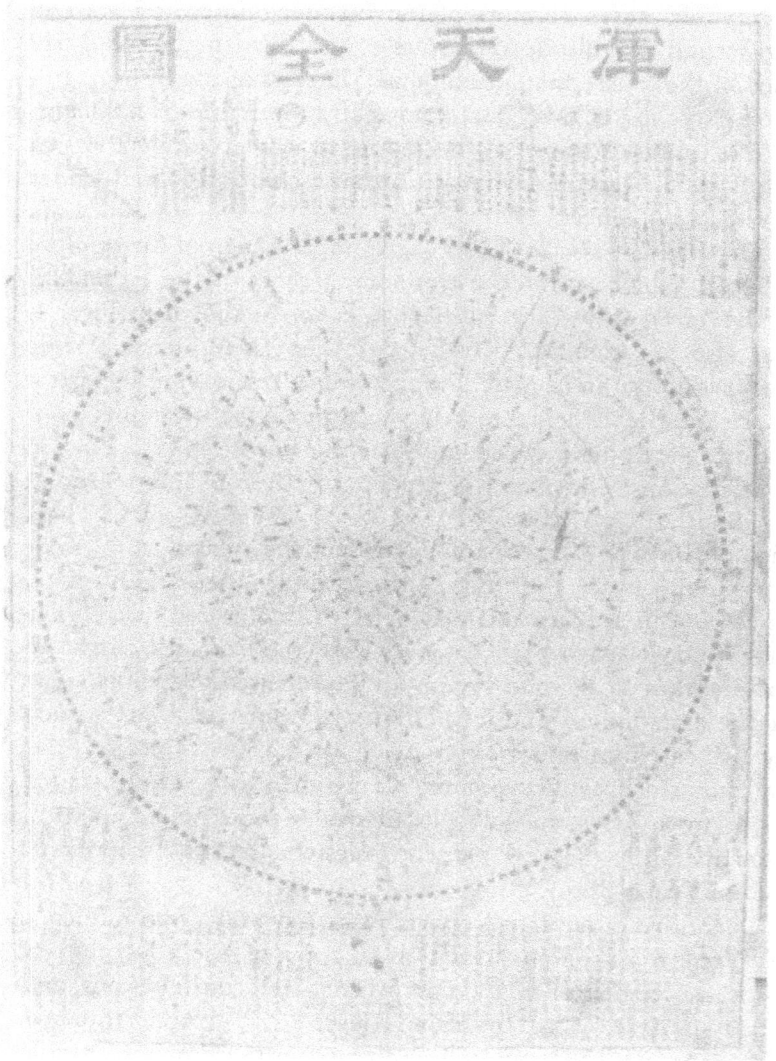

Fig. 1.4. The wood-block print of the Complete Map of the Celestial Sphere, *Honch'ŏn chŏndo,* late eighteenth century. Sungshin Women's Teachers College Museum, Seoul.

hemispheres on ecliptic coordinates); and a block-printed copy of the *Hwangdo ch'ong-sŏngdo* 黃道總星圖 (Star atlas of Kögler and Moggi of 1723).[50] A block-print of the same star map was made in 1834 by Kim Chŏng-ho 金正浩. Also available is the *Sŏnggyŏng* 星鏡 (Mirror of stars), compiled in two volumes by Nam Pyŏng-gil 南秉吉, who is said to have begun with the ancient astronomical literature of China and supplemented it by the new theories of Western astronomers. The *Sŏnggyŏng* depicts, besides the 28 lunar mansion constellations and other traditional Chinese constellations, the southern circumpolar constellations. Altogether, 1,317 stars are included, including 16 of the first magnitude, 51 of the second, 159 of the third, 349 of the fourth, 399 of the fifth, and 343 of the sixth.

The star maps of Korea are in the main derived from Ch'en Cho's astral map, *Hsing t'u* 星圖, and Wang Hsi-ming's 王希明 ancient mnemonic guide to the constellations, *Pu t'ien ko* 步天歌 (The song of the sky pacers), as well as Western star maps, such as those in the *I hsiang chih* 儀象志 (Descriptions of astronomical instruments) of Ferdinand Verbiest.[51] To obtain a fuller understanding of this point, we may cite a brief passage from a chapter of the *Chŭngbo munhŏn pigo* 增補文獻備考 (Comprehensive study of civilization, revised and expanded edition) dealing with fixed stars:[52]

According to *Han shu* "T'ien wên chih" 漢書天文志 (Treatise on astrology of the history of the Han Dynasty), fixed stars always occur north and south of the equator, comprised in 188 constellations and numbering 783. The same treatise of the *Chin shu* 晉書 (History of the Chin Dynasty) says that Ch'en Cho, Astronomer Royal of the Wu State, made first a list of the stars and the constellations of the Three Masters, namely, Shih Shen, Kan Te, and Wu Hsien. Ch'en Cho illustrated 283 constellations and 1,464 stars. Because *Pu t'ien ko* (The song of the sky pacers) was in agreement with Ch'en Cho as regards the number of stars, the later Astronomers-Royal used the *Pu t'ien ko* as their standard reference work. Ferdinand

50 Full translation of the inscriptions will be found in Needham and Lu, "A Korean Astronomical Screen," pp. 145–149.

51 Verbiest's *I hsiang chih* was reprinted in Korea in 1714.

52 *Munhŏn pigo*, ch. 1, pp. 23b–25a (slightly modified).

Verbiest maintains in his *I hsiang chih* that the number of stars that preserve their old names unchanged amounts to 259 constellations and 1,129 stars, or 24 constellations and 335 stars less than the numbers given in the *Pu t'ien ko*. Otherwise, there is an increase of 597 stars, and of the south circumpolar stars, 23 constellations and 150 stars. Through recent observations, it has come to our notice that the numbers of stars of various degrees shown in the *I hsiang chih* are not sufficient; therefore, we have made determinations of the stars one by one, and found that they amount to 300 constellations and that those stars with their old names make up a total of 277 constellations and 1,319 stars, or 18 constellations and 190 stars more than the numbers given in the *I hsiang chih,* while the numbers are closer to those given in the *Pu t'ien ko*. Since there is an increase of 1,614 constellations (above the famous Constant Number) while 23 stars and 150 stars around the South Pole are invisible from China, the number of fixed stars is determined at 3,033 stars comprised in 300 constellations, including those that are observable in the west and that have been observed since the early days.

The fixed stars were grouped into six classes according to their magnitude. A star of the first class was considered 68 times as large as the earth, a star of the second class 28 times, a third-class star 11 times, a fourth-class star 4.5 times, a fifth-class star $1\frac{1}{18}$ times, and a star of the sixth class half as large.

In 1861, Nam Pyŏng-gil, after a comparative study of the diagrams of fixed stars in the Chinese astronomical works and observational data obtained by the members of the Bureau of Astronomy in Seoul, compiled and published the *Sŏnggyŏng* (Mirror of stars), a catalog of the equatorial coordinates of fixed stars. He also utilized the theories of the Western astronomers to develop a theory about the varying distances of stars. He states: "The space in which stars are located should be discussed in terms of distance [from the earth], i.e., height. However, there has been no measurement of the distance in China or in the West. Perhaps this is because the variation in distance is negligible."

Also, mention may be made here of an astral chart in the form of a brass disk measuring 36.5 cm across, now preserved in the T'ongdo Temple 通度寺 and reputed to have been made by a nun named Sŏnjahwa 仙子花 at the Munsu Temple 文殊庵 on Mt. Samgak 三角山 north of Seoul.

Ancient Observatories in Korea

KYŎNGJU OBSERVATORY

To the northeast of the Panwŏl (Half-moon) Castle 半月城 in Kyŏngju 慶州 stands a bottle-shaped stone tower, stable and graceful in appearance. It is the Ch'ŏmsŏngdae 瞻星臺, the oldest known astronomical observatory extant in East Asia. The observatory, built in 647 (sixteenth year of Queen Sŏndŏk 善德女王),[53] is 9.108 meters high with a diameter of 4.93 meters at the base and 2.85 meters at the top. At a height of 4.16 meters from the stone base opens a square window, facing due south, which measures about one meter square.[54]

This observatory is presumed from its peculiar shape (see Fig.1.5) to have been built in accordance with the traditional round-heaven, square-earth 天圓地方 theory of ancient China. Its 27 stone layers are believed by some scholars to symbolize Queen Sŏndŏk, the twenty-seventh reigning monarch of the Silla Dynasty. The Ch'ŏmsŏngdae is thought by some to have supported an armillary sphere on the top, placed upon a wooden structure; but others believe that it was an observatory platform of the open-dome type.[55] The shape of the stone top seems to suggest that the eastern half of the interior space was paved with stone slabs while the western half had a wooden floor, but there is insufficient evidence to prove that an armillary sphere was permanently mounted on top. Also, the interior surfaces are made up of rough-hewn stone faces, generally too

[53]Ibid., ch. 2, p. 22a; *Tongguk yŏji Sŭngnam* (Geographical conspectus of the Eastern Kingdom [Korea]), ch. 21, p. 24b.

[54]"Ch'ŏmsŏngdae silch'ŭk pogo" (Report on the survey of the remains of the observatory of Silla), *Kogo misul*, 1963, *4*.5; 162–163; Hong Sa-jun, "Kyŏngju Ch'ŏmsŏngdae silch'ŭk chosŏ" (Report on the survey of the remains of the observatory in Kyŏngju), *Kogo misul*, 1965, *6*.3–4: 56–60.

[55]Wada Yūji, "Keishū senseidai no setsu" (A theory on the observatory of Silla), *Chōsen kodai kansoku kiroku chōsa hōkoku* (Report on the survey of the ancient record of observation in Korea; Seoul, 1917), pp. 144–151; Jeon, "Samguk mit t'ongil Sillaŭi ch'ŏnmun ŭigi," p. 19, n. 23.

Fig. 1.5. The astronomical observatory, the Ch'ŏmsŏngdae, built in 647 at Kyŏngju, the capital city of the Silla Dynasty. The observatory is 9.1 m high. It permitted the measurement of the sun's shadow, much like a gnomon; was an observatory platform of the open-dome type; and gave the standard points of reference for measurement of the four cardinal points and directions.

crude to support the belief that the interior was used for observational purposes as in an open-dome type observatory. For these reasons, it is difficult to believe that the Ch'ŏmsŏng-dae served at any time as a permanent astronomical observatory of either of the types mentioned above.

It is entirely probable, however, that in times of "unnatural" phenomena visible for some time, such as the appearance of a comet, an armillary sphere or a similar instrument was mounted on top for observation purposes. Also, the tower may well have served as an open-dome type observatory from time to time for determination of the positions of constellations for exact measurement of the solstitial and equinoctial points as well as the 24 fortnightly periods. This hypothesis seems to receive support from the fact that among the local people the tower is called *pidu* 比斗, meaning "comparison with the Dipper (as the standard point of reference)," and that the whole neighborhood is referred to as Pidu-kol 비두 (比斗) 골 or Pidugŏri 비두 (比斗) 거리.

This observatory, whose shape embodies Korean aesthetic preferences, is of particular interest in several respects. First of all, the tower permitted the measurement of the sun's shadow, much in the way of a gnomon, to ascertain the season. Besides, the window facing the south opens in such a way that sunlight is fully shed upon the interior bottom at noon on the vernal and autumnal equinoxes, whereas sunlight is completely absent at both solstices.

The Ch'ŏmsŏngdae was probably the center of astronomical activity of the Kingdom of Silla. It served as the meridian for astronomical observations throughout Silla and gave the standard points of reference for measurement of the four cardinal points and directions. The stone slab on top, square with extended diagonals, pointed to the eight cardinal points, and the window faced due south. The area surrounding the Ch'ŏmsŏngdae was presumably paved with flagstones to prevent the erection of houses and thus secure an open space in all directions to facilitate unobstructed observations.[56]

[56]Jeon, "Samguk mit t'ongil Sillaŭi ch'ŏnmun ŭigi," pp. 18–22.

The Ch'ŏmsŏngdae of Kyŏngju is presumed to have been built after the model of a Paekche observatory, for during the reign of Queen Sŏndŏk a large number of architects from Paekche were invited to build the Hwangnyong Temple 皇龍寺 and its famous nine-storied pagoda. Several years later, in 675, Paekche astronomers personally supervised the erection of an astronomical observatory in Japan.[57] The treatise on geography of the *Sejong sillok* (Veritable records of the King Sejong era), which was compiled in the fifteenth century, and the *Tongguk yŏji sŭngnam* 東國輿地勝覽 (Geographical conspectus of the Eastern Kingdom) of the sixteenth century record that there were ruins of an astronomical observatory in P'yŏngyang castle. This leads us to the presumption that Koguryŏ had a similar observatory. From all these facts, it is evident that there were then a number of astronomical observatories in Paekche, built according to the *Chou pi* system, and that under their influence were built both the Ch'ŏmsŏngdae of Silla and the astronomical observatory in Japan. In turn the observatories in Paekche and Silla probably influenced the reconstruction in T'ang China (723) of the Chou Kung Ts'e Ying T'ai 周公測景臺 (Chou Kung's Tower for the Measurement of the Sun's Shadow).[58]

THE KORYŎ OBSERVATORY

In Songdo 松都 (present-day Kaesŏng), to the west of the Manwŏltae 滿月臺, stands a stone structure known to be a remnant of the Ch'ŏmsŏngdae of the Koryŏ era. The structure consists of a stone slab about 3 m square, propped up on five stone pillars. On the four corners of the stone slab are holes about 15 cm in diameter which seem to have been used as sockets for setting up stone railings. Therefore, one may imagine a stone observatory of about 3 m on a side,[59] which had stone railings on top.

[57]See, for instance, *Meiji zen Nihon tenmongaku shi* (A history of Japanese astronomy before the Meiji era; Tokyo, 1960), p. 382.
[58]Jeon, "Samguk mit t'ongil Sillaŭi ch'ŏnmun ŭigi," pp. 18–22.
[59]Rufus, "Astronomy in Korea," p. 18.

There are no extant records of this observatory, and there is no way of determining its date of erection, its functions, or whether any instruments might have been set up on the stone framework. It is known, however, that since the early days of the Koryŏ Dynasty there were officials in charge of astronomical and meteorological observations, bearing such titles as *yŏngdaerang* 靈臺郎 (director of the observatory), *sasin* 司辰 (timekeeper), *saryŏk* 司曆 (calendar technician), and *kamhu* 監候 (meteorological technician).[60] The structure extant in Songdo may be presumed to be one of the *yŏngdae* 靈臺 (astronomical observatories) erected during the Koryŏ era.

Although a full set of observatory facilities was presumably in existence by the time of the reign of Munjong 文宗, most of them fell into disrepair in the course of the war with the Mongols, which lasted for thirty years beginning in 1231. Consequently, the astronomical phenomena often did not agree with the predictions of the current ephemerides. In 1281 the Susi calendar 授時曆 (*Shou-shih li*) was brought over from the Yuan court,[61] and the confused calendrical administration was put in order. In 1308 the existing T'aesaguk 太史局 and Sach'ŏndae 司天臺 were merged into the new astronomical office of Sŏun'gwan 書雲觀,[62] and this change was presumably accompanied by a large-scale reconstruction and renovation of the observatory facilities. It is probable that the remains of the Ch'ŏmsŏngdae in Songdo date back to this era.

It seems that observational instruments and a sundial were mounted upon the Koryŏ Ch'ŏmsŏngdae in Kaesŏng and that there was also a clepsydra inside. This is probable from the fact that the astronomical observatories erected in the Yi Dynasty era, which were in most respects faithful copies of the Koryŏ models, were equipped with these instruments.

60 *Koryŏsa*, ch. 76, p. 44a.

61 Ibid., ch. 29, pp. 3b, 31b.

62 Ibid., ch. 76, p. 44a. During 1047–1083, the Bureau of the Grand Astrologer (T'aesaguk) and the Office of the Observatory (Sach'ŏndae) had been established. In the Bureau of the Grand Astrologer were two professors of astrology (*yŏngdaerang, ling-t'ai lang*), two technicians (*sasin, ssu-ch'en*)

KANŬIDAE

The Yi Dynasty, inheriting the institutions of the Koryŏ Dynasty, established the Sŏun'gwan as its central royal astronomical and meteorological observatory, and the Kanŭidae 簡儀臺 (equatorial torquetrum observation platform) as a multi-purpose observatory. During the early Yi Dynasty era, there was a Sŏun'gwan both in the Kyŏngbok Palace 景福宮 and in Kwanghwa-bang 廣化坊 to the north. For the mounting of large-scale astronomical observation instruments, the manufacture of which began in 1432 (fourteenth year of Sejong), a stone platform measuring 6.3 m high, 9.7 m long, and 6.6 m wide was erected north of the Kyŏnghoeru 慶會樓 Pavilion in Kyŏngbok Palace. It was here that the Kanŭidae was set up, surrounded by stone railings, in 1434.[63]

In the Kanŭidae were mounted an armillary sphere, a celestial globe, a graduated scale, a direction-determining table, and other instruments. A great graduated scale about 8.24 m high was set up north of the observatory. Units were engraved on the surface of the blue stone scale so that examiners might read the length of the sun's shadow cast by the gnomon in the center.[64] This observatory was the largest to be built in East Asia after the giant Kuan Hsing T'ai 觀星臺 (Star Observation Platform) of the Yuan Dynasty, and from the spring of 1438 five officials of the Sŏun'gwan were assigned to it every night for continuous celestial observation.[65]

Around this time, another, smaller platform was erected at the Sŏun'gwan in Kwanghwa-bang in the north, and this is the

---

who ran the clepsydras, two technicians (*saryŏk, ssu-li*) who made calendars and published them, and two observers (*kamhu, chien-hou*).

[63]Jeon Sang-woon, "Sŏun'gwankwa kanŭidae" (The Bureau of Astronomy and the observatory in the Yi Dynasty), *Hyangt'o Sŏul*, 1964, no. 19, pp. 45–48.

[64]*Sejong sillok*, ch. 77, p. 9b. One of the gnomon observation records in the first month of 1550 appears in the *Myŏngjong sillok* as follows: "On the day of the full moon, the length of the moon's shadow was 2 *chang* 1 *ch'ŏk* 2 *p'un* by the great gnomon and that of the small gnomon was 4 *ch'ŏk* 1 *ch'on* 9 *pun*" (ch. 10, p. 2a).

[65]*Tongguk yŏji pigo* (A note on Korean geography), ch. 1, Bureau of Astronomy.

so-called Kwanch'ŏndae 觀天臺 (Celestial Observation Platform) extant today on the grounds of Hwimun High School in Seoul. This granite structure, 3.46 m in height and 2.4 × 2.5 m within the stone railings on top, was used to mount a sundial. Sŏng Chu-dŏk (1759–?) 成周悳, in his *Sŏun'gwanji* 書雲觀志 (Treatise on the Bureau of Astronomy) identified this observatory, also commonly referred to as a Ch'ŏmsŏngdae, as a miniature model of the Great Kanŭidae in the Kyŏngbok Palace, constructed, like the latter, in the tradition of the Ch'ŏmsŏngdae of the Koryŏ Dynasty.[66]

After a series of renovations and repairs, the Great Kanŭidae in the Kyŏngbok Palace, first erected in the era of King Sejong, came to be equipped with the best and the largest-scale instruments of the Far East. These were not shown even

Fig. 1.6. Remains of the Yi Dynasty observatory, the Kanŭidae, built in the seventeenth century at Ch'anggyŏng Palace. Remains of an almost identical observatory of the fifteenth century are preserved at Hwimun High School, Seoul.

[66] *Sŏun'gwanji* (Treatise on the Bureau of Astronomy), vol. 2.

to foreign envoys. This observatory, the pride of the Korean astronomers of succeeding generations, was completely destroyed during the Hideyoshi invasion and was never restored.

Later, in 1688, a celestial observatory was set up outside the Kŭmho Gate 金虎門 of the Ch'anggyŏng Palace 昌慶宮, modeled after the Sŏun'gwan of Kwanghwa-bang in the north, measuring 2.2 m in height and 2.4 × 2.3 m at the base.[67] This is preserved today in the Ch'anggyŏngwŏn zoological garden (see Fig. 1.6). Another observatory is said to have stood outside the Kaeyang Gate of Kyŏnghŭi Palace and in Changdŏk Palace (Fig. 1.7).[68] When Kyŏngbok Palace was expanded

Fig. 1.7. Ch'angdŏk Palace observatory, dating from the Yi Dynasty (probably the seventeenth century). From the *Tonggwŏl-to* (Painting of the East Palace), watercolor by an unknown artist of the eighteenth century. Koryŏ University Museum.

[67]Jeon, "Sŏun'gwankwa kanŭidae," p. 49.

[68]*Kunggwŏlchi* (Records on palaces), ch. 4, Kyŏnghŭi Palace.

during 1865–1872, an observatory (sundial platform) and buildings of the Kwansanggam 觀象監 (the later Bureau of Astronomy) were built. The equipment and instruments remaining in Kwanghwa-bang in the north were removed to this place. This reconstructed observatory of Kyŏngbok Palace, however, was torn down after the Japanese annexation of Korea in 1910.[69]

Because all these astronomical observatories of the Yi Dynasty era were located in the center of the capital, the vista was not sufficiently wide to permit unobstructed observation of astronomical phenomena near the horizon. Consequently in times of solar and lunar eclipses and other unusual astronomical phenomena, a provisional observatory was set up at the top of Mt. Samgak, which forms the background of the capital. There are some remains of this temporary observatory along the path leading to the top of Paegundae 白雲臺, one of the three major peaks of Paegun Mountain. Besides, Seoul's Namsan Hill and Mt. Mani 摩尼山 on Kanghwa Island 江華島 were utilized for the same purpose occasionally.[70]

HONG TAE-YONG'S PRIVATE OBSERVATORY

Besides these observatories of the Sŏun'gwan, private observatories of remarkable merit were erected in the late Yi Dynasty era (eighteenth century). The most outstanding of these is the Nongsugak 籠水閣 Pavilion of Hong Tae-yong (1731–1783) 洪大容. The existence of this private observatory, extraordinary under the feudal institutions of the Yi Dynasty when scientific facilities were expected to exist only as part of the central government apparatus, was known even to contemporary Chinese scholars and astronomers. It was equipped with, among other things, an armillary sphere, observational mechanical clock (*hujong* 候鐘), celestial sphere, sighting tube, and triangulation instrument (*kukoŭi* 勾股儀). The armillary sphere and striking clock were weight-driven. The celestial sphere was

[69]Wada, *Chōsen kodai kansoku kiroku chōsa hōkoku*, p. 169.
[70]*Munhŏn pigo*, ch. 2, pp. 21b, 10b–11a.

also automatic, driven by a waterwheel that needed a daily supply of 2 *chak* (almost exactly 2 cups) of water.[71]

Hong, who had been deeply impressed by the facilities of the observatory he saw in Peking, built his private observatory in the 1760s. Through astronomical observation and writing he contributed to the development of astronomy and mathematics in the late period of the Yi Dynasty. His book, *Chuhae suyong* 籌解需用 (Practical computation) is a textbook of applied mathematics, observational methods, and information about instruments, in which he synthesized mathematical knowledge of China and the West.

Sundials

SUNDIALS OF SILLA

In the Kyŏngju Museum are preserved fragments of a circular granite sundial of the Silla era (see Fig. 1.8).[72] About 33.4 cm in radius, this sundial lacks only the third that covers the hours *tzu* (rat) to *mao* (hare). It is believed to have been made in the sixth to seventh century. Its face shows the twenty-four seasonal divisions (of the year) by twenty-four characters indicating the corresponding directions, around which are eight symbolic inscriptions of the eight trigrams (*p'algoe* 八卦), representing the eight cardinal points. A gnomon was presumably fixed at the center of the plate face, mounted equatorially so that the gnomon points to the North Pole. It has many features in common with Chinese sundials of the Han through Yuan eras.[73]

The clear, fine weather with abundant sunshine that continues throughout the year in Korea encouraged use of the sundial from early times. The measurement of time in Korea is believed to have started somewhere around the beginning of the Christian era, in view of the relatively high standards of scientific knowledge attained by Nangnang 樂浪 under the influence of Han culture and the fact that Koguryŏ as early

[71] *Tamhŏnsŏ*, ch. 3, *Oejip* (part two).
[72] It was excavated in 1930 at a castle in Kyŏngju.
[73] See for instance, Needham, *SCC*, vol. 3, pp. 302–309.

Fig. 1.8. Remains of a Silla sundial. About 33.4 cm in radius and made of granite. Sixth to seventh century. Kyŏngju Museum.

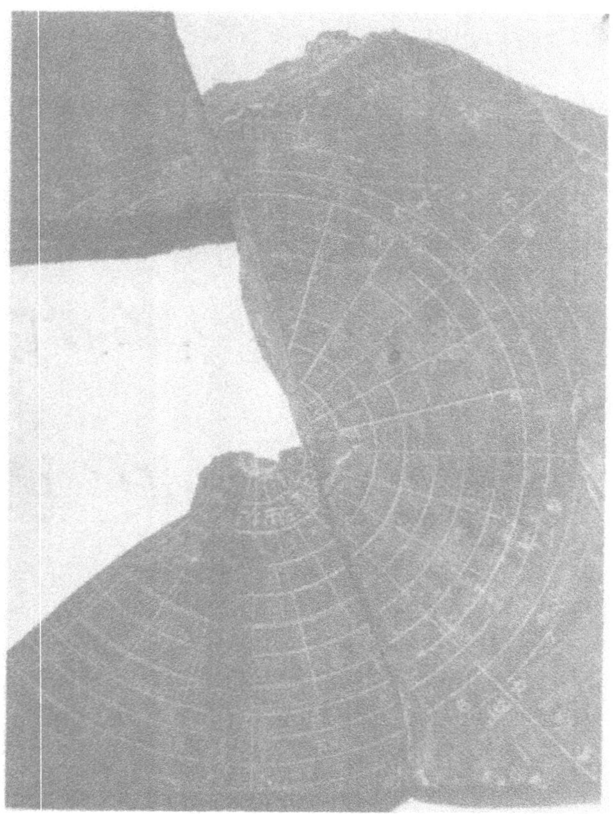

Fig. 1.9. Decimal circle sundial. This type probably antedates the sixteenth century. About 35 cm in radius and made of black marble. Iksan High School collection.

Fig. 1.10. Sundial and moondial preserved in Ch'anggyŏng Palace. It was destroyed during the Korean War, and now only the decimal circles and their stands remain. Photo from Rufus, "Astronomy in Korea."

as A.D. 62 sent to Han China for calendars.[74] The *ilcha* 日者 of Koguryŏ and the *ilgwan* 日官 of Paekche were both presumably court officials in charge of the custody and employment of the graduated gnomon scale and the sundial, and of time measurement generally.

The only change that took place in such sundials under the Yi Dynasty was the substitution of characters for the eight trigrams to indicate the eight cardinal points on the face of the sundial. The *paekkak-hwan* 百刻環 (decimal circle) sundial of the early Yi Dynasty era, such as those shown in Figs.1.9 and 1.10, is one of the most accurate sundials ever invented by mankind and furnishes invaluable information regarding the sundials of the era in general. It had a special water-level scale for accurate measurement of the length and position of the

[74]Jeon, "Samguk mit t'ongil Sillaŭi ch'ŏnmun ŭigi," pp. 13–14.

shadow of the gnomon, as did all other observational instruments of the Yi Dynasty.

SCAPHE SUNDIALS AND OTHER SUNDIALS OF THE SEJONG ERA

Official records regarding the manufacture of sundials first appear in the *Sejong sillok* 世宗實錄 (Veritable records of the King Sejong era). The sundials made in the Sejong era are of five kinds: *angbu ilgu* 仰釜日晷 (scaphe sundial), *hyŏnju ilgu* 懸珠日晷 (portable horizontal plumb-sundial), *ch'ŏnp'yŏng ilgu* 天平日晷 (portable horizontal water-level sundial), *chŏngnam ilgu* 定南日晷 (self-orienting armillary dial), and *kyup'yo* 圭表 (gnomon with graduated scale), and they constitute part of the various astronomical instruments made by Chŏng Ch'o 鄭招, Chang Yŏng-sil 蔣英實, Kim Pin 金鑌, Yi Ch'ŏn 李蕆, Kim Ton 金墩, and others in April of 1437 (nineteenth year of Sejong).[75]

Fig. 1.11. Korean scaphe sundial of the latter half of the seventeenth century. The latitude of Seoul is represented as "polar height in Seoul 37. 20 *do*." Made of bronze, diameter 35.2 cm. Ch'angdŏk Palace.

[75]*Sejong sillok*, ch. 77, p. 10a.

These sundials, the scaphe sundial in particular (Fig. 1.11), were based on astronomical instruments made by the Yuan astronomer Kuo Shou-ching 郭守敬, as described in the astrological chapter of the *Yuan shih* 元史 (History of the Yuan Dynasty).[76] The term *angbu* 仰釜 (upward-looking bowl) describes the hemispherical shape of the sundial. Within the bowl are marked 13 parallels, representing the 24 fortnightly periods distributed between the two solstices. A meridian is drawn at right angles, while an inclined gnomon is fixed so that the tip points to the North Pole.

Two scaphe sundials were ordered by King Sejong for the use of the general public and emplaced upon stone platforms on the Hyejŏng Bridge 惠政橋 and in the commercial district near the Royal Shrine (Chongmyo 宗廟).[77] These sundials, in fact, served as the first public clocks of Korea. The original sundials were lost during the Hideyoshi invasion but were restored in the latter half of the seventeenth century, during the reigns of Kings Hyŏnjong 顯宗 and Sukchong 肅宗. While the sundials of the Sejong era were put to public use, the restored instruments were installed within the royal court, with appropriate adornments such as the silver threads used for lines and characters on the dial face of black lacquer and bronze, which was mounted on four "Dragon Pillars." The division of the day into 96 rather than the customary 100 day parts indicates European influence in the reconstruction.

Several examples of this scaphe sundial were produced later, the most representative of which was a copy made in the Yŏngjŏng 英正 (英祖一正祖) era (eighteenth century) after the Hyŏnjong-Sukchong model, mounted on a beautifully carved stone platform. They are both characterized by statements of the latitude of Seoul (in Chinese degrees) at the time of their production: "Polar height in Seoul 37. 20 *do*" in the earlier one and in the later one, "Polar height in Seoul 37. 3915 *do*." This second figure is too precise to be based on direct

[76]Ibid., ch. 77, p. 10a; *Munhŏn pigo*, ch. 2, p. 25a; *Chega yŏksangjip*, vol. 2, ch. 14.

[77]*Sejong sillok*, ch. 77, pp. 10a ff.

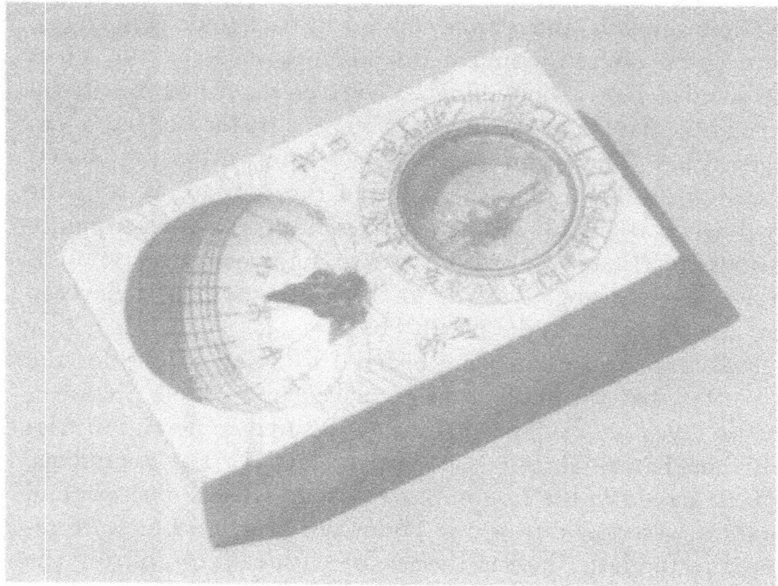

Fig. 1.12. Portable scaphe sundial with compass, made by Kang Kŏn, dated 1871. Its dimensions are 5.6 cm × 3.3 cm × 1.6 cm; the diameter of the bowl is 2.8 cm and that of the compass is 2.0 cm. Author's collection. It is almost identical to the sundial made by Kang Yun (probably Kang Kŏn's brother) and dated 1870, shown in Needham, *SCC*, vol. 3.

observation. The time plate of the sundials shows divisions of the day into 12 equal double hours (*shih*) and 96 quarters (*k'o*), covering the range from the hour of *yin* to the hour of *hsu*.

Another kind of sundial that has survived is a portable scaphe sundial the size of a matchbox. There are several extant models reputedly manufactured by Kang Yun 姜潤 and Kang Kŏn 姜健 in the latter half of the nineteenth century, and several others by anonymous hands. They all give the latitude of Seoul as "polar height 37. 3915 *do*" and also are equipped with a compass for their correct orientation (see Fig. 1.12). There are several other models believed to have been privately commissioned by provincial government offices or court mandarins for private use.[78]

[78]Jeon Sang-woon, "Yissi Chosŏnŭi sige chejak sogo" (A study of timekeepers in the Yi Dynasty), *Hyangt'o Sŏul*, 1963, no. 17, pp. 55–64.

Thus, the scaphe sundial, introduced into China by Kuo Shou-ching of the Yuan Dynasty but later discontinued there, passed to the Yi Dynasty and was preserved until modern times.

The *hyŏnju* 懸珠 and *ch'ŏnp'yŏng* 天平 sundials were both portable types provided with the means to maintain directional orientation and horizontal position with the use of a compass and a plumb-line weight hung by a rod to keep the gnomon perpendicular to the dial plate. A fine cord stretched diagonally plays the role of the gnomon and collapses upon the dial plate when the cover is closed. The only difference between the *ch'ŏnp'yŏng* sundial and the *hyŏnju* sundial was that instead of a plumb-line the latter had a small water level in the dial plate for maintaining the horizontal position.[79] The diagonal-thread gnomons seem to have come from the Arab world during the Middle Ages. The *chŏngnam* 定南 sundial, exceptionally accurate, combined features of the *hyŏnju*, and *ch'ŏnp'yŏng* sundials and the *kanŭi* 簡儀 (*chien-i*, equatorial torquetrum) in a horizontal circle and sighting tube.[80]

INTRODUCTION OF WESTERN SUNDIALS

While a few sundials of the Ming Dynasty were introduced into Korea during the reign of King Sejong, a larger number were brought from China after the Hideyoshi invasion. The most important of these was the *sinpŏp chip'yŏng ilgu* 新法地平日晷 (new model horizontal sundial), introduced in 1636.[81] Made by Li T'ien-ching 李天經 of Ming according to the *Shih-hsien li fa* 時憲曆法 of Adam Schall, mentioned earlier in this chapter, the "new model" sundial was a new version of the scaphe sundial on a plate model. This sundial may therefore be said to incorporate the traditions of both the Pelekinon sundial of ancient Greece and a similar one of Arabia, as well as of the scaphe sundials of Kuo Shou-ching. Another kind of horizontal sundial was made in Korea during the eighteenth

79*Sejong sillok*, ch. 77, pp. 10a ff.; *Munhŏn pigo*, ch. 2, p. 25a.
80Ibid.; Jeon, "Yissi Chosŏnŭi sige chejak sogo," pp. 55–64.
81*Munhŏn pigo*, ch. 3, p. 9b–10a.

century, a portable model with a compass and a triangular gnomon made of bronze.

Another special sundial was made in 1785, namely the *kanp'yŏng* and *hon'gae* sundial 簡平渾蓋日晷, a horizontal sundial of the plate type with coordinates.[82] We find a full explanation in the *Ŭigi chipsŏl* 儀器輯說 by Nam Pyŏng-ch'ŏl 南秉哲, in a chapter dealing with astronomical instruments and plotting the coordinates of the celestial sphere. The *kanp'yŏng* sundial could be described as the plane version of the scaphe sundial, said in the *Ŭigi chipsŏl* to have been based on the *Chien p'ing i shuo* 簡平儀說 (Description of a simple equatorial torquetrum) of Sabbatino de Ursis. The *hon'gae* sundial was made with the help of the *Hun kai t'ung hsien i shuo* 渾蓋通憲儀說 (On plotting the coordinates of the celestial sphere and vault) of the Ming Jesuits' associate Li Chih-tsao 李之藻. Twenty-six different kinds of measurement are possible with the use of these two sundials.[83] Meanwhile, new trends in Western-style astronomy gave rise to still another type of horizontal sundial. Made by Kang Yun in 1881,[84] its dial plate is in the form of a large semicircle of diurnal hours, and in it there is a small circle marked with the beginning points of each of the double hours (*shih*) and the 24 compass points. There is also a statement of the latitude of Seoul as "polar height 37. 3915 *do*," in seal characters. We may regard this sundial as one of mixed origins, originating in the West in antiquity and the Middle Ages and undergoing modifications under the indigenous traditions of China. It became the last sundial to be made in Korea according to the Chinese double-hour system. Made in portable

[82]Jeon, "Yissi Chosŏnŭi sige chejak sogo," pp. 70–72.

[83]For example, angular measures for sun, moon, and stars, ecliptic longitude of the sun, azimuth of the lunar mansions, angular extent of the lunar mansions, time of rising, setting, and meridian transits of stars, solar azimuth, solar latitude, solar noon azimuth, north polar elevation, projection of each of the twenty-four solar divisions, time of solar rising and setting, north azimuth of the twenty-four solar divisions, daylight time, time of solar transit of zenith in each of the twenty-four solar divisions and variation of noon shadow length with latitude, and so on. *Ŭigi chipsŏl* (Collected writings on astronomical instruments), vol. 2, pp. 19a–26b, 1a–8b.

[84]Jeon, "Yissi Chosŏnŭi sige chejak sogo," p. 73.

models, these sundials have foldable triangular gnomons and compasses.

Toward the close of the Yi Dynasty, Korea came to use the Arabic numerals and the daily 24-hour system (in lieu of the Chinese 12 double-hour system) under the influence of Western culture. Sundials, accordingly, came to reflect these changes, and the "modern" sundials had dial plates marked with 13 lines representing the hours from 6:00 to 12:00 A.M. and from 1:00 to 6:00 P.M., shown in Arabic numerals.[85] Some of these new sundials were made in portable models in the latter days of the Yi Dynasty. They were the last sundials ever made in Korea.

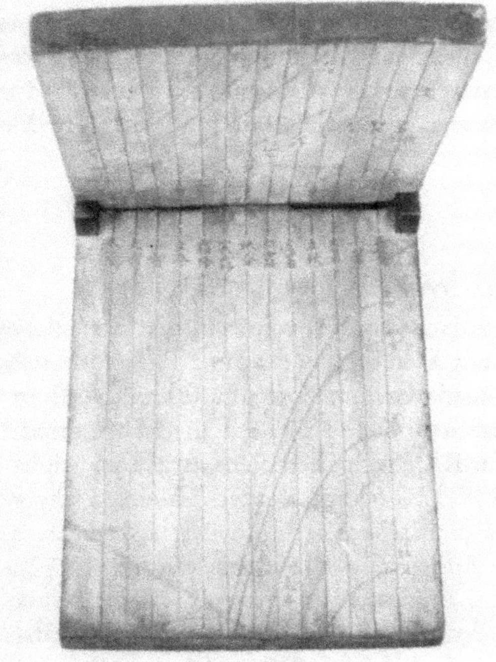

Fig. 1.13. Portable sundial of wooden plates without compass dated 1849. Its dimensions are 15.8 cm × 11.5 cm. Koryŏ University Museum.

[85]Ibid., pp. 73–75.

PORTABLE SUNDIALS WITHOUT COMPASS

A particularly interesting sundial made in Korea is a wooden portable instrument that gives its date of manufacture as 1849 (see Fig. 1.13). Measuring 15.8 cm × 11.5 cm, this sundial has two plates facing each other at right angles, in lieu of the conventional dial plate and hour marks. Inside the plates are parallel vertical lines indicating the seasons and obliquely crossing lines representing the hours. Its orientation is determined by adjusting the positions of the plates so that the end of the shadow cast by the standing plate is brought in line with the corresponding season on the opposite plate, giving indication of the time. Again we find inscribed the polar height at Seoul as 37. 3915 *do*, and the distance from the ecliptic longitude to the right ascension as 23. 27 *do*.[86]

Another type of sundial we have to mention here is the *yosŏn yang ch'ŏn-ch'ŏk* 曜仙量天尺, apparently a copy of the *liang t'ien ch'ih* 量天尺 (sky measuring scale) of Kuo Shou-ching, used in the Kuan Hsing T'ai, the star observation platform of thirteenth-century China.[87]

Clepsydras

FIRST REFERENCES TO CLEPSYDRAS

The usefulness of sundials is confined to sunny days; they have no function when it is cloudy or dark. Therefore the clepsydra came to be considered more useful and reliable than the sundial as a time indicator in the Three Kingdoms period. As was the case in China, Koreans called clepsydras by various names, such as *kangnu* 刻漏 (*k'o lou*, graduated leaker), *kyŏngnu* 更漏 or *nuho* 漏壺 (*lou hu*, drip-vessel). The first reference to the clepsydra in Korea appears in the *Samguk sagi* 三國史記 (History of the Three Kingdoms), written by Kim Pu-sik 金富軾 in 1145. According to this book, clepsydras were first made in 718, when the Nugakchŏn 漏刻典 (Clepsydra Board) was established and put in the charge of six professors and one

[86] Ibid., pp. 66–67.
[87] Ibid., pp. 69–70.

recorder.[88] The Silla clepsydras (Figs. 1.14 and 1.15) are believed to have been similar to the *fu lou* 浮漏 (inflow type with an indicator rod), typical of the Han and later eras of China.[89] From the hole of a reservoir vessel, water flows down into an inflow receiver, where the graduated indicator rod on a float ascends as the water within increases.

The clepsydras of Japan that were constructed in 671 under the direction and supervision of Paekche astronomers had four reservoir vessels,[90] a feature typical of the clepsydra of Lü Ts'ai 呂才 of the T'ang (first half of the seventh century).[91] We may speculate by analogy that all clepsydras used in Korea during the Three Kingdoms and Unified Silla eras were after the model of Lü Ts'ai.

The position of the *nugak-paksa* 漏刻博士 (clepsydra professor), in charge of clepsydras, is said to have been first created in 718 simultaneously with the establishment of the Nugakchŏn, but Paekche appears to have employed a similar official at least two centuries earlier. After Paekche's calendar professors first crossed over to Japan in 554, Japan created the positions of *rōkoku hakase* (clepsydra professor) and *reki hakase* 暦博士 (calendar professor) in 702.[92] These institutions are believed to be modelled on the systems of Sui and T'ang and therefore may be presumed to have been universally in use among the Three Kingdoms of Korea.

Silla in 749 assigned one astronomical professor and six clepsydra professors to the Nugakchŏn to ensure accurate time-keeping.[93]

There are no available records or data on clepsydras of the Koryŏ era, except that in the *Koryŏsa* 高麗史 (History of the

[88]*Samguk sagi*, ch. 8, seventeenth year of King Sŏngdŏk; ch. 38, *Chapchi* (various treatises), Clepsydra Board.
[89]Jeon, "Samguk mit t'ongil Sillaŭi ch'ŏnmun ŭigi," pp. 15–16.
[90]*Nihon kagaku gijutsushi* (A history of science and technology in Japan; Tokyo, 1962), ed. Yajima Suketoshi (Asahi Shinbunsha), p. 45.
[91]Yabuuchi Kiyoshi, "Chūgoku no tokei" (Timekeepers in ancient China), *Kagakushi kenkyū*, 1951, *19*: 20–22.
[92]Nakayama, *Japanese Astronomy*, pp. 17–19.
[93]*Samguk sagi*, ch. 9, eighth year of King Kyŏngdŏk.

Fig. 1.14. Remains of an ancient clepsydra preserved at Chŏndŭng Temple, Kangwha Island. This primitive water clock seems to have been made by the monks in the Silla Dynasty period. It is made of granite; the larger vessel is 39 cm in outer diameter and 30.5 cm in height, and the smaller one is 43 cm in outer diameter, 26 cm in height.

Fig. 1.15. Clepsydra preserved at Chŏndŭng Temple (see Fig. 1.14.)

Koryŏ Dynasty), the "Paekkwanji 百官志" (Treatise on government posts) mentions titles of officials in charge of the clepsydra, such as *sŏlhojŏng* 挈壺正 and *changnu* 掌漏. Furthermore, it is stated that the work of the Sŏun'gwan included astronomy, the calendar, meteorology, and clepsydra technology.[94] There seems little doubt, therefore, that these officials were put in charge of the construction and management of clepsydras.

FIRST STANDARD CLOCK INSTALLED IN SEOUL

After the Yi Dynasty moved the capital to Hansŏng 漢城 (present-day Seoul), there arose the need to install a new standard clock, and accordingly a *kyŏngnu* 更漏 (night clepsydra) was installed in 1398 (seventh year of T'aejo) in the center of Seoul.[95] Together with the new clepsydra a belfry was erected to ring the standard time throughout the royal capital. Therefore, the neighborhood of the belfry (*chongnu* 鐘樓) and the clepsydra was called Chongno 鐘路, or Bell Street.

The *kyŏngnu*, which was the first clepsydra to be used in the Yi Dynasty, was probably of the same model as that in use in the latter days of Koryŏ, belonging to the same category as the polyvascular inflow clepsydra made at Canton in 1316 during the Yuan era by Tu Tzu-sheng 杜子盛 and Hsi Yunhsing 洗運行. The time was announced by ringing the bell 28 times at the beginning of the night watch, for the 28 constellations, and 33 times at the end of the night watch (after the "33rd heaven," the Indra heaven). The former signal was called *in'gyŏng* 人定, serving as the signal for closing the castle gates, and the latter *para* 罷漏, a signal for opening the castle gates again.[96]

In the sixth year of Sejong (1424), a clepsydra made of bronze after the Chinese model was installed in Kyŏngbok Palace in May.[97]

[94] *Koryŏsa*, ch. 76, p. 45a.
[95] *T'aejo sillok*, ch. 14. p. 7a.
[96] *T'aejong sillok*, ch. 27, pp. 46ab.
[97] *Sejong sillok*, ch. 24, p. 10b.

CLEPSYDRA WITH AUTOMATIC TIME-SIGNAL APPARATUS

It was an earnest wish of both King Sejong and his clockmakers to have a clepsydra with an automatic time-signal apparatus. To this end, the king installed Chang Yŏng-sil 蔣英實, formerly a slave in Tongnae 東萊 prefecture, in the office of *pyŏlchwa* 別座 (assistant section chief) in the Sangŭi-wŏn 尙衣院 (Royal Clothing Office) and commissioned him to undertake this task in cooperation with the astronomer Kim

Fig. 1.16. Remains of the clepsydra with automatic time-signal apparatus (*chagyŏngnu*). It was built in 1434 and reconstructed in its present form in 1536. It consisted of three reservoirs and two inflow receivers. The two inflow vessels had an outer diameter of 37 cm and a height of 196 cm. In 1653, when the Shih-hsien calendar was adopted, the automatic striking apparatus was removed (photo from Takabayashi, *Tokei hattatsu shi,* taken in the 1920s at Ch'anggyŏng Palace). The clepsydra is now at Tŏksu Palace, but the pipes and stone stands were lost during the Korean War.

Pin 金銚. After two years' labor, the new device was finally completed in June of 1434.[98] The automatic clepsydra, called *chagyŏngnu* 自擊漏, was installed in the Poru Pavilion erected in the southern part of the Kyŏngbok Palace grounds and was formally inaugurated as the official standard clock of the Yi Dynasty.

The *chagyŏngnu*, shown in Fig. 1.16, shows signs of influence from the palace clock made by order of Emperor Shun 順帝 of the Yuan Dynasty and the automatic time signal device of the Arab clepsydra.[99] The new clepsydra was constructed in such a way that the bell struck the time in an entirely automatic way, in perfect agreement with the irregular night-watch system. It was thus a cause of constant wonder and admiration to all those who viewed the functioning of the device.

According to the *Porugak-ki* 報漏閣記 (Description of the chiming clepsydra) by Kim Ton 金墩, and also the inscriptions made by Kim Pin 金銚, the clepsydra had a total of 4 reservoir vessels, 2 inflow receivers, 12 indicator rods, and an automatic time-ringing device, which worked roughly in the following manner:[100]

. . . As water drips into the water-receiving scoop, the float indicator rod ascends. When it reaches a certain point a copper plate reacts to cause a small ball to roll down through a copper pipe. It makes the impact plate rebound, in turn driving another, larger ball into a short pipe at the base and activating a mechanical spoon which strikes the Hour God 時神 on the arm, thus resulting in the ringing of the bell. . . . Since all parts of the machine are hidden within an enclosure, all that can be seen from without is the wooden man in the mandarin costume.

The enormous size of the clepsydra (its water-receiving scoop was 11 *ch'ŏk*, 2 *ch'on* high and 1 *ch'ŏk*, 8 *ch'on* in diameter) and the ingenious mechanisms incorporated therein attest to the

98Ibid., ch. 64, p. 44b.

99See, for instance, Joseph Needham, Wang Ling, and D. J. Price, *Heavenly Clockwork: the Great Astronomical Clocks of Medieval China* (Cambridge, England, 1960), pp. 28–94.

100*Sejong sillok*, ch. 65, pp. 1a–3b; *Tongguk yŏji sŭngnam*, ch. 1, pp. 22a–27a; *Munhŏn pigo*, ch. 2, pp. 27b–31b.

genius of Chang Yŏng-sil, the designer of the *chagyŏngnu*. This device, it is obvious from records in the *Sejong sillok* and other extant literature, made other innovations possible.[101]

In 1455 (third year of Tanjong) the *chagyŏngnu*, after twenty-one years of service, stopped operation in February, as its inventor, Chang Yŏng-sil, and his coinventor, Kim Pin, both died.[102] It was not until fourteen years later, in 1469, that the automatic clepsydra was once again in working order.[103] In November of 1505, the automatic clepsydra was removed to Ch'angdŏk Palace 昌德宮.[104]

The completion of the Ch'angdŏk Palace gave rise to a call for the construction of a new automatic clepsydra—a call that was intensified as the old clepsydra failed to keep correct time during the reign of Sŏngjong 成宗. In September of 1534 (twenty-ninth year of Chungjong), or a hundred years after the installation of the first automatic clepsydra, work was started on both the reconstruction of the *chagyŏngnu* of the Poru Pavilion and the construction of a new *chagyŏngnu*, at the suggestion of two court officials,[105] Yu Chŏn 柳耶 and Ch'oe Se-jŏng 崔世鄭. The work was completed in June of 1536 after two years' endeavor.[106]

The *chagyŏngnu* of Yu Chŏn and Ch'oe Se-jŏng was put into use on August 20 of the same year. It was constructed on the same plan as that of Chang Yŏng-sil but modified so that the bell would ring at every fifth part of each night watch.[107]

It consisted of three reservoirs and two water-receiving scoops. Of the three reservoirs, one was made of bronze, 70 cm high and 13.5 cm in diameter, while the other two were made of stoneware, each measuring 46 cm in maximum diameter and 40.5 cm in height. The two inflow vessels had an outer diameter of 37 cm and a height of 196 cm each, receiving water through a pipe

[101] Ibid.
[102] *Tanjong sillok*, ch. 13. See Jeon, "Yissi Chosŏnŭi sige chejak sogo," p. 90.
[103] *Yejong sillok*, ch. 8. See Jeon, "Yissi Chosŏnŭi sige chejak sogo," p. 90.
[104] *Yŏnsan'gun ilgi*, ch. 60. See Jeon, "Yissi Chosŏnŭi sige chejak sogo," p. 90.
[105] *Chungjong sillok*, ch. 78, p. 4b.
[106] Ibid., ch. 82, pp. 12b, 31b, 32a.
[107] Ibid.

2.5 cm in diameter.[108] The old clepsydra, which continued in use for some time, was destroyed by fire during the Hideyoshi invasion. The new clepsydra, installed at the Porugak 報漏閣 (Chiming Clepsydra Pavilion) of Ch'angdŏk Palace, continued in use, surviving many subsequent repairs and renovations.

Meanwhile, a new calendar system from the Ch'ing court of China, the Sihŏn (Shih-hsien) Calendar 時憲曆, was officially adopted in 1653 (fourth year of Hyojong), replacing by a system of 96 quarter hours the conventional system of 100 divisions (*k'o*) of the day. Consequently, the new clepsydra lost its usefulness because of discrepancies with the new system of 96 quarters per day. With the automatic striking apparatus removed, only the vessels survived in use. They are preserved today in Tŏksu Palace 德壽宮.[109]

THE JADE CLEPSYDRA

Chang Yŏng-sil, the former slave who was advanced to the rank of Grand Protector-General (*taehogun* 大護軍) thanks to the patronage of King Sejong, in gratitude constructed and submitted to the monarch as a gift the *ongnu* 玉漏, a jade clepsydra serving both as a celestial clock and an automatic water clock. Completed in February of 1438 (twentieth year of Sejong) and installed at the Hŭmgyŏnggak 欽敬閣 (Pavilion of Respectful Veneration) of the Kyŏngbok Palace, the device is described in the *Hŭmgyŏnggak-ki* 欽敬閣記 written by Kim Ton, as follows:[110]

A papier-mâché "mountain" about 7 *ch'ŏk* high is set up within the Hŭmgyŏng Pavilion with the timekeeping wheel of the *ongnu*, which is set to rotate by means of falling water. A golden image of the sun, made out of a cannonball, is provided to move across the middle of the mountain surrounded by multicolored clouds. The sun makes its daily revolution, appearing at the rim of the mountain at dawn and hiding behind it at dusk. The inclination of the sun varies depending upon the polar distance while its rise and set follow that of the true

108Jeon, "Yissi Chosŏnŭi sige chejak sogo," pp. 91–94.
109Ibid., pp. 95–97.
110*Sejong sillok*, ch. 80, pp. 5a–6a.

sun for each of the fortnightly periods concerned. Below the imitation sun stand four female immortals placed at the four cardinal points, each with a golden bell in hand. At the beginning of each of the three morning double hours ranging from *yin* to *ch'en*, the immortal in the east jingles the bell, followed by the next immortal to the west for the next three double hours, and similarly by those to the north and the south by turns. At the four cardinal points on the ground stand the Four Gods 四神, each facing the central mountain. The first of the Four Gods, the blue Dragon, faces north at the double hour of *yin*, south at *mao*, and again west at *ssu*. Next, the Red Bird faces east, and so on. At the southern foot of this mountain stands a high platform where an "hour jack" (*ssu ch'en* 司辰) stands turning his back to the mountain, while three warriors, all in armor, are arranged in such a way that the first one, carrying an iron hammer, stands in the east facing west; the second one, carrying a drumstick, stands near the west facing east; and the third one, carrying a gong stick, also stands in the west facing east. At each double hour, the hour jack turns to face the bell man and they strike their respective instruments at each night watch. On the level ground are Twelve Gods 十二神 occupying their respective positions and behind them are holes. At the hour of *tzu* the hole behind the Rat opens while an immortal with a time tablet (*shih p'ai* 時牌) comes out and the Rat stands still. The work is done by the Ox at the double hour of *ch'ou* with a similar performance, and so on down the successive double hours. At the south point is located another platform carrying an "advisory vessel" (*i ch'i* 欹器), which lies on its side when empty, stands upright if half-full with water, and falls over again if filled to the brim. All these are performed completely automatically without any help from anyone. Around the mountain is an enclosure with paintings of rural scenery in the four seasons, and wood carvings of men, birds, and plants displaying the labors undertaken by the people in the different seasons.

The *ongnu*, the Jade Clepsydra, incorporated various features of ancient instruments. One of its sources is the armillary sphere of which mention is made in the *Hou Han shu* 後漢書 (History of the Later Han Dynasty) in its biography of Chang Heng. The clepsydra also derives from an armillary sphere of which records appear in the astronomy chapter of the *Chin shu* 晉書 (History of the Chin Dynasty), and from a third, which was made by

Liang Ling-tsan 梁令瓚 in the eighth century, and of which record appears in the Treatise on Astrology of the *T'ang shu* 唐書 (History of the T'ang Dynasty). Presumably, reference was also made in its manufacture to other ancient records such as the Treatise on Astrology of the *Sung shih* 宋史 (History of the Sung Dynasty),[111] in which it is said that Chang Ssu-hsun 張思訓 of the Sung Dynasty in 979 made a clepsydra with a device for telling the time, and the *Hsin i hsiang fa yao* 新儀象法要 (New description of an armillary clock), in which is described the time-measuring device rotating an armillary sphere made by Su Sung 蘇頌 in the eleventh century. The time-telling device of the hour jack and the three warriors, as well as the female immortals and the Twelve Gods, was no doubt an imitation of apparatus in use in Shun Ti's palace clock and also in the Arab world in the Middle Ages.[112]

We may therefore characterize the *ongnu* as an astronomical clock worked by an escapement similar to those of Chinese astronomical clocks, in combination with the time-ringing apparatus of Shun Ti's palace clock and of the Arabic clepsydras carrying various figures (such as the female immortals, and so on), with the further addition of a model of the sun. According to Kim Ton's description of Hŭmgyŏng Pavilion, the operation of the *ongnu* was entirely automatic, whereas the clepsydras of Chinese and Arab origin required manual attention for the winding-up process. We owe this thorough and complete automation to Chang Yŏng-sil, the slave turned royal clockmaker.

Chang Yŏng-sil's original *ongnu* was lost in the conflagration of Kyŏngbok Palace early in the reign of Myŏngjong, but a copy was made by Pak Min-hŏn 朴民獻, Pak Yŏng 朴泳, and others in August of 1554.[113]

CLEPSYDRAS OF THE LATE YI DYNASTY

As mentioned above, the clepsydras used during the reign

[111] Yabuuchi, "Chūgoku no tokei," pp. 20 ff.
[112] Needham, Wang, and Price, *Heavenly Clockwork*, pp. 160 ff.
[113] *Myŏngjong sillok*, ch. 15, ch. 17, pp. 21ab.

of King Chungjong 中宗 were without the automatic striking device. These simpler vessels continued in use until the end of the Yi Dynasty, with indicator rods that pointed to the hour marks in accordance with the new system of 96 quarters daily. Of the latter clepsydras, the clepsydra chapters of *Kukcho yŏksanggo* 國朝曆象考 (Compendium of calendrical science and astronomy in the Yi Dynasty) and *Chŭngbo munhŏn pigo* 增補文獻備考 read as follows:[114]

The *nuho* 漏壺 consists of three vessels of different sizes. The largest has a circumference of 12 *ch'ŏk*, corresponding to the circumference of the *ya-ch'ŏnji* 夜天池 (upper reservoir for nocturnal use). The second is the *il-ch'ŏnji* 日天池 (reservoir for diurnal use) while the last and the smallest corresponds to the *p'ing shui hu* 平水壺 (constant-level tank) of early days. Besides, two inflow receiver tubes of 1 *ch'ŏk*, 2 *ch'on* in diameter and 6 *ch'ŏk*, 8 *ch'on* in length each serve day and night by turns. . . . Each of the 24 indicator rods corresponding to the 24 fortnightly periods has a length of 6 *ch'ŏk*, 2 *ch'on* and is divided into 96 quarters in conformity with the new Shih Hsien calendar system instead of the 100 quarters of the old days. Each quarter is subdivided into 15 *p'un*. Therefore, a double hour consists of 8 quarters. The receiver tube has a "turtle float" made of copper plate fitting into the inner diameter of the receiver tube. On the back of the turtle float there is a hole of square shape for fixing an indicator rod that shows the time as the float ascends.

The clepsydra now preserved in the Tŏksu Palace was the standard clock of the Yi Dynasty from the reign of Hyojong 孝宗 until the end of the dynasty. The time report was performed in the following manner: The clepsydra official announced the night watches and the half-double hours to the soldier in charge of night watches, who in turn conveyed the information to the belfry guard, who struck the bell. As the bell was rung, the first signal-fire post on Namsan Hill reported by torch flame the beginning of the night watch to all the citizenry in and out of the castle.

The great bell now housed in the Posingak 普信閣 on Chongno Street was cast in 1467 (thirteenth year of Sejo), and was in

[114]*Kukcho yŏksanggo* (Compendium of calendrical science and astronomy in the Yi Dynasty), vol. 2, ch. 4, Clepsydra; *Munhŏn pigo*, ch. 3, p. 11b.

service from the time of the Hideyoshi invasion (1592) to the end of the Yi Dynasty in 1910.

THE PORTABLE CLEPSYDRA

While clepsydras were satisfactory for use in the capital and other main cities, they were of no service in mobile military camps. This called for a portable version of the clepsydra (*haengnu* 行漏). The earliest of these, as far as we know, was produced in 1437, and the first models were intended for dispatch to border stations in Hamgil and P'yŏngan provinces in the north. The typical model was made up of a reservoir vessel and an inflow receiving vessel, connected by a siphon provided to diminish error arising from irregular water pressure. They seem to have originated from the device described in *Kuan shu k'o lou t'u* 官術刻漏圖 (Illustrated treatise on standard clepsydra technique), published in the Sung Dynasty,[115] showing a leather bag or reservoir vessel hung up at the end of a long stick, and a wooden receiving vessel with indicator rod connected with the bag by a pipe.[116]

Incense Clocks

INCENSE STICKS

Another means of measuring time is the incense stick, or joss stick, widely used in Buddhist and Taoist temples. From an early period attempts were made to measure time so that the various rituals and ceremonies might be punctually performed. The ancients of course observed the stars at night and made use of sundials and clepsydras; but these were of no use when the sky was clouded or when no water was available. In the course of their rites, the Buddhist monks learned that incense sticks burn at a constant rate. They eventually found incense sticks to be more reliable for accurate measurement of time than incense powder. Writings about the use of incense sticks for timekeeping

[115]*Sejong sillok*, ch. 77, pp. 10ab, 38a.
[116]Needham, *SCC*, vol. 3, p. 326, plate XLVII.

are found in the *Ŭigi chipsŏl* 儀器輯說 (Collected writings on astronomical instruments) of Nam Pyŏng-ch'ŏl 南秉哲.[117]

Although there is ample evidence about the use of incense sticks for the measurement of time in the Silla and Koryo eras, we have no information on further developments with regard to the use of the incense clock in the Yi Dynasty. In China and Japan, incense clocks consisted of a tray with channels in which powder burned, winding its way through the strokes of a stylized character such as *hsi* 囍 (happiness). Many such clocks are extant today.[118]

According to the tales of old Buddhist monks, mushrooms grown on oak trees were the raw material for incense powder in Korea. They were first boiled in lye, dried in the shade, and ground into powder. The powder was packed into the stylized template of the incense tray. These accounts suggest similarities to the incense clocks formerly in use in China and Japan. In Korean Buddhist temples, the *nojŏn* 爐殿 (tray room) was usually off the *taeungjŏn* 大雄殿 (main hall). Incense sticks, often referred to as *mansuhyang* 萬壽香 (longevity incense), are said to have been burned continually throughout the Yi Dynasty era. Even now, the *mansuhyang* is in use at Buddhist temples in Korea.

THE JADE LAMP

Outdoor lanterns, called *changmyŏngdŭng* 長明燈 (eternal lamp), hang in many Buddhist temples of Korea. They are also called *oktŭngjan* 玉燈盞 (jade lamp), that is, a lamp made of white precious stone. These lanterns generally use sesame or soybean oil as fuel. The time is measured by the amount of oil consumed.

Armillary Spheres and Clocks

ARMILLARY SPHERES OF THE SEJONG ERA

The armillary sphere was the basic astronomical instrument of ancient and medieval times in East Asia. First manufactured

[117] *Ŭigi chipsŏl*, vol. 2, pp. 31 ff.
[118] Silvio A. Bedini, "Scent of Time," *Transactions of the American Philosophical Society*, 1963, *53*.5: 5 ff.

in China around the second century B.C.,[119] the instrument is believed to have been brought into Korea sometime between the Three Kingdoms era and the age of Unified Silla.[120] Instruments similar in principle to the armillary sphere seem to have continued in use during the Koryŏ Dynasty era.

However, it is not until the era of King Sejong of the Yi Dynasty that the first reliable record of the manufacture of an armillary sphere appears. The *Sejong sillok* mentions an armillary sphere made through the cooperative effort of Chŏng Ch'o 鄭招, Pak Yŏn 朴堧, and Kim Chin 金鎭 in June of the fifteenth year of King Sejong (1433).[121] In August of that year, another armillary sphere was made by Chŏng Ch'o, Yi Ch'ŏn 李蕆, Chŏng In-ji 鄭麟趾, and Kim Pin 金鑌.[122] However, the *Sillok* makes no further mention of the two armillary spheres said to have been made in that year, probably as part of a program of constructing astronomical instruments that had been started in the previous year. The program was presumably supervised and directed by Chŏng In-ji and Chŏng Ch'o on the basis of ancient literature and records then available, while the actual work of construction was in charge of Yi Ch'ŏn, Chang Yŏng-sil 蔣英實, and others. They first fabricated a wooden instrument (*mokkanŭi* 木簡儀) and measured the latitude of Hanyang (Seoul), and then according to the results of the measurement constructed the various accessory instruments, which were completed in April of 1437.[123] Among these various instruments were, according to the brief account appearing in the *Sejong sillok*,[124] several armillary spheres made of bronze according to the designs given in the records of Yuan. The armillary spheres were made of lacquered wood, and the celestial globe, covered by lacquered cloth, was round like a cannonball with a circumference of 10 *ch'ŏk*, 8 *ch'on*, 6 *pun*. Latitudes and longitudes were drawn at right angles on the surface, with the

[119]Needham, *SCC*, vol. 3, p. 340.
[120]Wada Yūji, "Keishū senseidai no setsu," pp. 144 ff.
[121]*Sejong sillok*, ch. 60, p. 38b.
[122]Ibid., ch. 61, p. 24a.
[123]*Munhŏn pigo*, ch. 2, pp. 22b–23a.
[124]*Sejong sillok*, ch. 77, pp. 9b ff.

equator crossing the ecliptic at an angle of 24 *do*. On the lacquered cloth surface were drawn the various fixed stars. The armillary sphere rotated once per solar day, with a daily recession of 1 *do* to coincide with the annual motion of the sun in the opposite sense.

For the construction of these astronomical instruments, King Sejong dispatched a number of scholars to the Ming court for the study of mathematics in 1422 and 1431.[125] The king also ordered a restudy of the various ancient classics on astronomy preserved in the Sŏun'gwan since the Sung and Yuan eras, such as Chang Heng's *Hun i* 渾儀. Such studies were assigned to Chŏng In-ji, Chŏng Ch'o, and others, and their results were incorporated by Yi Sun-ji 李純之 in his *Chega yŏksangjip* 諸家曆象集 in 1445.

The resulting astronomical clock was in the tradition of the great astronomical clock of the eleventh century, constructed by Su Sung of Sung China. While Su Sung's astronomical clock was probably for observational use, the one constructed in Korea in the Sejong era was a demonstrational armillary clock. That the armillary clock of the Sejong era was successfully fabricated after only one year's documentary research attests to the generally high standard of astronomical and academic learning of that age.

The armillary sphere of the Sejong era was one of the most outstanding astronomical instruments made throughout the Yi Dynasty period, and from this era the waterwheel-driven astronomical clock served as the standard time-measuring instrument of the Yi Dynasty. This particular armillary clock was repaired once in May of 1526[126] and remained in use until it was replaced by a new armillary clock installed in the Hongmun'gwan 弘文館 in January of 1549.[127] The latter, however, was essentially a replica of the old model. Both these armillary

[125] *Yŏllyŏsil kisul* (Narrative of Yŏllyŏsil = Yi Kŭng-ik), *pyŏlchip* (separate chapters), ch. 15, Astronomy. The scholars sent to Ming first were Yun Sa-ung, Ch'oe Ch'ŏn-hyŏng and Chang Yŏng-sil.

[126] *Chungjong sillok*, ch. 57, p. 5b.

[127] *Munhŏn pigo*, ch. 2, p. 33b.

clocks, however, were destroyed during the Hideyoshi invasion. Armillary clocks were not included in the post-invasion reconstruction of astronomical instruments.[128]

THE ARMILLARY CLOCKS OF YI MIN-CH'ŎL AND SONG I-YŎNG

The introduction of Western astronomy via the Ch'ing Dynasty in the era of King Injo 仁祖 (1623–1649), combined with the introduction of the Jesuit Shih-hsien Calendar by Kim Yuk 金堉 during the reign of King Hyojong 孝宗, served as a great stimulus for astronomical studies. There was an urgent need of new astronomical instruments to support the universal enforcement of the new calendrical system.[129] Thus, in March of 1652 (third year of Hyojong), the calculations made by Kim Sang-pŏm 金尙范, a member of the Bureau of Astronomy, on the basis of the new calendrical system, were checked against literature freshly brought over from Peking. In January of the following year the Shih-hsien Calendar was finally promulgated, and in May the clepsydras of the Poru Pavilion 報漏閣 were reconstructed to suit the new calendrical system.[130]

In view of the need of astronomical clocks for the enforcement of the new calendrical system, King Hyojong ordered Hong Ch'ŏ-yun 洪處尹 to construct a new astronomical clock. Because the armillary clock made by him was not satisfactory, the king ordered Ch'oe Yu-ji 崔攸之, sheriff of Kimje 金堤 County, to fabricate another model. This was the waterwheel-driven armillary clock completed in May of 1657, known as *sŏn'gi okhyŏng* 璇璣玉衡.[131]

Ch'oe Yu-ji's armillary clock, of which numerous copies were made, was used at a number of clepsydra bureaus. In March of 1664, a major remodeling was carried out by Yi Min-ch'ŏl and Song I-yŏng.[132] The repairs were apparently not satisfactory, for in October of 1669 Yi Min-ch'ŏl 李敏哲 and Song

[128]Jeon, "Sŏn'gi okhyŏnge tachayŏ," p. 4.
[129]Ibid., pp. 4–5.
[130]*Hyojong sillok*, ch. 8; *Munhŏn pigo*, ch. 1, pp. 5b–6b.
[131]*Hyojong sillok*, ch. 18, p. 46a; *Munhŏn pigo*, ch. 3, pp. 1b–2a.
[132]*Munhŏn pigo*, ch. 3, p. 2a.

I-yŏng 宋以穎 produced their own models.[133] While Yi Min-ch'ŏl's armillary clock was of the waterwheel-driven type, in the tradition inherited from the Sejong era, the model of Song I-yŏng was made on a smaller scale, incorporating the principles of the Western striking clock. In *Chŭngbo munhŏn pigo* appear structural descriptions of these devices.[134]

Fig. 1.17. Seoul armillary clock of Song I-yŏng and Yi Min-ch'ŏl. It was constructed in 1664–1669 and is housed in a wooden case about 120 cm × 98 cm × 52.3 cm. The armillary sphere is about 40 cm in diameter. It incorporates numerous special features of the Chinese, Arabic, and European traditions of clockmaking. Koryŏ University Museum.

[133]*Hyŏnjong sillok*, ch. 17, pp. 35ab.
[134]*Munhŏn pigo*, ch. 3, pp. 2a–3a.

The armillary clock of Song I-yŏng (see Fig. 1.17) is housed in a wooden case about 120 cm long, 98 cm high, and 52.3 cm wide. The armillary clock itself is about 40 cm in diameter while the diameter of the globe in the center is 8.9 cm. This device consists of two main components: the clock mechanism and the armillary sphere.

The clock mechanism is activated by the movement of two lead weights, one of which moves the time wheel, adjustable by a simple escapement attached to the pendulum. The hours, inscribed on a wheel rotating on a perpendicular axis resting on the disk, are observable from the window in the clock. The 12 double hours are marked on the wheel. The other plumb activates the automatic ringing device. A number of metal balls, placed atop a tube by the paddles fixed to the rotating wheel, roll down to activate the hammer which in turn strikes the main bell.[135]

The armillary sphere is attached to the clock device. The sphere itself consists of an outer nest (*yukhapŭi, liu ho i* 六合儀), middle nest (*samjinŭi, san ch'en i* 三辰儀), and an earth model which was adjusted to make precisely one rotation each day. The position of the sun is given on the ecliptic circle, which is marked with the 24 fortnightly periods of the year, 28 lunar mansions (*hsiu* 宿), and 360 degrees on both side of the band. The motion of the moon is indicated on the lunar circle, which is divided by pegs to mark the 28 mansions.[136]

The armillary clock of Song I-yŏng combined the principles incorporated in both the armillary spheres of the Sejong era (1433), which in turn reflected the tradition of the waterwheel-driven astronomical clock of Su Sung (1090), and the Western striking clock, introduced via China in 1663. In Song I-yŏng's clock, water as the motive power was replaced by lead weights, while the Chinese armillary clock was installed indoors and used for the measurement of time. Song I-yŏng's demonstrational armillary clock combined the outstanding features of East Asian astronomical clocks of that age with Western in-

[135]Jeon, "Sŏn'gi okhyŏnge taehayŏ," pp. 7–8.
[136]*Munhŏn pigo*, ch. 3, pp. 3ab, 6b–7b.

novations. He also fully exploited Arabic inventions and the original work of Chang Yŏng-sil of the Sejong era.[137]

The *sŏn'gi okhyŏng* of Song I-yŏng underwent repairs at the hand of Yi Chin-jŏng 李鎭精 in 1687.[138] The armillary clock was acquired, under unknown circumstances, by Kim Sŏng-su 金性洙, the founder of Koryŏ University of Seoul, who donated it to the university museum, where it is preserved today.

In July of 1687, King Sukchong 肅宗 caused the *sŏn'gi okhyŏng* constructed by Yi Min-ch'ŏl in the era of King Hyŏn-jong 顯宗 to be repaired. The clock was then housed in the newly built Chejŏng Pavilion 齊政閣 south of the Hŭijŏng Hall 熙政堂 within the Ch'angdŏk Palace 昌德宮 grounds in May of the following year.[139] In July of 1704, An Chung- t'ae 安重泰, Yi Si-hwa 李時華, and others made a copy slightly larger than the original model.[140]

In March of 1732 (eighth year of Yŏngjo), the copies of the armillary clock made in the era of Sukchong were remodeled by An Chung-t'ae and others to correct the errors caused by the lapse of time. The repairs completed, the machine was installed in the Kyujŏng Pavilion 揆政閣 within the Kyŏnghŭi Palace 慶熙宮.[141] The wooden plaque hung over the doorway of the pavilion, written by the hand of King Yŏngjo 英祖, is today extant in Kyŏngbok Palace.

In the *Chŭngbo munhŏn pigo* there appears a detailed description of the structure and workings of the armillary clock in the Kyujŏng Pavilion.[142] Since the original model was by the hand of Yi Min-ch'ŏl in 1669, the descriptions were similar to those for the original model. It is mentioned, among other things, that there were 24 metal balls approximately the size of a pigeon's egg. In August of 1777 (first year of Chŏngjo), the armillary clock housed in the Chejŏng Pavilion was repaired by

[137]Jeon, "Sŏn'gi okhyŏnge taehayŏ," p. 8; Needham, Wang, and Price, *Heavenly Clockwork*, pp. 161–163.
[138]*Sŏun'gwanji*, vol. 2, ch. 3.
[139]*Sukchong sillok*, ch. 19, p. 13a.
[140]Ibid., ch. 39, p. 65b.
[141]*Yŏngjo sillok*, ch. 32, p. 10a.
[142]*Munhŏn pigo*, ch. 3, pp. 6b–7b.

Yi Tŏk-sŏng 李德星, Kim Kye-t'aek 金啓宅, and others under the supervision of Sŏ Ho-su 徐浩修.[143]

Between 1760 and 1761, Na Kyŏng-jŏk 羅景績 and An Ch'ŏ-in 安處仁 collaborated with the support of Hong Tae-yong 洪大容 in the construction of a weight-driven armillary clock,[144] and an astronomical clock was made between 1809 and 1830 by Kang I-jung 姜彝重 and Kang I-o 姜彝五 in accordance with the principles incorporated in Song I-yŏng's mechanism.[145] Several other models were made, both privately and under royal commissions, for training purposes. Among the extant models are a wooden armillary clock (from the sixteenth century), reputed to have been manufactured by Yi Hwang 李滉 of the Tosan Academy 陶山書院, and a bronze armillary sphere preserved in the Ch'anggyŏng Palace.

Astronomical Instruments of the Yi Dynasty

THE KANŬI

Together with the sundial, the clepsydra, and the armillary clock, the *kanŭi* 簡儀 (*chien-i*, equatorial torquetrum) was one of the most noteworthy astronomical instruments installed in the observatories of the Yi Dynasty. In the Yi Dynasty era, an astronomical observatory was commonly referred to as a *kanŭi-dae* 簡儀臺, or *kanŭi* platform. The *kanŭi* system, perfected in 1437, was constructed in accordance with the system of Kuo Shou-ching as described in the *Yuan shih* (History of the Yuan Dynasty).[146] According to extant literature, the greater *kanŭi* was similar to the gigantic observatory instrument made by Kuo Shou-ching of Yuan in 1270,[147] while the lesser *kanŭi* was a portable version.[148] The *kanŭi* was in fact a skeletal miniature

[143]*Sŏun'gwanji*, vol. 2, ch. 3.
[144]*Tamhŏnso*, ch. 8, *Oejip*.
[145]*Oju yŏnmun changjŏn san'go* (Collected works of Oju = Yi Kyu-kyŏng), ch. 13.
[146]*Sejong sillok*, ch. 77, p. 9a; *Munhŏn pigo*, ch. 2, pp. 23ab.
[147]*Sejong sillok*, ch. 77, p. 9a; *Munhŏn pigo*, ch. 2, pp. 23ab; *Sŏun' gwanji*, vol. 2, ch. 3.
[148]See, for instance, Needham, *SCC*, vol. 3, pp. 369–372.

armillary sphere. Whereas in China the armillary sphere was developed for astronomical observation outdoors, in Korea of the Yi Dynasty the same instrument was installed indoors for use as an astronomical clock while the *kanŭi* was used chiefly for observational purposes.[149]

The *kanŭidae*, or the observational platform, usually held, besides the *kanŭi* itself, the *chŏngbangan* 正方案, a square dais carrying the azimuth circle.[150]

THE ILSŎNG CHŎNGSIŬI

The *ilsŏng chŏngsiŭi* 日星定時儀 (instrument for determining the time by the sun and stars) was a bronze instrument 2 *ch'ŏk* in diameter for measuring the solar and sidereal hours, and was made up essentially of a circular disk paralleling the equator, equipped with three circles or rings. The outermost ring, for measuring the heavenly circumference, had within it a solar ring marked with the 12 double hours and 100 quarters, the latter being subdivided into six equal parts each. Innermost was the stellar ring, marked like the solar ring.[151]

The instrument was in fact an abridged version of the *kanŭi*, from which the equatorial ring and the 100-quarter ring were detached and combined with the solar and hourly rings devised by Kuo Shou-ching. The *ilsŏng chŏngsiŭi* was also made in a portable model, which was called the *so* (small) *chŏngsiŭi* 小定時儀.

ASTROLABE AND ANGLE MEASUREMENT INSTRUMENT

In October of 1525, Yi Sun 李純 made a kind of astrolabe called *mongyun* 日輪 from the diagrams in the *Ko hsiang hsin shu* 革象新書, recently imported from China.[152] On the front and back sides of a brass disk 33.9 cm in diameter were carved the fixed stars south and north of the ecliptic. The disk was marked in the circumference and fixed with a rotating arm for observation.

[149] *Sejong sillok*, ch. 77, pp. 9b ff.; *Munhŏn pigo*, ch. 2, p. 24a.
[150] *Sejong sillok*, ch. 77, pp. 9b ff.
[151] Ibid.; *Munhŏn pigo*, ch. 2, pp. 25b–27b.
[152] *Munhŏn pigo*, ch. 2, pp. 33ab.

Among the various astronomical instruments made in the latter part of the Yi Dynasty era was the *yangdoŭi* 量度儀 (angle measurement instrument), invented by Nam Sang-gil 南相吉 in the 1850s. Trigonometry had been introduced by then to facilitate astronomical calculations, but the complicated calculations could be dispensed with if one made use of tables worked out by Mei Wen-ting 梅文鼎 (1633–1721), which, however, required a separate table for each degree of arc. Nam Sang-gil therefore proceeded to invent a new instrument to supersede the Chinese tables. Nam in 1855 published his *Yangdoŭi tosŏl* 量度儀圖說 (Illustrated description of the angle measurement instrument).[153]

According to the *Yangdoŭi tosŏl*, the angle measurement instrument (*yangdoŭi*, [shown in Fig.1.18]) consists of eight parts: a *pangp'an* 方版 (square wooden board), a *chao-taekwŏn* 子午大圈 (round meridian ring graduated in 360 degrees),

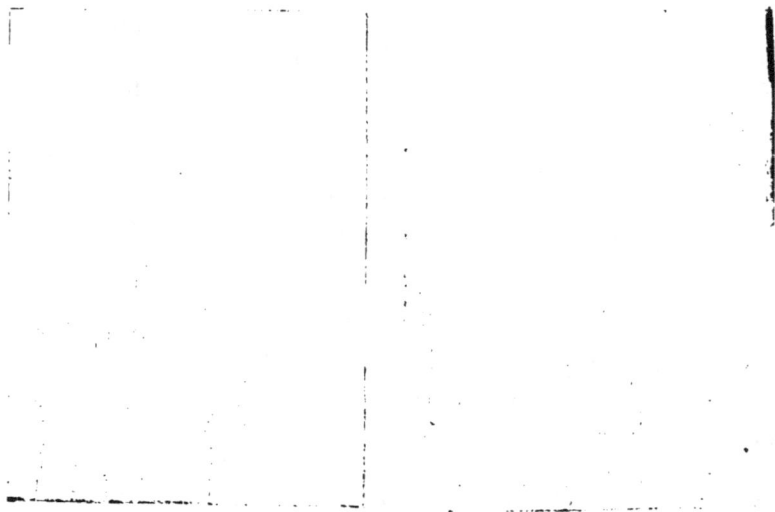

Fig. 1.18. *Yangdoŭi* (angle measurement instrument), invented by Nam Sang-gil in the 1850s. The text shown in the photograph is from his book, *Yangdoŭi tosŏl* (Illustrated description of the angle measurement instrument).

[153] *Yangdoŭi tosŏl* (Illustrated description of the angle measurement instrument), pp. la ff.

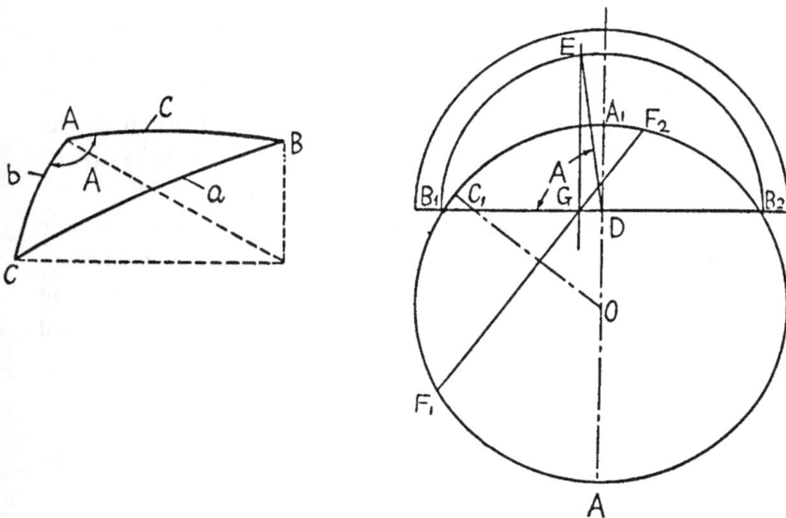

Fig. 1.19. Diagram of the principle employed by the *yangdoŭi*.

a *hwangdo-yukwŏn* 黃道游圈 (movable ecliptic ring), a *yangdo-p'an* 量度版 (semicircular protractor), a *punwisŏn* 分緯線 (line of division into equal parts), a *yanggak-chiksŏn* 量角直線 (straight ruler for measuring angles), and two *yangho-koksŏn* 量弧曲線 (curved rulers for measuring arcs).

The *yangdoŭi* was in essence a kind of slide rule for use in astronomical calculations. It worked on the following principle: Let us consider, for example, the problem of getting angle A in a spherical triangle when its three sides a, b, c are known (see Fig.1.19). Let A_1 be the upper end point of the vertical diameter of the disk *chao-taekwŏn*. A point C_1 is taken on the circumference of the disk at the distance $AC = b$ from A_1, then C_1 is joined to the center O. Points B_1 and B_2 are taken on the circle such that $A_1B_2 = A_1B_2 = c$. Then the protractor is placed on the disk so that the diameter of the protractor is on points B_1 and B_2 and is symmetric about the vertical line A_1O. Thus a concentric semicircle passing through B_1 and B_2 is determined on the protractor. Points F_1 and F_2 are taken respectively at the distance $BC = a$ from C_1, and G is the intersection of the chord F_1F_2 and the diameter of the protractor. If the straight ruler *yanggak-chiksŏn* is placed vertically on the protractor so that it passes through G, E is determined as the intersection of the straight ruler and the semicircle B_1B_2 on the protractor. Now the angle EDB_1 is the desired angle A.

In a similar way, we can get side a of a spherical triangle if the two sides b and c and the enclosed angle A are known. The cosine law and sine law are completely worked out using the same equipment, and thus the calculations of spherical trigonometry become simple.

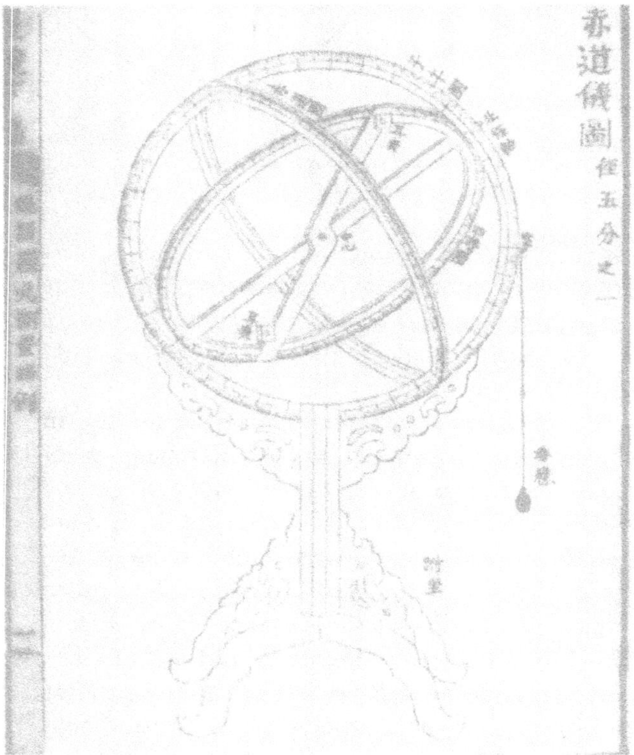

Fig. 1.20. Korean equatorial armillary sphere in the Chinese tradition, described in Nam Pyŏng-ch'ŏl's *Sŏnggyŏng*. This was one of the basic observational instruments of the Korean Astronomical Bureau in the latter Yi Dynasty.

Nam Pyŏng-ch'ŏl's *Ŭigi chipsŏl* includes descriptions of such other astronomical instruments as the *kujin ch'ŏnch'uryŏngŭi* 勾陳天樞令儀 and *yanggyŏng kyuirŭi* 兩景揆日儀, both of which appeared in the latter part of the Yi Dynasty,[154] but their actual designs as well as their specific uses remain unknown today (Fig.1.20).

TELESCOPES

As far back as the age of Silla, Koreans used a spherical burning-lens called *hwaju* 火珠 (*huo chu,* fire-pearl) which, however, had no image-forming applications. There is well-corroborated evidence to show that the Western telescope was brought into Korea in the middle of the Yi Dynasty period by way of China. The first record is for July of 1631 when Chŏng Tu-wŏn 鄭斗源 is said to have brought in from Ming China the *ch'ŏlligyŏng* 千里鏡 (far-seeing optical instrument) as well as a "Western gun" and *chamyŏngjong* 自鳴鐘 (striking clock).[155] There is no mention of whether the telescope was ever used for astronomical observation. It appears certain, at any event, that no attempt was ever made throughout the Yi Dynasty era to make telescopes or to use the imported instruments for astronomical observing in Korea. In July 1744, the astronomer Kim T'ae-sŏ 金兌瑞 is said to have imported from Ch'ing China a "giant" *ch'ŏlligyŏng*.[156] Throughout the *Yijo sillok,* there appear only four mentions of the *ch'ŏlligyŏng,* or telescope. Nam Pyŏng-ch'ŏl's *Ŭigi chipsŏl* contains descriptions of ten kinds of astronomical instruments used in the later Yi Dynasty period; the telescope is not among them.

Calendar and Time Measurement

CALENDAR

The development of calendrical science in Korea had its start with the introduction of the Chinese art in the Three

[154] *Ŭigi chipsŏl,* vol. 2.
[155] *Munhŏn pigo,* ch. 1, pp. 4b–5a.
[156] *Yŏngjo sillok,* ch. 59, p. 38a.

Kingdoms period. Paekche is said to have introduced the Chinese Yuan-chia calendar 元嘉曆 in the fifth century.[157] In 554 Paekche was asked by Japan to send Kodŏk Wangson 固德王孫, doctor of calendrical sciences, as an advisor, and in 602 Kwallŭk 觀勒 carried books on the calendar and astrological literature to Japan. Japan began the use of the Yuan-chia calendar in 604. It is clear that actual use of the Yuan-chia calendar in Paekche preceded that date.[158]

In 624 Koguryŏ sent to T'ang for calendrical books and probably obtained the Wu-yin ephemerides 戊寅曆 in that year. Silla in 674 adopted the Lin-te calendrical system 麟德曆 of China, according to records appearing in the *Samguk sagi* 三國史記.[159]

Between the late eighth century and the early ninth century, Silla substituted the Hsuan-ming calendar 宣明曆, which was subsequently inherited by the Koryŏ Dynasty.[160] Koryŏ continued the use of the obsolete Hsuan-ming calendar in spite of its errors.[161]

Finally, in March of 1062 Koryŏ compiled a number of new calendars, including the Sipchŏng calendar 十精曆 of Kim Sŏng-t'aek 金成澤, the Ch'iryo calendar 七曜曆 of Yi In-hyŏn 李仁顯, the Kyŏnhaeng calendar 見行曆 of Han Wi-haeng, 韓爲行 the Tun'gap calendar 遁甲曆 of Yang Wŏn-ho 梁元虎, and the T'aeil calendar 太一曆 of Kim Chŏng 金正,[162] but none of these was an original attempt.

Koryŏ's efforts to reduce somewhat the resulting errors produced the calendar of 1218, the Sinch'an calendar 新撰曆 of Kim Tŏng-myŏng 金德明, the *Koryŏ Sasŏngyosŏ* 高麗師星曜書, and the *Koryŏ Illyŏk* 高麗日曆.[163]

In 1281 Kublai Khan of Yuan sent to Koryŏ, his vassal state, Kuo Shou-ching's Shou-shih calendar, which had been

[157]Hong, *Chosŏn kwahaksa*, p. 54.
[158]Nakayama, *Japanese Astronomy*, pp. 9, 69.
[159]*Samguk sagi*, ch. 29, pp. 12a–15a; *Munhŏn pigo*, ch. 1, pp. 1b–2a.
[160]*Munhŏn pigo*, ch. 1, pp. 2a–3a.
[161]Ibid.
[162]*Koryŏsa*, ch. 7, p. 1a; Rufus, "Astronomy in Korea," p. 19.
[163]Hong, *Chosŏn kwahaksa*, pp. 102–103.

invented in the previous year.[164] To prepare for its adoption, Koryŏ sent Ch'oe Sŏng-ji 崔誠之 to the Yuan court, and in 1298 (twenty-fourth year of Ch'ungyŏlwang), Crown Prince Ch'ungsŏn 忠宣 was sent to the Yuan court to study the Shou-shih calendar personally. The calendrical sciences were then taught to Kang Po 姜保, who wrote a book of calendrical calculations called *Susiryŏk ch'ŏppŏp ipsŏng* 授時曆捷法立成 (A ready reckoner for Shou-shih Calendar calculations), which appeared in 1343.[165] The *Susiryŏk ch'ŏppŏp ipsŏng* is a mathematical table containing values worked out at certain intervals. It was designed for expediting repeated calculations or the vast work of the continuous addition, subtraction, multiplication, and division of certain numbers in preparing the Shou-shih calendar. This table, which Kang calculated himself, is an invaluable source for studying the theory and techniques of Korean and Chinese calendrical science.[166] It was not until the reign of King Ch'ungsŏn (1309–1313) that the Shou-shih calendar was finally put into effect in Koryŏ, but because the Koreans' knowledge of the calendrical calculations was deficient (calculations were made in part according to the obsolete Hsuan-ming calendar) errors and miscalculations resulted.[167]

After the Ming adopted Liu Chi's Ta-t'ung calendar 大統曆, King Kongmin of Koryŏ, in the nineteenth year of his reign (1370), sent Sŏng Chun-dŭk 成准得 to Ming for a copy of the Ta-t'ung calendar, which was adopted by the Koryŏ court. However, since the Ta-t'ung calendar was but a revised version of the Shou-shih calendar, it contained errors inherent in the original Shou-shih calendar as well as others which arose from calculations made in the Koryŏ capital.

It was not until the reign of Sejong of the Yi Dynasty that the first successful attempt at local adaptation of the Chinese

[164] *Koryŏsa*, ch. 29, pp. 3b, 31b.
[165] Ibid., ch. 52, pp. 17b; ch. 108, p. 12b.
[166] *Susiryŏk ch'ŏppŏp ipsŏng* (A simplified method for Shou-shih calendar calculation), pp. 1a ff.
[167] *Munhŏn pigo*, ch. 1, pp. 2a–3a.

calendars was made.[168] By the order of King Sejong, corrections were made in February of 1423 in the Hsuan-ming calendar of the T'ang, the Shou-shih calendar of the Yuan, and other calendars of Chinese origin.[169] For the first time, calculations were successfully made for eclipses and planetary phenomena. In the fifteenth year of his reign, King Sejong ordered the *Ch'ilchŏngsan* 七政算 (Calculation of the motions of the seven governors) to be compiled by his scholars, who included Chŏng In-ji 鄭麟趾, Chŏng Ch'o 鄭招, Chŏng Hŭm-ji 鄭欽之, Kim Tam 金淡, and Yi Sun-ji 李純之.[170] Completed in 1442 and published two years later, the book consisted of an Inner Part (*Naep'yŏn* 內篇) comprising two chapters in two volumes and an Outer Part (*Oep'yŏn* 外篇) of three chapters in five volumes. The Inner Part corrects errors in both the Shou-shih and Ta-t'ung calendars, while the Outer Part treats of the Arabic calendars (see Fig. 1.21).

The characteristic of this calendrical system is that it was prepared for the latitude of Seoul on the basis of observations made there. The calendar (Inner Part) takes the mean length of a year as 365.2425 days and that of a month as 29.530593 days, from the Shou-shih calendar. It further shows very precise differences between the mean values of a year or of a month and the daily true values.

In describing the parallactic movements of the sun, moon, and other planets, efforts were made to deal empirically with the equation of center and angular acceleration. In the theory of eclipses, Chinese values for north-south and east-west differences in parallax were adopted. Thus the Yi Dynasty astronomers in the first half of the fifteenth century not only had a thorough grasp of basic Chinese principles and theories of the calculation of calendars, but also made good use of their instruments.

[168]Ibid.

[169]*Sejong sillok*, ch. 19, p. 13b.

[170]*Ch'ilchŏngsan Naep'yŏn* (Inner Part of the Calculation of the motions of the seven governors), vol. 1, p. 1a. The Seven Governors are the sun, moon, and five classical planets.

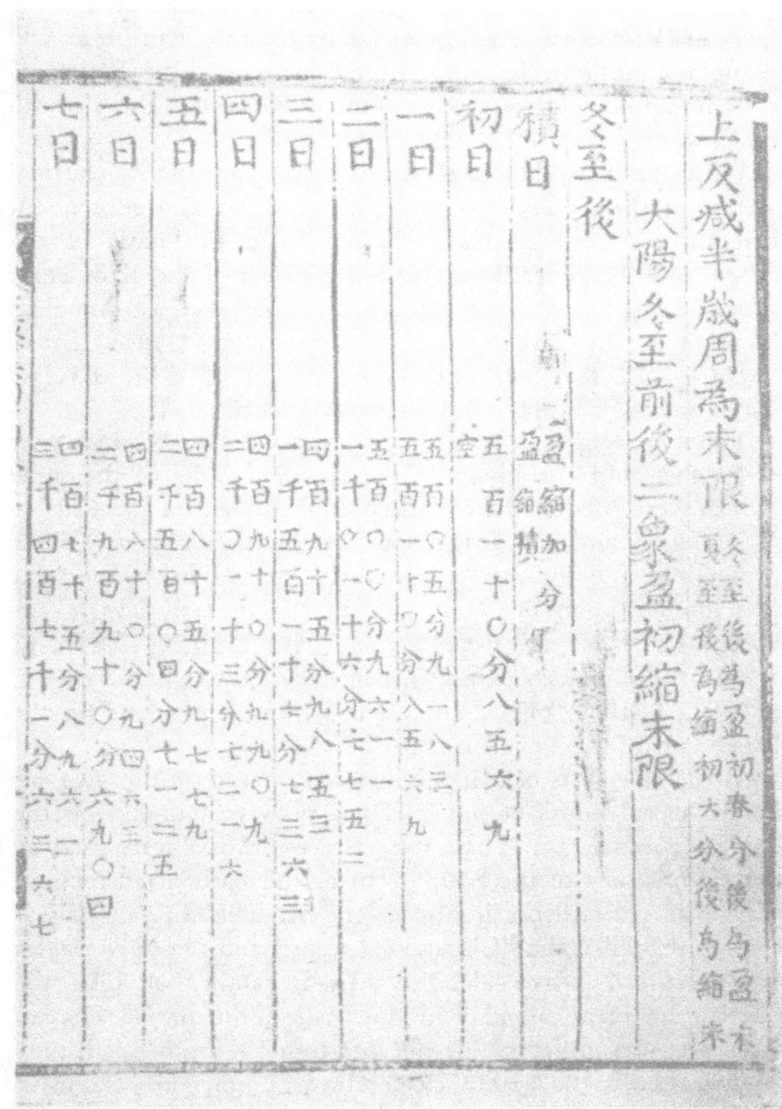

上反減半歲周為末限

大陽冬至前後二象盈初縮末限

冬至後

積日

初日

一日

二日

三日

四日

五日

六日

七日

Fig. 1.21. A page from the *Ch'ilchŏngsan naep'yŏn*, one of the basic calendrical treatises of the Yi Dynasty, completed in 1442 and published two years later.

It is especially notable that the Outer Part uses an Islamic form of the theory of epicycles for its treatment of lunar and planetary motions. In other words, aspects of the Ptolemaic system had been introduced into the astronomy of the fifteenth century.

With the compilation of the *Ch'ilchŏngsan,* the calendrical system of the Yi Dynasty was put in order, and subsequent calendrical computations were made on the basis of the chapters of the Inner Part, chiefly by Yi Sun-ji and Kim Tam, who compiled from these calculations a series of astronomical handbooks, the *Ta-t'ung li il t'onggwe* 大統曆日通軌, *Kyosik t'onggwe* 交食通軌, *Wŏlsŏng orŭngbŏm* 月星五陵犯, *Chungsu Ta Ming li* 重修大明曆, and *Kyŏngo wŏllyŏk* 庚午月曆.

The solar eclipse in early August of 1447 (twenty-ninth year of Sejong) and the lunar eclipse in the middle of the same month were predicted from calculations based on the *Chungsu Ta Ming li* and the Inner Part of the *Ch'ilchŏngsan.* These predictions and the results of actual observations of the solar and lunar eclipses are contained in the reports *Chungsu taemyŏngyŏk chŏngmyo ilsik wŏlsik karyŏng* 重修大明曆丁卯日食月食假令 and the *Ch'ilchŏngsan naep'yŏn chŏngmyonyŏn kyosik karyŏng* 七政算內篇丁卯年交食假令, the last of which helps make up. the Inner Part of the *Ch'ilchŏngsan.*

As the calendars of Ming China were based on the *Ta-t'ung li,* so the calendars of the Yi Dynasty were founded on the *Ch'ilchŏngsan naep'yŏn.* These two reference works were used by the astronomers of the Sŏun'gwan as the basis of corrections whenever errors arose in the calendars. The *Ta-t'ung li* was commonly called the "Chinese calendar" and the *Ch'ilchŏngsan naep'yŏn* the "native calendar." In December of 1504 the publication of a calendar in the *han'gŭl,* the native Korean alphabet, was projected for the first time.[171] From this it may be inferred that the use of the calendar was sufficiently extensive in Korea at that time to justify the publication of a vernacular version.

[171] *Yŏnsan'gun ilgi,* ch. 56, p. 27a.

The astronomy and mathematics of the Yi Dynasty, however, gradually declined after the reign of King Sŏngjong 成宗, until a rapid deterioration set in following the disastrous Hideyoshi invasion in the reign of King Sŏnjo 宣祖. Whenever discrepancies arose between the Chinese calendar used by the troops of Ming who aided Korea during the Japanese invasion and Korea's own calendars, the former prevailed. Inevitably, the Ta-t'ung, or the "Chinese" calendar, obtained a position of superiority over the native calendar, obstructing the development of a calendar based on local observations and calculations. Even before this confusion was completely dissolved, the Manchu dynasty of Ch'ing took over in China and proclaimed the Shih-hsien calendar,[172] which was accepted by the Yi Dynasty court after its capitulation to Ch'ing.

In 1644 Kim Yuk 金堉 on a trip to Yenching (Peking), heard of the compilation of the Shih-hsien calendrical system by T'ang Jo-wang (Adam Schall von Bell). Upon his return home, Kim made a strong report to Chief Astronomer Kim Sang-pŏm 金尙范 and others in favor of converting to the new Chinese calendar because the calendrical system of Kuo Shou-ching, after a lapse of 365 years, had developed a large number of errors and discrepancies.[173]

During his visit to Yenching, Kim Yuk had taken along two calendrical astronomers to have them study with Adam Schall, but their entry was barred. Therefore Kim Yuk merely bought a copy of the Shih-hsien calendar, but when, upon his return to Seoul, Kim Sang-pŏm and others studied it, they achieved only imperfect understanding. In 1651, Kim Sang-pŏm was again dispatched to China and, with the aid of well-placed bribes, succeeded in obtaining knowledge of the new calendar system.[174] Unfortunately, however, the mission failed to obtain information on calculations about the movements of the planets according to the new calendar, and Kim

172 Nakayama, *Japanese Astronomy*, p. 166.
173 *Munhŏn pigo*, ch. 1, pp. 5ab.
174 Ibid., ch. 1, pp. 5a–6b.

Sang-pŏm, who was again sent to China in the following year to make up this deficiency, died on the trip.[175]

In 1705 Court Astronomer Hŏ Wŏn 許遠 was sent to the Ch'ing court to learn the Shih-hsien calendar, and in 1708 the Shih-hsien system in its entirety, including planetary calculations, was put into use in Korea. In 1710 Hŏ Wŏn issued his *Sech'o yuhwi* 細草類彙, a guide to details of the computational techniques learned from the Ch'ing court astronomers.

The new calendar, while accurate to the minute as to the seasons and lunations, failed to predict accurately the time of solar and lunar eclipses. To resolve the problem, the Yi Dynasty court again sent Hŏ Wŏn to the Ch'ing court.[176] He returned from Yenching in April of 1715, bringing with him calendrical books, measuring instruments, and clocks, among other things.

According to the *Chŭngbo munhŏn pigo,* use of the "newly revised" Shih-hsien calendar 新修時憲七政法 was started in 1725, chiefly on the basis of the *Li hsiang k'ao ch'eng* 曆象考成, compiled by Mei Ku-ch'eng 梅穀成 and others and published in 1723.[177] With this, the Korean astronomers could reach a reasonable degree of accuracy in calculations on solar and lunar eclipses and planetary movements for the first time. Efforts to compile increasingly accurate calendars were continued, and Korean astronomers were dispatched to the Ch'ing court again and again to learn new developments and to import new calendrical literature in 1730 and 1735.[178] Later, beginning in 1738, Tai Chin-hsien 戴進賢 (Ignatius Kögler), Hsu Mou-te 徐懋德 (Andrew Pereira), and others in China revised the *Li hsiang k'ao ch'eng* on the basis of new observations by Cassini and Flamsteed. Calculations concerning the solar and lunar movements were made according to Cassini's methods and those concerning planetary motions by the methods devised by

[175]Ibid., ch. 1, p. 6a.
[176]Ibid., ch. 1, pp. 6b–7a.
[177]Ibid., ch. 1, pp. 7ab.
[178]Ibid., ch. 1, pp. 7b–8a.

Mei Ku-ch'eng.[179] After the 1742 publication of the *Li hsiang k'ao ch'eng hou pien,* the new techniques contained therein were orally transmitted by Kögler and Pereira with the aid of translators to the Korean astronomer An Kuk-pin 安國賓, who went to Yenching in 1741.[180] In 1744, another Korean astronomer, Kim T'ae-sŏ 金兌瑞, learned the new calculations from Kögler, and in 1745 Kim and the translator An Myŏng-sŏl 安命說 took to Korea a copy of the revised edition (*hsin fa* 新法) of the *Li hsiang k'ao ch'eng.* From then on, it was possible for the court astronomers of the Yi Dynasty to calculate solar movements according to Kepler's theory of elliptic motion.[181] However, imperfections persisted, so that it was impossible to enter with accuracy in the calendars the hours of sunrise and sunset, the precise change of seasons, and so on.

The use of a combination of techniques from the Shih-hsien calendar and the *Ch'ilchŏng paekchung* 七政百中曆, first compiled and made public in 1772, was discontinued in 1781 as a result of the publication in the previous year of the Paek-chung calendar 百中曆, which combined elements of the Ta-t'ung and Shih-hsien calendars. This new calendar formed the basis for the compilation in the following year (1782) of the Ch'ŏnse (Thousand-year) calendar 千歲曆, covering the 100 years from the first year of King Chŏngjo 正祖 (1777).[182] At the end of each decade, the Ch'ŏnse calendar added a new decade at the other end so as to perpetuate itself. This seems to indicate the solution of several problems unsolved since the reign of Yŏngjo 英祖.[183] Thus, in 1783, the hours of inception of the seasons were accurately determined by the use of Seoul's "latitude" (polar height) of 37. 3915 *do*, and in 1792 the precise latitudes of the principal towns and countries in the eight provinces were determined.[184] The astronomers of the Kwansanggam (Bureau of Astronomy) published the *Ch'ilchŏng*

179Ibid., ch. 1, pp. 8a–9a.
180Ibid., ch. 1, 9a.
181Ibid., ch. 1, p. 9b.
182Ibid.
183*Ch'ŏnseryŏk,* pp. la ff.
184*Munhŏn pigo,* ch. 1, pp. 10ab.

sech'o 七政細草, describing the techniques of computing motions of the *ch'ilchŏng*—the "seven governing celestial bodies," that is, the sun, moon, and Five Planets.

In November 1894 it was decided to use the Western solar calendar,[185] with occasional references to the conventional Shih-hsien calendar. Thus, while the solar calendar was official, the Shih-hsien was the basis for observation of the change of seasons, births, funerals, and other rites and events. In 1904, the Ch'ŏnse calendar was revised into a Manse (Ten-thousand-year) calendar 萬歲曆.[186]

Nam Pyŏng-gil 南秉吉 wrote the *Sihŏn kiyo* 時憲紀要 (published in 1860), a textbook for teaching the methods of calculation for operation of the Shih-hsien system. Part 1 contained examples of calculations on the ecliptic and the equator, the latitude and the meridian, fundamentals of the calendar, the sidereal year, and the solar, lunar, and planetary motions. Part 2 dealt chiefly with solar and lunar eclipses. The *Sihŏn kiyo* was the only Yi Dynasty textbook on calendrical and astronomical calculations until a modern astronomy textbook was translated by Chŏng Yŏng-t'aek 鄭永澤 in 1908.

Lastly, we must take note of the *Myŏngnamnu munjip* 明南樓文集 (Collected works of Ch'oe Han-gi), in several hundred chapters, and *Chigu chŏnyo* 地球典要 (A handbook of world geography), in 13 chapters, by Ch'oe Han-gi 崔漢綺 (1803–1879) who greatly contributed to the transplanting to Korea of Western astronomy and geography toward the close of the Yi Dynasty. Ch'oe Han-gi was influenced by the Western cosmology of the Jesuit scholars in the Ch'ing court and developed his own theory of the Four Elements (*Sawŏn* 四元).

SYSTEM OF HOURS

The system of hours (*sije* 時制) was an integral part of every calendar system. Since the beginning of mathematical astronomy, the system of hours in Korea has been subject to the same changes and modifications that took place in China itself.

[185]Ibid., ch. 1, pp. 11a–12b.
[186]Ibid., ch. 1, p. 12b.

The most basic of these changes was the replacement in 1653 (fourth year of Hyojong) of the system of 100 divisions of the day (*kak* 刻) with one of 96 quarter-hours.[187] Under both systems, the day was divided into 12 double hours, named Rat, Ox, Tiger, Hare, Dragon, Serpent, Horse, Sheep, Ape, Cock, Dog, and Swine.

Each double hour was equally divided into *ch'o* 初 (first half) and *chŏng* 正 (latter half) and again subdivided into *kak* (quarters), each of which had 60 *pun* 分. Under the 100-*kak* system, one double hour had 8 *kak* and 20 *pun*, while each half of the double hour was divided into 4 *kak* and 10 *pun*. The fraction of 10 *pun* was assigned to the first part of both *ch'o* and *chŏng;* therefore, for both *ch'o* and *chŏng*, the first *kak* was made up of 10 *pun*, followed by four regular *kak* of 60 *pun* each. Each double hour, thus, had a total of 500 *pun*. Midnight, therefore, did not fall at the beginning of the hour of Rat but halfway through it; likewise, noon fell at the start of *chŏng* of the hour of Horse.

In conjunction with the system of 12 double hours and 100 *kak* per day, another system dividing the night into five watches (*kyŏng* 更) was in force. Of necessity, the latter system kept seasonal hours 不定時法, starting at dusk 昏, that is, 2.5 *kak* after sunset and ending at dawn 旦, or 2.5 *kak* before sunrise the next morning.[188] Each *kyŏng* was subdivided into five *chŏm* 點, so that each night was made up of five *kyŏng* and 25 *chŏm*. Since the time of sunset and sunrise is subject to seasonal variation, the length of each *kyŏng* and *chŏm* varied according to the changes of season. According to the Shou-shih calendar, the length of the day at the summer solstice was 62 *kak*, while at the winter solstice, day was made up of 38 *kak* and night of 62 *kak*. At the summer solstice, the night of 38 *kak* was divided into five equal parts so that 7.6 *kak* made one *kyŏng*. On the other hand, at the winter solstice, the length of one *kyŏng* was as great as 12.4 *kak*, or 1.6 times the length at the summer sol-

[187] Ibid., ch. 3, p. 11b.

[188] *Nusu t'ongŭi* (Manual for the operation of the clepsydra), pp. 1a ff; *Munhŏn pigo*, ch. 3, pp. 11b–16a.

stice.[189] The system of temporal night hours was not as inconvenient as it seems, since 1 *kyŏng,* 1 *chŏm* was always 2.5 *kak* after sunset, and 5 *kyŏng,* 5 *chŏm* was always 2.5 *kak* before sunrise. The clepsydra, which was the standard instrument for measurement of time at night, usually employed a number of indicator rods marking the lengths of the *kyŏng* and the *chŏm.*[190]

The basically Western system of 96 *kak* a day, brought into use simultaneously with the Shih-hsien calendar, assigned 8 *kak* of 15 *pun* to each double hour, while the double hour was also divided as before into two equal parts of *ch'o* and *chŏng.*[191] One *kak* corresponded to 15 minutes.

The system of night watches did not change much. While the 100-*kak* system put *hon* at 2.5 *kak* after sunset and *tan* at 2.5 *kak* before sunrise, the 96-*kak* system set *hon* and *tan* at the hours when the sun was 18 *tu* below the horizon. This, therefore, was subject to seasonal variation; the shortest time, at the vernal and autumnal equinox, was 6 *kak* and 1 *pun.* At both the vernal and autumnal equinox, the lengths of day and night are equal at 48 *kak,* but at the winter solstice, day is 37 *kak* and 9 *pun* and night 58 *kak* and 6 *pun,* while at the summer solstice, these lengths are reversed.[192]

The *Nusu t'ongŭi* 漏籌通儀 (Manual for operation of the clepsydra) contains illustrated instructions for the use of the indicator rods of the clepsydra (see Fig.1.22). The book, compiled by An Kuk-pin and others in 1754 (thirtieth year of Yŏngjo), states that 24 such indicator rods were used. In the edition compiled by Kim Yŏng 金泳, the use of the rods is detailed as in the following examples (names of the fortnightly periods may be found in Table 1.1):

First rod: Used first day of winter solstice to day before the

[189]See *Nusu t'ongŭi* and *Sinpŏp nusu t'ongŭi* (Manual for the operation of the clepsydra by a new system).

[190]*Kukcho yŏksanggo,* vol. 2, ch. 4, Clepsydra; *Munhŏn pigo,* ch. 3, p. 11b.

[191]*Nusu t'ongŭi*; *Sinpŏp nusu t'ongŭi.*

[192]Ibid.; Hashimoto Manpei, "Taiyō no Shutsunyū ni mirareru Nihon chūsei no jikoku seido" (On the Japanese medieval time system), *Kagakushi kenkyū,* 1955, *36*: 8–12.

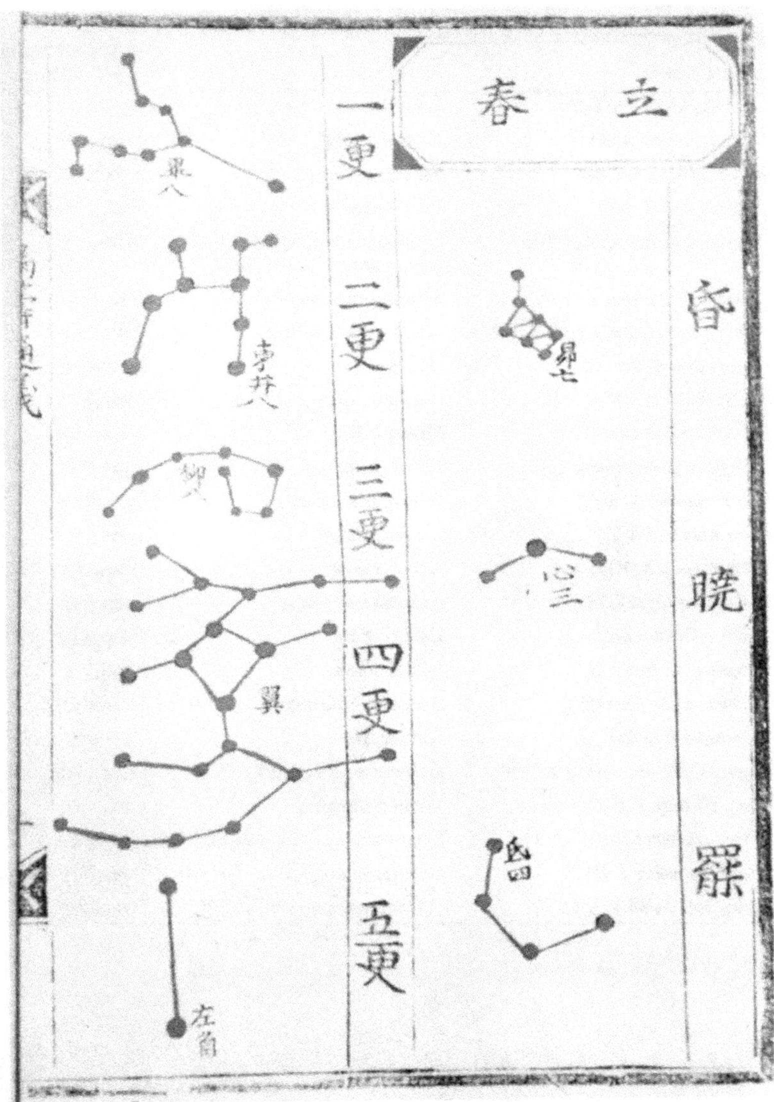

Fig. 1.22. A page from the *Nusu t'onggi* (Manual for operation of the clepsydra), published in 1754. The page shows the position of the stars for standardizing the hours on the day designated Beginning of Spring.

Table 1.1 The Twenty-Four Fortnightly Periods (*chŏlgi, chieh ch'i*)

Solar Term	Translation	Approximate Solar Date
Sohan (*hsiao-han* 小寒)	Lesser cold	Jan. 5
Taehan (*ta-han* 大寒)	Greater cold	Jan. 20
Ipch'un (*li-ch'un* 立春)	Spring begins	Feb. 4
Usu (*yü-shui* 雨水)	Rainwater	Feb. 19
Kyŏngch'ip (*ching-chih* 驚蟄)	Awakening of hibernating creatures	Mar. 6
Ch'unbun (*ch'un-fen* 春分)	Spring equinox	Mar. 21
Ch'ŏngmyŏng (*ch'ing-ming* 清明)	Clear and bright	Apr. 5
Kogu (*ku-yü* 穀雨)	Grain rains	Apr. 20
Ipha (*li-hsia* 立夏)	Summer begins	May 6
Soman (*hsiao-man* 小滿)	Grain fills	May 21
Mangjong (*mang-chung* 芒種)	Grain in ear	June 6
Haji (*hsia-chih* 夏至)	Summer solstice	June 21
Sosŏ (*hsiao shu* 小暑)	Lesser heat	July 7
Taesŏ (*ta-shu* 大暑)	Greater heat	July 23
Ipch'u (*li-ch'in* 立秋)	Autumn begins	Aug. 8
Ch'ŏsŏ (*ch'u-shu* 處暑)	End of heat	Aug. 23
Paengno (*pai-lu* 白露)	White dews	Sept. 8
Ch'ubun (*ch'iu-fen* 秋分)	Autumn equinox	Sept. 23
Hallo (*han-lu* 寒露)	Cold dews	Oct. 8
Sanggang (*shuang-chiang* 霜降)	Descent of hoarfrost	Oct. 23
Iptong (*li-tung* 立冬)	Winter begins	Nov. 7
Sosŏl (*hsia-hsueh* 小雪)	Lesser snow	Nov. 22
Taesŏl (*ta-hsueh* 大雪)	Greater snow	Dec. 7
Tongji (*tung-chih* 冬至)	Winter solstice	Dec. 22

fortnight Lesser Cold; day after the fortnight Greater Snow to day before winter solstice

1 *kyŏng*, 1 *chŏm:* Cock *chŏng, ch'ogak* (first quarter hour)
2 *kyŏng*, 3 *chŏm:* Swine *ch'o*, 1 *kak*
3 *kyŏng*, 2 *chŏm:* Rat *ch'o, ch'ogak*
4 *kyŏng*, 1 *chŏm:* Ox *ch'o, ch'ogak*
5 *kyŏng*, 1 *chŏm:* Tiger *ch'o*, 1 *kak*

Second rod: First day of Lesser Cold to 11 days after Lesser Cold; 4 days after Lesser Cold to first day of Greater Snow

1 *kyŏng*, 1 *chŏm:* Cock *chŏng*, 1 *kak*
2 *kyŏng*, 3 *chŏm:* Swine *ch'o*, 1 *kak*
3 *kyŏng*, 2 *chŏm:* Rat *ch'o, ch'ogak*
4 *kyŏng*, 1 *chŏm:* Ox *ch'o, ch'ogak*
5 *kyŏng*, 1 *chŏm:* Tiger *ch'o*, 1 *kak*

Twenty-fourth rod: First day of Grain in Ear to day before summer solstice; first day of summer solstice to first day of Lesser Heat

1 *kyŏng*, 1 *chŏm:* Dog *chŏng*, 3 *kak*
2 *kyŏng*, 5 *chŏm:* Rat *ch'o, ch'ogak*
3 *kyŏng*, 1 *chŏm:* Rat *ch'o*, 1 *kak*
4 *kyŏng*, 3 *chŏm:* Ox *ch'o, ch'ogak*
5 *kyŏng*, 1 *chŏm:* Ox *ch'o*, 3 *kak*

Another detailed description of this sytem of *kyŏng-chŏm* 更點 and *si-kak* 時刻 is given in the *Chŭngbo munhŏn pigo.* During the reign of King Chŏngjo, a revised edition of the *Nusu t'ongŭi* was briefly employed, which divided each of the 24 *ki* 氣 (fortnights) into 72 *hu* 候 and used 37 indicator rods. According to this system:

First rod: First *hu* of winter solstice

1 *kyŏng*, 1 *chŏm:* Cock *chŏng*, 2 *kak*, 12 *pun*
2 *chŏm:* Dog *ch'o, ch'ogak*, 7 *pun*

3 *chŏm:* Dog *ch'o,* 2 *kak,* 2 *pun*
4 *chŏm:* Dog *ch'o,* 3 *kak,* 11 *pun*
5 *chŏm:* Dog *chŏng,* 1 *kak,* 6 *pun*
2 *kyŏng,* 1 *chŏm:* Dog *chŏng,* 3 *kak,* 1 *pun*
2 *chŏm:* Swine *ch'o, ch'ogak,* 11 *pun*
4 *chŏm:* Swine *chŏng, ch'ogak*
3 *kyŏng,* 1 *chŏm:* Swine *chŏng,* 3 *kak,* 5 *pun*
2 *chŏm:* Rat *ch'o,* 1 *kak*
5 *chŏm:* Rat *chŏng,* 2 *kak*
4 *kyŏng,* 1 *chŏm:* Rat *chŏng,* 3 *kak,* 10 *pun*
3 *chŏm:* Ox *ch'o,* 3 *kak*
5 *chŏm:* Ox *chŏng,* 2 *kak,* 4 *pun*
5 *kyŏng,* 1 *chŏm:* Ox *chŏng,* 3 *kak,* 14 *pun*
2 *chŏm:* Tiger *ch'o,* 1 *kak,* 9 *pun*
5 *chŏm:* Tiger *chŏng,* 2 *kak,* 8 *pun*

Second rod: Second *hu* of winter solstice, third *hu* of Greater Snow

1 *kyŏng,* 1 *chŏm:* Cock *chŏng,* 2 *kak,* 12 *pun*
2 *chŏm:* Dog *ch'o, ch'ogak,* 7 *pun*
3 *chŏm:* Dog *ch'o,* 2 *kak,* 2 *pun*

The hours of sunset and sunrise from the winter solstice to the summer solstice in Seoul were as follows (cf. *Chŭngbo munhŏn pigo,* ch. 2, "Sangwigo" (chapter on astronomy and meteorology).[193]
Winter solstice: Sunrise at Serpent *ch'o,* 1 *kak,* 3 *pun,* sunset at Ape *chŏng,* 2 *kak,* 12 *pun*
Vernal equinox: Sunrise at Hare *chŏng, ch'ogak,* sunset at Cock *chŏng, ch'ogak*
Summer solstice: Sunrise at Tiger *chŏng,* 2 *kak,* 12 *pun,* sunset at Dog *ch'o,* 1 *kak,* 3 *pun.*
The method of ringing the bell to tell the hours at night varied in different ages, but the system established in 1469 (first year of Yejong) was generally followed throughout the Yi Dynasty. According to this system, the beginning of the night

[193]*Munhŏn pigo,* ch. 2, pp. 11a–13a.

was signaled by the ringing of the great bell at 1 *kyŏng,* 3 *chŏm.* The bell was rung 28 times, symbolizing the number of lunar mansions. After that, the hours were told by the use of the drum and the gong, the former telling the number of *kyŏng* and the latter the number of *chŏm.* Thus, at 1 *kyŏng,* 4 *chŏm,* the striking of the drum once was followed by four strikes of the gong, the whole process being repeated five times. The end of the night was signaled by the ringing of the great bell 33 times according to the number of the Indra Heaven.[194] Beginning in 1884, a gun was fired at the Kŭmch'ŏn Bridge 禁川橋 both at noon and at the start and close of the nightly curfew.[195] Finally, the length of time it takes for the sun to travel the distance between the horizon and the point 18 *do* below it, both before sunrise and after sunset, was measured at Seoul in different seasons.[196]

Observation of Celestial Phenomena

SOLAR AND LUNAR ECLIPSES

The succeeding dynasties of Korea were extremely sensitive to the occurrence of meteorological and astronomical phenomena, especially solar and lunar eclipses. The prediction of solar and lunar eclipses was a part of the monarchic ritual which augmented the dignity of the rulers in the eyes of the common people. Many, indeed, are the royal astronomers and meteorologists who lost their positions—or even lives—because of failure to predict such occurrences correctly.

The first Korean records of solar eclipses occur in the *Samguk sagi* 三國史記. According to this work, 29 eclipses are recorded throughout the Silla kingdom period of 965 years (April 54 B.C.– January 911), 11 eclipses for Koguryŏ between March 114 and December 559 (445 years), and 26 for Paekche between July 19, 14 B.C. and July 592 (606 years). Many scholars, however, doubt the independence of the records for the early

[194]Ibid., ch. 3, pp. 18ab.
[195]Ibid., ch. 3, p. 18b.
[196]Ibid., ch. 2, pp. 13a–14a.

part of the Silla era and regard them as merely copies of records in earlier Chinese literature.[197]

The author of the *Samguk sagi* made such records of solar eclipses during the Silla era because of their political as well as their astronomical significance. It is immediately apparent from comparing the records of solar eclipses for the three kingdoms that the number on record for the kingdom of Koguryŏ is disproportionately small and that for Silla disproportionately large. The first date of observation in Koguryŏ, March of 114 (sixty-second year of T'aejo), is some 168 years after the first record for Silla and 128 years after the first record for Paekche.

Of far greater importance are the records for the 475 years of the Koryŏ Kingdom which appear in the Treatise on Astrology of the *Koryŏsa* 高麗史. The astronomical science of Koryŏ, inheriting the tradition of Silla, has left a total of 132 records of observations (including 13 occasions on which it is said actual observations were prevented by bad weather). As a whole, this compares with the records left by the Arab astronomers of the Middle Ages. However, the astronomers of Koryŏ were unfortunately unable to make precise predictions of solar and lunar motions. There are six separate records of punishment of court astronomers for failure to make accurate predictions of solar and lunar eclipses.[198] Such failure mainly resulted from defects in the calendrical system. Following the example of China, the astronomers of Koryŏ, in their excess of concern for empirical approaches to prediction, failed to delve into the theoretical principles governing the motions of heavenly bodies.

For the Yi Dynasty, according to the meteorology and astronomy chapter of the *Chŭngbo munhŏn pigo,* a total of 190 solar eclipses is on record, including 19 during the reign of King Kojong, the last monarch of the dynasty.[199]

[197]Iijima Tadao, "Sankoku shiki no nisshoku kiji ni tsuite" (On the records of the solar eclipse in the *Samguk sagi*), *Tōyō gakuhō*, 1926, *15*.3:410 ff. On the ritual function of eclipses, see Sivin, "Cosmos and Computation," and other sources cited there.

[198]*Munhŏn pigo,* ch. 4, pp. 1b–10a.

[199]Ibid., ch. 1. pp. 3ab; ch. 2, pp. 20ab.

From the records of solar and lunar eclipses appearing in the *Yijo sillok*, we can make reasonable guesses at the methods of observation in use then. In October of 1808, "There was a solar eclipse on the first. [The eclipse] lasted from Serpent *ch'o* to Horse *ch'o*. The obscuration commenced in the northwest, reaching its maximum extent due north, and ending in the northeast." There is also record of the lunar eclipse that occurred in March of 1819 (nineteenth year of Sunjo): "A total lunar eclipse occurred on the sixteenth. It lasted from the first half of the double hour of Dog to the early part of the double hour of Swine. Obscuration commenced due east, reaching total eclipse at the third *kak* of the second part of the double hour of Dog, and ending due west." According to the regulations concerning the observation of solar and lunar eclipses laid down in the *Sŏun'gwanji*, records were made of the hours, directions, and extent of eclipses.

Recording and observation of solar and lunar eclipses was a very important part of the functions of the Sŏun'gwan in the Yi Dynasty era, and any court astronomer who failed to make accurate predictions of eclipses was dismissed or punished. Especial efforts were made during the reign of King Sejong to ensure accuracy in the prediction of eclipses, and to this end an accurate measurement of the latitude of Seoul was ordered. In order to facilitate observations of solar and lunar eclipses, officials of the Sŏun'gwan were dispatched to such points as Mt. Samgak and the Diamond Mountains, and an observatory was set up atop Mt. Mani on Kanghwa Island.

Because accurate prediction of solar and lunar eclipses was mandatory, King Sejo, in December of the third year after his accession (1437), ordered Yi Sun-ji 李純之 and Kim Sŏk-che 金石悌 to popularize the methods of computation of solar and lunar eclipses first brought into use during the reign of Sejong.[200] These scholars put the computation formulas into mnemonic verses in the *Kyosik ch'ubopŏp* 交食推步法 (Method of calculation of eclipses). This book was later to be used as a text for civil service examinations.

After the Shih-hsien calendrical system of China was brought

[200] *Sejo sillok*, ch. 10, p. 29a.

into use during the reign of King Hyojong, the hours and directions of solar and lunar eclipses were calculated by computing the time differential between Yenching (Peking) and Seoul.[201] "The time of maximum obscuration in the home country is obtained by adding 42 *pun* to the time of maximum obscuration in Yenching." Illustrated reports on eclipses were first made in the early part of the Yi Dynasty era,[202] while from the middle part it was mandatory to make predictions of each solar and lunar eclipse three months in advance.[203]

OBSERVATION OF PLANETS, COMETS, AND NOVAS

Next to solar and lunar eclipses, the greatest importance was attached in ancient times to the movements of the Five Planets, namely, Mercury, Venus, Mars, Jupiter, and Saturn. Among the innumerable stars that embellish the night sky, the planets, because of their complex motions, attracted attention in antiquity. The movements of the planets were incorporated into the art of astrology, which interpreted their motions in relation to the rise and fall of states and rulers.

Of no less significance was the appearance of comets, novas, and shooting stars (meteors), generally regarded as evil augurs. Among the planetary phenomena observed with special interest were the eclipse of a planet by the moon or vice versa; occultation of one planet by another; conjunction of two planets within the same constellation; occultation of a fixed star by a planet; and daytime visibility of a star. In all the hundreds of records of these phenomena, Venus was involved.[204]

The *Samguk sagi* carries a total of 57 records of the appearances of comets and novas, 10 of them for Koguryŏ, 15 for Paekche, and 32 for Silla. As in the case of solar eclipses, many records of Silla, especially for the early part, were no doubt copies of records appearing in ancient Chinese literature. But they soon became independent. This we know because some of

[201] *Munhŏn pigo*, ch. 2, pp. 15a–16a.
[202] *Munjong sillok*, ch. 8, p. 1a.
[203] *Kyŏngjong sillok*, ch. 12, p. 8b.
[204] *Munhŏn pigo*, ch. 4, p. 10b; ch. 6, p. 19a.

the records dated after the unification of Korea by Silla lack counterparts in China. Some of these can be verified by observations made in Europe or Japan. The fact that every comet observed in Japan, China, and Europe is also recorded by Silla attests to the reliability of astronomical observations made in that kingdom, especially after the erection of the Silla Ch'ŏmsŏngdae. Of the total of 57 records given for the three kingdoms, four are believed to have been duplicated, and the remaining 50 observations, covering the period from 49 B.C. to 908, represent the most outstanding body of astronomical observations outside China during this period of about ten centuries.[205]

Such fullness of observation is also characteristic of the records mentioned in the Treatise on Astrology of the *Koryŏsa*. The Koryŏ astronomers have left records of 87 observations, 20 of novas and the remaining 67 of comets.[206] One of the comets was observed for 72 days from the time of first appearance on July 2, 1264 to its disappearance on September 14, while another, which first appeared in the east on February 26, 1374 and disappeared 45 days later, was recorded as measuring more than 10 *ch'ŏk* (ca. 2.2 m). The method by which such linear measurements were taken is unknown so we can suggest no angular equivalent.[207]

Throughout the Yi Dynasty there are a total of 103 comets on record.[208] In case of the discovery of a comet, it was mandatory to submit a report in quadruplicate on the date and time of its appearance, its position and movement, its size and color, the

[205]Tamura, *Tōyōjin no kagaku to gijutsu*, pp. 127–142; Wada, *Chōsen kodai kansoku kiroku chōsa hōkoku*, pp. 182–183; Ho Peng Yoke, "Ancient and Mediaeval Observations of Comets and Novae in Chinese Sources," *Vistas in Astronomy*, 1962, *5*: 127–225, especially corrections of Tamura on p. 214.

[206]Pak Tong-hyŏn reported, in his article "Koryŏsa ch'ŏnmunjic kiroktoen hyesŏnge kwanhaesŏ" (Records of comets in the treatise on astronomy of the *Koryŏsa*), *Han'guk Munhwa Yŏn'guwŏn Nonch'ong*, 1960, *2*.1 : 195, that there were a total of 94 records of observation found in the *Koryŏsa*.

[207]*Koryŏsa*, Treatise on Astronomy; *Munhŏn pigo*, ch. 6, p. 25b. N. Sivin suggests that this "linear" measure is actually a decimal measure of angular breadth, with *ch'ŏk* as degrees and *pun* as tenths. This is consistent with early Chinese practice. See Needham, *SCC*, vol. 3, p. 3.

[208]Ibid., ch. 6, pp. 21a–22b, 26b–31a.

length of its tail, and the time of its disappearance, together with a sketch of its position and a notation of the observer's name.[209] The copies of this report were submitted to the Sŭngjŏngwŏn 承政院 (Royal Secretariat), Tanghu 堂後, Sigangwŏn 侍講院 (Tutorial Office) and the Naegak 內閣 (Royal Library) while the Bureau of Astronomy made entries in the *Kwansanggam ilgi* 觀象監日記 (Diary of the Bureau of Astronomy) and the *Ch'ŏnbyŏn tŭngnok* 天變謄錄 (Register of heavenly portents).[210]

Most of these records were lost with the passage of time, but several of them were recovered by Wada Yūji 和田雄治, formerly with the Inch'ŏn Meteorological Station (1910), from the archives of Kyŏngbok Palace.[211] These extant records cover the periods of January 1661, October–December 1664, January–February 1667, October 1695, February 1702, September, 1723, and March and December 1759. The record for October–December of 1664 mentions a large comet that was observed from October 10 until the early part of January in the following year.[212] Throughout this period observations were made daily except on eight cloudy days. This is regarded as the most complete and exhaustive record to be made of a comet in the annals of astronomy (see Fig 1.23).[213] The *Ch'ŏnbyŏn tŭngnok* of the Yi Dynasty Bureau of Astronomy, also painstakingly unearthed by Wada, was in the custody of the library of the Inch'ŏn Meteorological Station until Korea's liberation in 1945 but was lost in the midst of the ensuing confusion. One story alleges its loss by fire during the Korean War of 1950–1953.

There is also a book that gave astrological interpretations for appearances of comets and sunspots. The *Yŏktae yosŏngnok* 歷代妖星錄 (Record of stars of ill omen in successive ages), by Kim Ing-nyŏm 金益廉 (1622–1674), records evil happenings

[209]*Sŏun'gwanji*, vol. 1, ch. 1.
[210]Ibid.
[211]Wada, *Chōsen kodai kansoku kiroku chōsa hōkoku*, pp. 173–176.
[212]*Munhŏn pigo*, ch. 6, pp. 29ab.
[213]Wada, *Chōsen kodai kansoku kiroku chōsa hōkoku*, pp. 173–176.

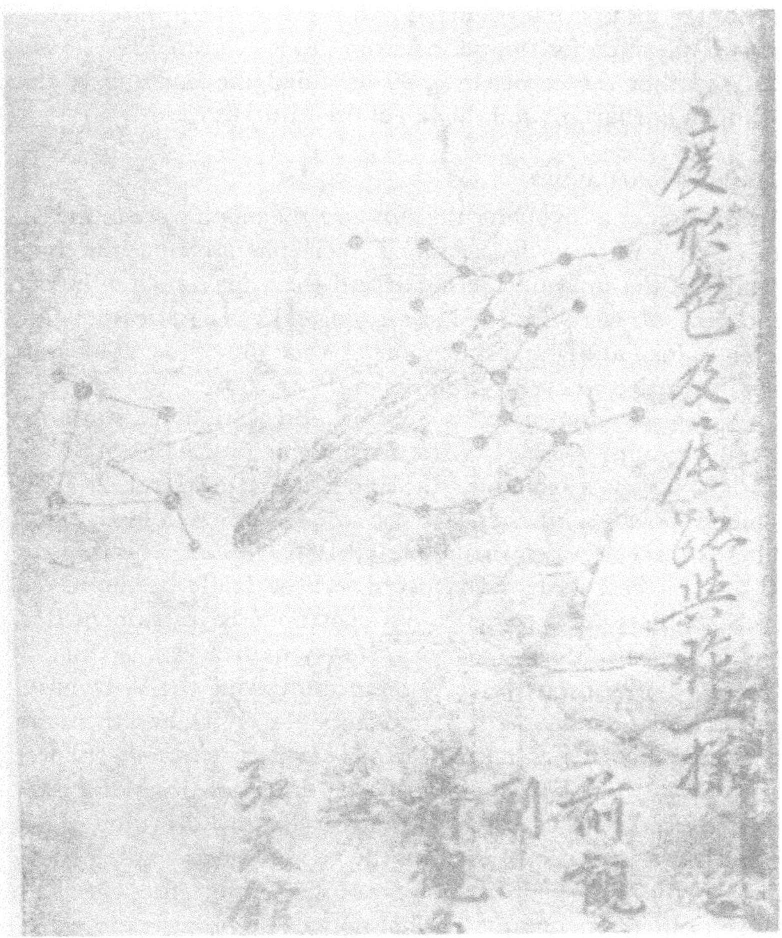

Fig. 1.23. Observation record (or report) of a comet. MS drawing of a comet passing between the lunar mansions I and Chen, on the night of 28 October 1664, from the records of the Korean Astronomical Bureau (from W. C. Rufus, "Astronomy in Korea," and Needham, *SCC*, vol. 3). The annotation at the side says that the shape and color of the comet and its tail were the same as on the previous night. The notes of observers' official posts and names are not shown.

in human affairs that occurred as a result of the appearance of stars of ill omen for the period from 281 B.C. until 1590.[214] He asserted that the comet of 1590 heralded the invasion of the Manchu barbarians and the fall of the Ming Empire.

SUNSPOTS AND HALOES

All phenomena involving the sun and the moon were minutely observed and recorded. Among such phenomena, the reddening of the sun and the moon and the appearance of haloes and sunspots were regarded as astrologically insignificant, while solar haloes and lunar phenomena thought to be analogous were regarded as events of ill omen.[215]

Of special interest to us are the observations of sunspots. The astronomy chapter of the *Koryŏsa* carries a total of 34 records of sunspots including one under the date March 2, 1151, which reports "a black spot in the sun, the size of a hen's egg." These observations cover the period 1024–1383.[216] It is difficult to believe these observations to have been made by the naked eye, but there is no evidence on record of any instrument that might have been used, except a smoky quartz 烏水晶 disk.[217]

Records of sunspot observations occur during the Yi Dynasty period also. One bears the date October 20, 1402, and there are a total of nine entries in the Meteorology and Astrology chapter of the *Chŭngbo munhŏn pigo*.[218] The number of observations is thus markedly smaller than in the Koryŏ era. According to the *Sŏun'gwanji*, the occurrence of the sunspot was explained in terms of the *ki* 氣 (Chinese *ch'i*) concept, in some senses energetic and in others pneumatic.[219] One noticeable point is the periodicity of sunspot observations appearing in the astronomy chapter of the *Koryŏsa*. Between March of 1151 and August of

[214] *Yŏktae yosŏngnok* (Record of stars of ill omen in successive ages), pp. 1 ff.
[215] *Sŏun'gwanji*, vol. 1, ch. 1.
[216] *Koryŏsa*, ch. 19, pp. 16ab; *Munhŏn pigo*, ch. 7, pp. 1a–4a.
[217] *Oju sŏjong pangmul kobyŏn* (Oju's book on the investigation of phenomena; Seoul, 1956), "Eyeglass," p. 1142.
[218] *Munhŏn pigo*, ch. 7, pp. 2b–3a.
[219] *Sŏun'gwanji*, vol. 1, ch. 1. This was an extremely conventional Sino-Korean mode of explanation.

1278,[220] sunspot observations were made at intervals of 8 to 20 years each, closely corresponding to the average periodicity of 7.3–17.1 years established in modern astronomy.

While haloes properly belong to the sphere of meteorological phenomena, they are treated here because their significance was analogous to that of the celestial events already discussed. Haloes were classified into eight kinds according to their appearances. According to the *Sŏun'gwanji*, the complete halo was called *hong* 虹 (rainbow), a partial lateral arc of the 46° halo *i* 珥 (earring), the partial upper arc of the 22° halo *kwan* 冠 (crown), the upper tangent arc of the 22° halo *pae* 背 (back), partial lateral arc of Hall's halo *kyŏng* 璚 (ring), infralateral tangent arcs of the 46° halo *kŭk* 戟 (halberd), and partial lower arcs of circumscribing oval halo *ri* 履 (shoe).[221]

These heavenly occurrences were observed with especial attention because they were thought to herald evil happenings or disaster. The "Chaeigo" 災異考 (Observations of natural disasters and strange events), written in the reign of King Hyojong (1650–1659), records early calamities consequent upon strange phenomena in the heavens, which were no doubt copied from the *Yijo sillok*. The book, pages of which are shown in Figs. 1.24 and 1.25 , covers the 24-year period from 1624 (second year of Injo) to 1655 (sixth year of Hyojong). There are, besides, records of hundreds of meteors, meteorites, and meteor showers.[222]

[220]The dates of these observations were
1151, third month, second, twelfth, and thirteenth day
1160, eighth month, twenty-eighth day
1171, ninth month, twentieth day, and tenth month, seventeenth day
1183, eleventh month, eighteenth day
1185, first, second, third, and tenth months
1200, eighth month, tenth day
1201, third month, second day
1202, eighth month, fifth day
1204, first month, first day
1258, eighth month, sixteenth and seventeenth days
1278, eighth month, twelfth day.

[221]*Sŏun'gwanji*, vol. 1, ch. 1; Needham, *SCC*, vol. 3, pp. 473–476.

[222]*Munhŏn pigo*, ch. 7, pp. 17a–37b.

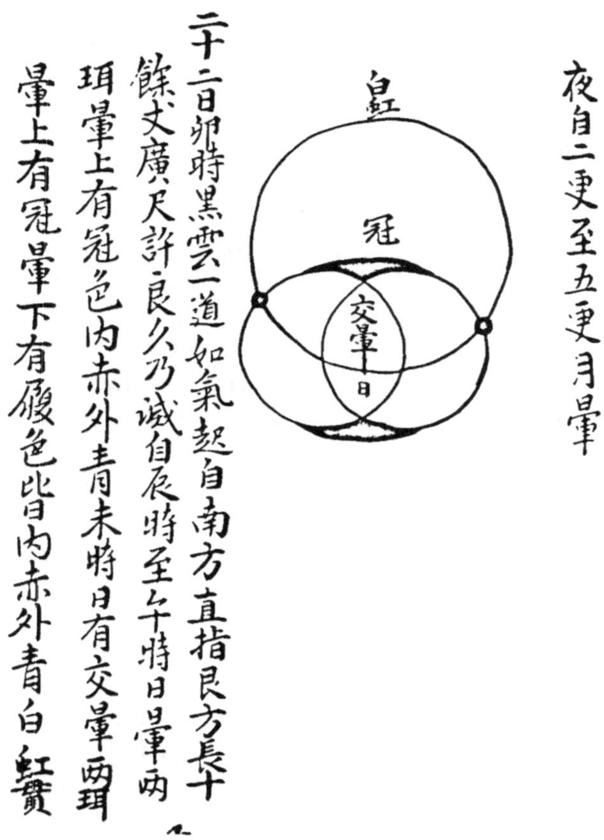

夜自二更至五更月暈
冠
交暈
日
二十二日卯時黑雲一道如氣起自南方直指艮方長十
餘丈廣尺許良久乃滅自辰時至午時日暈兩
珥暈上有冠色內赤外青未時日有交暈兩珥
暈上有冠暈下有履色皆內赤外青白虹貫

Fig. 1.24. Halo drawn in the "Chaeigo" MS.

MEASUREMENT OF POLAR ALTITUDE

The time of appearance of a heavenly body is determined by the polar altitude of the place of observation and the declination of the heavenly body. The observatories of the succeeding dynasties, therefore, made it a point first to measure the accurate polar altitude of the point of observation. In the Treatise on Astronomy of the *Yuan shih*, the polar altitude of the Koryŏ capital of Songdo is listed as 38¼ *tu*,[223] and the *Yijo sillok*

[223] *Yuan shih* (History of the Yuan Dynasty), ch. 48, pp. 12b ff.

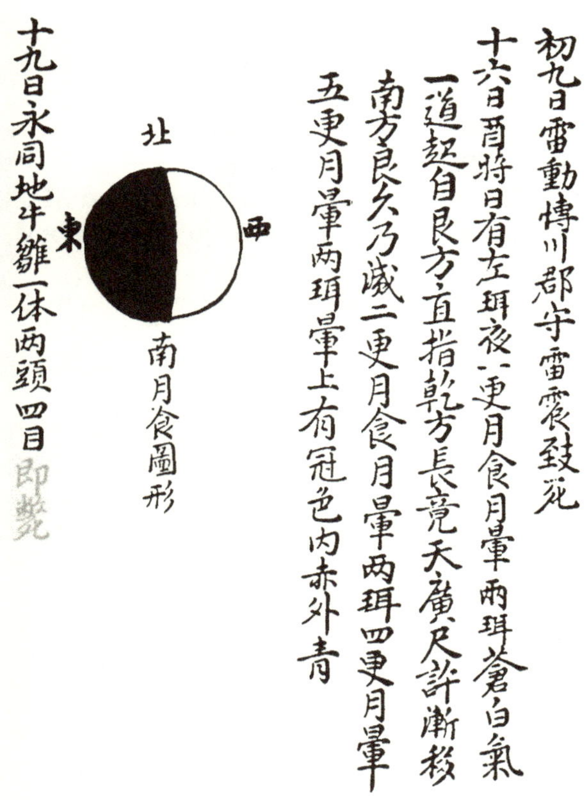

初九日雷動傳川郡守雷震致死
十六日酉時日有左珥夜一更月食月暈兩珥蒼白氣
一道起自艮方直指乾方長竟天廣尺許漸移
南方良久乃滅二更月食月暈兩珥四更月暈
五更月暈兩珥暈上有冠色内赤外青
北
東
西
南
月食圖形
十九日永同地牛雛一体兩頭四目即斃

Fig. 1. 25. Lunar eclipse drawn in the "Chaeigo" MS.

records an observation made in the reign of King Sejong setting that of Hansŏng (Seoul) at slightly over 38 *do*.[224] The most accurate measurement of all was that made by Mu K'o-teng 穆克登 and five other Chinese astronomers in 1713 (thirty-ninth year of Sukchong), which put Seoul's latitude at 37. 3915 *do*.[225] Since until the era of King Sejong, ancient astronomy put down the circumference of the heaven as 365¼ *do*, conversion of this

[224]*Sejong sillok*, ch. 77, pp. 9b ff.
[225]*Munhŏn pigo*, ch. 2, pp. 10ab.

value to the scale of 360° gives a reading of 37° 41′, which is the latitude of Seoul as determined by modern methods.

Besides making measurements from Mt. Samgak 三角山 in Seoul, the court astronomers of the Sejong era made observational excursions to Mt. Paektu 白頭山 on the Korean-Manchurian border, Mt. Mani 摩尼山 on Kanghwa Island, and Mt. Halla 漢拏山 on the southern island of Cheju 濟州島, but unfortunately their records are not extant.[226]

The latitudes of the principal cities of Korea were measured in 1792 (sixteenth year of Chŏngjo), with the following results:[227]

Seoul, 37.3915 *do*
Taegu 大邱, 35.21 *do*
Haeju 海州, 38.15 *do*
Hamhŭng 咸興, 40.57 *do*
Kongju 公州, 36.6 *do*
Chŏnju 全州, 35.15 *do*
Wŏnju 原州, 37.6 *do*
P'yŏngyang 平壤, 39.33 *do*

THE SYSTEM AND INSTITUTIONS OF OBSERVATION

Beginning in the period of the Three Kingdoms, institutions for astronomical observation were set up in imitation of those of China. The Nugakchŏn 漏刻典 of Paekche and Silla is among those recorded. In Paekche there were such officials as professor of timekeeping (*nugak-paksa* 漏刻博士), professor of calendar making (*yŏk-paksa* 曆博士), and timekeeper (*ilgwan* 日官) as early as the sixth century. Silla established the Nugakchŏn with six professors (*paksa* 博士) and one technician (*sa* 司) in 718 and included a professor of astrology (*ch'ŏnmun-paksa* 天文博士) in 749. It is probable that Koguryŏ also had such a system, though no record is extant. In Koguryŏ, however, there was a timekeeper (*ilcha* 日者) who is thought to have been in charge of measuring and regulating time.

[226]Ibid.

[227]*Toritop'yo* (Table of the distances between various cities and towns), unpaginated.

The government organization of early Koryŏ was derived mainly from Unified Silla, adding some elements from T'ang China. In Koryŏ, the T'aebokkam 太卜監 (Bureau of Divination) and T'aesaguk 太史局 (Bureau of the Grand Astrologer) were established to take care of astronomy, the calendar, meteorological observation, and timekeeping. It is thus believed that institutions like the T'aesaguk were created in the late period of the Unified Silla. In 1023, the Bureau of Divination was transformed into the Sach'ŏndae 司天臺 (Office of the Observatory) and later developed into Sŏun'gwan 書雲觀 (Bureau of Astronomy) in 1308. There were 20 officials: the director (*chejŏm* 提點), deputy director (*yŏng* 令), second deputy director (*chŏng* 正), third deputy director (*pujŏng* 副正), fourth deputy director (*sŭng* 丞), two recorders (*chubu* 主簿), two clepsydra technicians (*changnu* 掌漏), three solar observers (*siil* 視日), three observers (*kamhu* 監候), two timekeepers (*sasin* 司辰). The rank of the director and deputy director was Senior 3, and of the timekeeper was Junior 9, on the scale of eighteen official ranks.

The institutions of the late period of Koryŏ were succeeded by those of the Yi Dynasty. According to the organization enforced in July 1392 (the first year of King T'aejo), the Sŏun'gwan was made responsible for astrology, geomancy, and the calendar. Its officials consisted of 34 members ranging from the rank of Senior 3 to Junior 9. The number of members of the Sŏun'gwan was greatly increased during the reign of King Sejong (1419–1450). The Sŏun'gwan, which preceded the era of King Sejo, was renamed the Kwansanggam 觀象監 (Bureau of Astronomy) in the reformation of government organizations in January 1466 (the twelfth year of King Sejo), and it took charge of astronomy, geography, the calendar, meteorological observation, and timekeeping. According to the *Kyŏngguk taejon*, the chief state councillor was in charge of the Bureau of Astronomy and two officials of ministerial rank seconded him. There were sixty-five officials: the director (*chŏng*), deputy director (*pujŏng*), second deputy director (*ch'ŏmjŏng* 僉正), two chief clerks (*p'an'gwan* 判官), two recorders (*chubu*), education officer (professor) of astronomy, education officer of geography, two chief technicians (*chikchang* 直長), two secretaries (*pongsa* 奉事), three deputy secretaries (*pubongsa* 副奉事), four preceptors (*hundo* 訓導), sixteen students (*ch'ŏnmun sŭptokkwan* 天文習讀官), and thirty timekeepers (*kŭmnu* 禁漏).

Up to the end of the Yi Dynasty, the Kwansanggam, an important institution of the government, contributed to the development of astronomy, meteorology, and geography. It functioned as an administrative office, research institute, observatory, and educational institute. In the Yi Dynasty the office of chief of the Kwansanggam was concurrently held by the prime minister and those of *sŏun-chŏng* 書雲正 (director) and *pujŏng* (assistant director), specialized positions, were filled by men of the relatively high ranks senior 3 and the junior 3 respectively.

A rotation system of observation seems to have been established during the reign of King Sejong. According to the record for June of the thirteenth year in the *Sejong sillok*, it was proposed that a certain number of members alternate on duty every two hours and that the observations of each member be reported every morning on opening the door of the Kwansanggam.

One of the factors that enabled the steady development of astronomy and meteorology in Korea in spite of incessant political turmoil and invasions from outside is that these sciences were sponsored and protected by the government organization.

2

Meteorology

Emergence of Agricultural Meteorology

METEOROLOGY IN THE YI DYNASTY

Meteorology in the modern sense had its start in Korea in the early Yi Dynasty era (fifteenth century) when a scientific method was devised for the measurement of rainfall. This invention was part of the effort to control and understand nature's influence on agricultural production.

At a time when the modern art of mechanical irrigation of paddies and farms was unknown, the amount and distribution of natural rainfall had an all-important bearing on the success of farming. Whether or not there was adequate rainfall at the appropriate time determined whether there was to be a bumper crop or a famine. The Yi Dynasty elaborated on the heritage of meteorological studies and researches from the eras of the Three Kingdoms and Unified Silla and further developed the system of the Sŏun'gwan, the Astronomical Bureau of the Koryŏ Dynasty. The practical genius of the Yi Dynasty also succeeded in bringing the fruits of meteorological studies and researches into direct application to problems of agriculture. This so-called agricultural approach to meteorology distinguished the Yi Dynasty era from the preceding ages.

From the beginning, therefore, the meteorologists of the Yi Dynasty concerned themselves primarily with quantitative measurements and observations of rainfall, frost, fog, and similar meteorological phenomena that affected the growth of crops. In this, as in the case of astronomy, purely theoretical or academic problems were relatively neglected. Thus, in forecasting rainfall or temperature changes, reliance was placed on sheer augury. While the meteorologists of the Yi Dynasty were extremely meticulous and accurate in the measurement of rainfall, they apparently paid scant attention to changes of temperature, probably because the latter factor little affected agricultural crops.

A similar trend is observable in their development of the theory of tidal changes. They linked such meteorological phenomena as solar or lunar haloes to the occurrence of auspicious or evil events in human society. Some such phenomena were taken as omens of great catastrophes on earth. Thus, meteorology throughout the Yi Dynasty was much swayed by astrological superstitions, which interfered with scientific investigation at the same time that they provided a structure that made it possible.

INVENTION OF THE RAIN GAUGE AND WATERMARK

A method of measurement of rainfall, presumed to have been devised first in the Koryŏ era, was in use during the early Yi Dynasty.[1] The depth of rainfall was gauged, and the statistics thus obtained were consolidated by the provincial governors and reported to the Board of Taxation (Hojo), which periodically recorded the figures. At first this practice was not conducted at regular intervals but only during the farming seasons or in times of severe drought when the method was used to measure the aridity of the soil.[2]

The first recorded mention of this measurement appears in the *Sejong sillok* (Veritable records of the King Sejong era), bearing the date May 3, 1423: "It rained tonight, penetrating earth to the depth of approximately 1 *ch'on* (about 2.13 cm)."[3] The next mention appears under the dates of April 1 and May 3, 1425: "Drought. Ordered the provinces and counties to measure the depth of moisture penetration of the soil and report."[4] This method, probably derived from the practice of measuring the height or depth of snowfall on the ground in

[1]It is recorded in the *Koryŏsa* (History of the Koryŏ Dynasty) that there was a heavy rain in July 1113 and the depth of water on the flat ground was over 1 *ch'ŏk* (ca. 22–30 cm). *Munhŏn pigo* (Comprehensive study of civilization), ch. 9, p. 8a.

[2]Jeon Sang-woon, "Richō jidai ni okeru Kōuryō sokuteihō ni tsuite" (On the scientific measurement of precipitation in the Yi Dynasty), *Kagakushi kenkyū*, 1963, *66*: 50.

[3]*Sejong sillok*, ch. 20, p. 13a.

[4]*Sejong sillok*, ch. 28, p. 5b.

winter,[5] was inaccurate since the depth of moisture penetration would differ according to the type and dryness of the soil.[6]

Again according to the *Sejong sillok*, heavy rains and severe droughts alternated around the year 1441, further rendering impracticable the conventional methods of measurement.[7] Such a fact was reported from the provinces to the Board of Taxation. The difficulties encountered then obviously led to the invention of the more scientific method of collecting rainwater in a vessel specially shaped for the purpose of accurate measurement.

The statistical, comparative measurement of precipitation, in an effort to alleviate the difficulties caused by the uneven seasonal fluctuation of rainfall, such as the decrease of agricultural production and government revenues, provided a convenient index and aid for improvement of agricultural production.

In spite of the pressing need for an accurate rain gauge, its invention could not have happened without the solid heritage of scientific and technological researches going back to the beginning of the Yi Dynasty, the national policy of encouragement of science during the reign of King Sejong, and the accumulated records of quantitative measurements of rainfall that were available by the mid-fifteenth century.

The *Sejong sillok*, under the date of August 18, twenty-third year of the reign of King Sejong (A.D. 1441), mentions the epoch-making event in the agricultural meteorology of the Yi Dynasty:[8]

The Minister of Taxation informed [His Majesty] that according to reports from the provincial governors on the amount of rainfall, the conventional method of measurement was unable to distinguish between the differences in the depth of rainwater on the ground when it was parched and when it was

[5]The *Samguk sagi* (History of the Three Kingdoms) records that 5 *ch'ŏk* of snow accumulated in the capital of Koguryŏ in December 116. *Munhŏn pigo*, ch. 9, p. 16b.

[6]*Sejong sillok*, ch. 93, pp. 22ab.

[7]Jeon, "Richō jidai ni okeru Kōuryō sokuteihō ni tsuite," p. 50.

[8]*Sejong sillok*, ch. 93. pp. 22ab.

soaked. [The Minister of Taxation] therefore recommended that the Sŏun'gwan should be instructed to prepare a pedestal or a base and place thereon a vessel of iron 2 *ch'ŏk* (42.5 cm) in depth and 8 *ch'on* (17.0 cm) in diameter to measure the amount of rainfall. Further, a slab of stone should be immersed in water west of the Majŏn Bridge to set thereupon two stone pillars embracing between them a wooden post graduated in *ch'ŏk*, *ch'on*, and *p'un* (that is, feet, inches, and tenths of inches) to measure the depth of rainfall. It was also recommended that upon a rock on the bank of the Han River there should be set up a marker graduated in *ch'ŏk, ch'on*, and *p'un* whereby the ferry-master should gauge the depth of the river and report it to the Board of Taxation. Also, officials responsible for various replicas of the cast [iron] vessel should be instructed to make ceramic or earthenware replicas of the cast [iron] vessel within the royal court and place the same in their respective courtyards for the measurement of the depth of [rain] water and report the same to the [provincial] governors. The recommendations were graciously adopted and put into practice.

In these words *Sejong sillok* succinctly explains the reasons that necessitated the making of the rain gauge, which replaced the conventional methods of precipitation measurement with a scientific device. Thus was invented the first rain gauge in the world, made of an iron cylinder with a depth of 2 *ch'ŏk* (42.5 cm) and a diameter of 8 *ch'on* (17.0 cm). The fact that the amount of rainfall could be measured by gauging the water level of rivers did not escape scientists of the Sejong era, who invented appropriate instruments of measurement. One of these watermarks was set up in the Ch'ŏnggyech'ŏn stream flowing through the middle of Seoul, just west of the Majŏn bridge, and another on a rock near the embankment of the Han River, which skirts the capital.

This article, however, fails to name the inventor of this device. According to hearsay, it was King Munjong 文宗, who devised the method of measuring the amount of rainfall by collecting rainwater in a cylindrical vessel.[9] We learn from the quotations from the *Sejong sillok* that the rain gauge and the watermark were based on research and preparatory study at

[9]Ibid., ch. 92, p. 25a.

the Sŏun'gwan, and that the rain gauge was in the charge of the Sŏun'gwan, even though the watermarks in the Ch'ŏnggyech'ŏn were in the charge of the Board of Taxation. The present writer is inclined to believe therefore that both the rain gauge and the watermark were a result of joint researches by the numerous scholars who belonged to the Sŏun'gwan.

As already stated, the invention of the rain gauge was no doubt a product of efforts in a predominantly agricultural country to overcome the natural condition of uneven fluctuations of precipitation. It seems fairly certain, as we shall see below, that most of this effort was motivated by the magic-ritualistic objective of securing succor from either drought or flood by appealing to heaven.[10] Thus medieval aspects can be noted in this eminently scientific invention.

SYSTEM OF RAINFALL MEASUREMENT

Although the rain gauge was invented in 1441, the need for improvement in several respects was felt when it was first put to practical application at the outset of the rainy season in the following year. It was not until May 8 of that year, therefore, that the final designs of the rain gauge were set and the device was officially designated *ch'ŭgugi* 測雨器 (rain-measuring instrument).

The *Sejong sillok* contains an entry describing this historic event:[11]

The Minister of Taxation informed [His Majesty] that in the capital, a *ch'ŭgugi* made of cast iron, measuring 1 *ch'ŏk*, 5 *ch'on* in length and 7 *ch'on* in diameter, was placed upon a pedestal or base set up at the Sŏun'gwan, and after the rain stopped, a functionary of the Sŏun'gwan measured the depth of the collected water, and at the same time kept the records of the date and time of rainfall, and recorded the depth of water in *ch'ŏk*, *ch'on*, and *p'un* by the Yi yardstick (*chuch'ŏk* 周尺). In the provinces, a similar device was established at each of the provincial, county, or township offices for the measurement of the amount of rainfall.

[10]Hong I-sŏp, *Chosŏn kwahaksa* (A history of Korean science; Seoul, 1946), pp. 156, 157.
[11]*Sejong sillok*, ch. 96, pp. 7ab.

The system, as improved and perfected by the measure of May 8, 1442, was a most scientific means for measuring rainfall. In fact, it shows no essential difference from the modern measuring instruments currently in use throughout the world except for the use of a separate measuring instrument (such as a ruler), which, by being immersed in the water, increased its volume and thus caused a slight discrepancy in the reading.

The changes effected in 1442 were as follows: (1) the depth of the rain gauge decreased considerably in relation to its diameter; (2) the measurement of the amount of rainfall was to be made after the rain had stopped; (3) the time of the start of rainfall and the time the sky cleared were to be recorded; (4) the depth was to be measured by yardstick down to the nearest *p'un*; and (5) the Sŏun'gwan was to report in detail on the readings of the rain gauge.

The measure used was a standard ruler of 21.27 cm.[12] From the length of this measuring stick, it is apparent that the rain gauge used in May 1442 must have measured 32 cm in length and 15 cm in diameter. The measurements referred to are believed to denote, respectively, the internal height and diameter of the vessel.

The amount of rainfall thus measured was recorded at the Sŏun'gwan in an entry that gave the time of the rain, the date, and the water depth in the rain gauge in *ch'ŏk* and *ch'on*.[13]

In the Yi Dynasty, records were kept not only of the depth of rainfall but also of the intensity: whether it was drizzle 微雨, very light rain 細雨, light rain 小雨, moderate rain 下雨, considerable rain 灑雨, squall 驟雨, heavy rain 大雨, or rainstorm 瀑雨.[14]

Refinement of the Method of Rainfall Measurement

STAGNATION DURING FOREIGN INCURSIONS

The modern method of measurement of rainfall, perfected

[12]See Chapter 3, p. 134.
[13]*Sŏun'gwanji* (Treatise on the Bureau of Astronomy), vol. 1, ch. 1.
[14]See for instance, Tamura Sennosuke, "Richō kishōgaku seiritsu no

during the reign of King Sejong through the invention of the rain gauge and the watermark, continued in use during the reign of King Sŏngjong 成宗, but a period of stagnation seems to have set in afterward. Thus, in the *Chungjong sillok* 中宗實錄 appears a notice that "it rained tonight, to the depth of 2 *p'un*."[15] and again that "there have been occasional rains since yesterday, and a reading of a depth of 5 *p'un* has been obtained."[16] It is evident that both these entries are not strictly in accordance with the procedures prescribed in May 1442.

A more extreme case appears in the *Sŏnjo sillok* 宣祖實錄 in 1586: "The depth of rainwater was 1 *ch'on*, 1 *p'un* by the cloth measure."[17] This entry indicates neither the time nor the length of rainfall, and the use of the cloth measure indicates that various different measures were in use in gauging the depth of rainwater collected in the rain gauge. However, until the time of the Hideyoshi invasion (1592), the system established in the era of King Sejong more or less continued in use. One notable fact is that most such entries were made between April and July, that is, during the farming season. This leads to the surmise that the measurement of rainfall was made systematically only during the farming season from spring to autumn.

Observations and records after the Hideyoshi invasion are to be found in the "Kiuje tŭngnok 祈雨祭謄錄" (Complete records of the prayer service for rain), which kept a minute record of rainfall in Seoul from 1636 to 1889.[18] All records previous to 1770 are based on watermark readings, and there appears in the *Injo sillok* 仁祖實錄 in 1648 an entry indicating that the report on the watermark reading for the heavy rains of the previous day had not been filed.[19] For May 18, 1743, appears the entry: "Heavy rains. Measurement made by the

kyakkanteki jōken" (Objective conditions for the formation of meteorology in the Yi Dynasty), *Kagakushi kenkyū*, 1958, *48*: 7–10.

[15]*Chungjong sillok*, ch. 68, p. 53a.

[16]Ibid., ch. 98, p. 41b.

[17]*Sŏnjo sillok*, ch. 20, p. 1a.

[18]Wada Yūji, *Chōsen kodai kansoku kiroku chōsa hōkoku* (Report on the survey of the ancient records of observation in Korea; Seoul, 1917) pp. 163–165.

[19]*Injo sillok*, ch. 49, p. 3a.

watermark."[20] It seems evident from these records that from the time of the Hideyoshi invasion to 1770, measurements of the amount of rainfall were made exclusively by means of the watermark. It appears also that during the era of King Injo, watermarks were set up at two places in Seoul, for in the *Injo sillok* appears mention of a "central watermark" and "southern watermark," together with mentions of watermark readings in the Han River.[21]

NEW START UNDER KING YŎNGJO

The rain gauges made during the reign of King Sejong were all destroyed or lost during the Japanese and Manchu invasions, rendering impossible the measuring of rainfall throughout the country. Only the watermarks of the Ch'ŏnggyech'ŏn stream flowing through the middle of Seoul remained as a means of measurement. There was stagnancy in both science and scholarship, and misery was universal. But the gradual resuscitation of national life, and particularly the external stimulation of cultural influences from Ch'ing China, led to the adoption of the Jesuit Shih-hsien Calendar during the reign of King Hyojong; the making of an astronomical clock; a new measurement of the altitude of the polestar at Seoul (a measure of the city's latitude); the printing of a new edition of the *Ch'ŏnsang yŏlch'a punyajido* in the reign of King Sukchong; and the full flowering of the science of astronomy under King Yŏngjo. The meteorological studies incorporated in the *Tongguk munhŏn pigo* 東國文獻備考 (Handbook of the history of Eastern Kingdom civilization) in the eighteenth century are important landmarks in the history of astronomical studies in Korea.

The long-forgotten utility and scientific principle of the rain gauge were realized afresh from the perusal of the appropriate entries in the *Sejong sillok,* and King Yŏngjo finally determined to set up rain gauges throughout the country and accordingly issued an edict to this effect:[22]

[20] *Yŏngjo sillok*, ch. 58, p. 2b.
[21] Wada, *Chōsen kodai kansoku kiroku chōsa hōkoku*, pp. 28–29.
[22] *Munhŏn pigo*, ch. 3, pp. 10b–11b.

> We have heard of the entries in the *Yijo sillok* about the rain gauge but until this day have never had occasion to read them in person. We have ordered reports of watermark readings to be sent in even when it is not the time for rain prayer. This device, in particular, is so clever and easy to use that we have decided to have many of these made by the Sŏun'gwan according to the existing specifications for distribution throughout the eight provinces. If all the incoming reports are compared and studied, it will be possible to learn the situation in detail with utmost accuracy. Instruct the Office of Weights and Measures (T'akji 度支) to make two rain gauges and install one in the Ch'angdŏk Palace and the other in the Kyŏnghŭi Palace. Also within the palace grounds is a streamer which since days of old has been of use in showing the direction of wind. A platform of stone has been set up inside the Toje Gate of Ch'angdŏk Palace and another inside the Sŏhwa Gate of Kyŏnghŭi Palace to fly such a streamer on a bamboo pole. A rain gauge shall be installed within both the Ch'angdŏk and Kyŏnghŭi Palaces, in accordance with His Majesty's [King Sejong's] wish to measure every wind and every rain—a wish that even this day shall not be disregarded or neglected. Since propitious wind and auspicious rain are of utmost importance to the realm, this royal order is of particular significance.

In another edict, King Yŏngjo issued the following instructions regarding the measurement of rainfall:

> We have also heard that the rain gauge of which mention is made in the *Sillok* was set up on a stone pedestal or stand. Inasmuch as this also was in accordance with His Majesty's wish, it is my desire now that in setting up rain gauges in the two palaces and the Sŏun'gwan, each of them shall be erected upon a stand of stone, measuring 1 *ch'ŏk* in height and 8 *ch'on* in width by the cloth measure, and put into a basin that shall be carved to the depth of 1 *ch'on*. Also, use the measure newly manufactured in the 57th year of the current sexagenary cycle.

Thus, on May 1 in the forty-sixth year of the reign of King Yŏngjo (1770), bronze rain gauges were made exactly in accordance with the specifications established during the reign of King Sejong.[23] The rain gauge stand was 46 cm high and 37 cm wide, and had a hole or basin 16 cm in diameter carved to the depth of 4.3 cm (see Fig. 2.1). Both on the front and rear

[23] *Yŏngjo sillok*, ch. 114, pp. 19ab.

Fig. 2.1. Korean rain gauge. Made in 1770, of bronze, exactly in accordance with the specifications established in 1442. The rain gauges made during the reigns of King Sejong and Yŏngjo were almost all destroyed or lost; the only known remaining example is now in the Science Museum, London. A stone stand of the fifteenth century is preserved by Maedong Elementary School and one of the eighteenth century by the Korean Meteorological Observatory in Seoul.

sides of the stone stand was inscribed the name *ch'ŭgudae* 測雨臺 (rain-measuring stand), while to the left on the rear side were inscribed the letters *Kŏllyung Kyŏngsin Owŏl Cho* 乾隆庚申五月造 (manufactured in the fifth month, sexagenary year 57 in the reign of the Ch'ien-lung Emperor [in China] = 1770). The measure used in these gauges was newly made in accordance with the standard measures adopted in 1740, which were based on the weights and measures specified in the *Kyŏngguk taejŏn* 經國大典 (National code) and incorporated features of the cloth measure of the Sejong era. Thus the system of rainfall measurement was revived after an interval of close to two centuries.

INCREASED FREQUENCY OF OBSERVATIONS, AND THE RAIN GAUGE IN KING CHŎNGJO'S ERA

According to the *Sŏun'gwanji* 書雲觀志 (Treatise on the Bureau of Astronomy), the measurement of rainfall under the reign of King Sejong was made after each rain only from May to September, but from the edicts of King Yŏngjo it is apparent that under the latter monarch's reign measurements and reports were made even outside the rice-growing season. From 1770 on, in fact, the practice was extended throughout the year. Measurements were made twice daily, once from late afternoon to sundown, recorded at sundown, and again from early evening to early dawn, recorded after gate-opening. From the fifteenth year of Chŏngjo (1791), observations were made thrice daily—from dawn to early noon, from noon to curfew (around 10:00 P.M.), and from curfew to dawn. In case of a heavy rain, measurement was made more frequently, sometimes every hour. Four copies of each report were prepared for submission to the Sŭngjŏngwŏn 承政院 (Royal Secretariat), Tanghu 堂後, Sigangwŏn 侍講院, and Naegak 內閣. The daily totals, added up by the Sŏung'wan, were forwarded to the Royal Secretariat for entry and dissemination on the *Ilsŏngnok* 日省錄 (Daily Records for Royal Introspection), *Tanghu ilgi* 堂後日記 (Officer's diary of the Royal Secretariat) and *Choji* 朝紙 (the official gazette).[24] From the twenty-fourth year of King Chŏngjo

[24]*Sŏun'gwanji*, vol. 1, ch. 1.

(1800), monthly records were also kept at the suggestion of the Kwansanggam 觀象監 (Bureau of Astronomy).[25] The statistics on the amount of rainfall for 8 years beginning in 1792 are reported in the *Chŏngjo sillok* 正祖實錄.[26] This source compares the amount of rainfall in May 1799 with that of the same month in the previous year, which shows that the monthly amount of rainfall was also known.

In 1782, a rain gauge was installed in the courtyard in the Ch'angdŏk Palace during a severe drought that lasted from June to July. The marble stand on which the rain gauge was placed has an inscription explaining the significance of the measuring device:[27]

> In the twenty-fourth year of the reign of King Sejong, a rain gauge was made of copper [iron] to the size of 1 *ch'ŏk*, 5 *ch'on* in height and 7 *ch'on* in diameter, one being installed in each province, county, and township to measure the depth of rain-water after each rain. In the forty-sixth year of the reign of the preceding king [Yŏngjo], copies of the old model were cast and installed in the Ch'angdŏk and Kyŏnghŭi Palaces and in the eight provinces. Although the vessels may be small in size, they are heavy in weight and import since they embody the efforts made under the two sacred eras to combat both floods and droughts. In the summer of the sixth year [of King Chŏng-jo], a severe drought visited the whole 500 *ri* of Kyŏnggi Province and the paddies were parched and cracked. There-upon, the king blamed his holy self and widely sought advice. Having had a cloth screen set upon the altar, His Majesty alit from the sedan chair to offer a prayer in person, until his royal attire was wet with evening dews. Also, convicts were amnestied,

[25] Wada, *Chōsen kodai kansoku kiroku chōsa hōkoku*, p. 32.

[26] *Chŏngjo sillok*, ch. 51. According to the record, annual precipitation was as follows:

1791	8.59 *ch'ŏk*
1792	7.19 *ch'ŏk*
1793	4.49 *ch'ŏk*
1794	5.80 *ch'ŏk*
1795	4.22 *ch'ŏk*
1796	6.85 *ch'ŏk*
1797	4.56 *ch'ŏk*
1798	5.56 *ch'ŏk*

The precipitation for the fifth month of 1798 was over 1 *ch'ŏk*.

[27] See Jeon, "Richō jidai ni okeru Kōuryō sokuteihō ni tsuite," pp. 54–55.

and the people and scholars, and women and children within the capital looked up in grateful wonder, and tearfully spoke in these words: "Since His Majesty has such great apprehension for the well-being of his subejcts, rain is bound to come; even if no rain shall fall, his subjects should feel more grateful than if rain came." That night, there was a great rain, and the depth reached 1 *ch'on*, 2 *p'un*. This was entirely due to His Majesty's virtue. His Majesty was worried that the rain was not sufficient; so he had a rain gauge installed in the courtyard of the Ch'angdŏk Palace and measurement taken, and it was found sufficient. Thereupon, His Majesty called upon his courtiers Sim Yŏm-jo 沈念祖 and Chŏng Chi-gŏm 鄭志儉 to record his exceeding joy. They are his most loyal subjects. They know that when then there is no rain, the apprehension of His Majesty on behalf of his subjects cannot be measured by the same measure as the apprehension of the subjects, and that when there is rain, His Majesty's joy must far surpass that of his subjects. This rain gauge embodies such apprehension and such joy on the part of His Majesty. Respectfully, subject Sim Yŏm-jo, subject Chŏng Chi-gŏm.

There is a copy of this rain gauge, extant on the grounds of the Inch'ŏn Meteorological Station, which was made in February 1811, and another in the Kongju Museum, which was made in 1837 and bears the inscription:

Kŭmyŏng Ch'ŭgugi Togwang Chŏngyu-je 錦營測雨器道光丁酉製 (Rain-measuring Instrument in Kongju, made in the thirty-fourth sexagenary year of Tao-Kuang).

IMPROVEMENT OF THE WATERMARK

Because the watermark erected west of the Majŏn Bridge in 1442 was, as mentioned early in this chapter, half made of wood, it must have required repair or replacement before long. Later, in the era of King Sŏngjong, the watermark began to be made of stone, and the model that remains today in the Changch'ung Park of Seoul emerged.

This watermark, made before the mid-seventeenth century, is a hexagonal pyramid of granite about 3 m in height and 20 cm in width mounted on a square base. On top is a capital in the lotus flower design, while the square base is buried underground. On the stone marker are carved degrees at 1-*ch'ŏk* intervals from 1 to 10 *ch'ŏk*, and at the levels of 3 *ch'ŏk*, 6 *ch'ŏk* and 9 *ch'ŏk* are inscribed circles to denote, respectively, drought, average, and high water.[28]

Compared with the watermark of the Sejong era, this device lacked finer gradations for *ch'on* and *p'un*, but one point of improvement was the marking of low and high levels of water. If was first erected near the Sup'yo 水標 (Watermark) Bridge, formerly the Majŏn Bridge, in Seoul. When the Ch'ŏnggyech'ŏn was covered up in a road expansion project, the watermark, together with the Sup'yo Bridge, was removed to its present location in the Changch'ung Park.

FORECASTING OF RAIN

The foretelling of rain has always been of great importance to farmers; therefore, since ancient times, rain forecasting has been practiced by means of sorcery. In various books dealing with agriculture, mention is made of the methods of foretelling rain first described in the *Sasi ch'anyo* 四時纂要 (A synopsis of agriculture in the four seasons), 1424–1483. *Chŭngbo sallim kyŏngje* 增補山林經濟 (The agricultural handbook, revised and enlarged), by Hong Man-sŏn 洪萬選, describes several methods of forecasting rain:

(1) Forecasting weather from observation of the sun and moon: (*a*) A halo around the sun heralds rain. If the halo "tends" southward, it will be clear; if it tends northward, there will be rain. If the halo tends both ways, the wind and rain will soon cease. (*b*) If the globe of the sun is clear at the time of sunrise or sunset, the weather will be clear. Even if the sun is partly hidden by thin clouds, and if the globe is clear and brilliant, the weather will be fine. If the contrary, there will be rain or wind. (*c*) If the sun is brilliant just before sundown, it will be clear, and if the sky reddens to the color of rouge, there will be no rain but some wind. (*d*) Generally speaking, if just before sunrise, east and southeast appear red and the air is clear, the weather will be fine, and if there are black clouds and the air is murky, or if purple-black clouds surround the sun, there will be rain that day. If the wind is strong, there will be rain in the morning, and if the wind is weak, there will be rain in the

[28]Ibid., p. 56.

afternoon. The same forecast can be made from observation of the sun setting in the west in the afternoon. (*e*) If the moon has a halo, there will be wind. (*f*) If the moon is red, the weather will be dry; if the clouds floating near the moon are white, there will be wind; and if black clouds hover below the new moon, there will be rain the following day.

(2) Forecasting weather from observation of the stars: (*a*) If immediately after rain, one or two stars can be observed in spite of cloudy weather, there will be a clear sky that night. (*b*) If a long rain suddenly ceases in the afternoon and the skies are full of stars that night, there will be rain the next day.

(3) Forecasting weather from wind and rain: (*a*) If in the spring there is a south wind or if in the summer there is a north wind, there is bound to be rain. If in the winter a south wind blows three or four days, there will be rain. (*b*) If there is wind from the northeast, there will be rain that will not cease quickly. If there is a strong east wind, there will be rain. If a strong wind is accompanied by scudding clouds, there will be rain without fail. (*c*) If there is rain just before dawn, the weather will be clear that day. If rain drops cause foam on the water, the rain will not cease soon. (*d*) If rain is mixed with snow, the weather will not clear easily. A sudden and vigorous rain is a harbinger of fine weather.

(4) Forecasting weather from cloud movement: (*a*) If clouds move eastward, the weather will be clear, and if they move westward, it will rain. Northward movement heralds clear weather and clouds moving south will bring rain. (*b*) If a cloud descends on the fields, spreading around like a fog, it is called *p'unghwa* 風花 (wind-flower). The phenomenon is caused by wind. (*c*) If clouds arrive from the southeast, the rain will cease shortly. (*d*) If cloud banks arise from the southwest, there will be a lot of rain, and the same phenomenon happening in the northwest heralds great winds followed by a downpour, which, however, will stop soon. (*e*) Cirrocumulus clouds herald no rain or wind. (*f*) Even if an autumn sky has clouds, if there is no wind there will be no rain. If the winter sky is suddenly covered by carp-clouds in late evening, there will still be no rain. (*g*) If black clouds arise and gradually cover the whole sky, there will be a heavy rain.

The source mentions other methods of foretelling the weather from observations of mist, rainbows, thunderclaps, lightning, frost, snow, ice, fog, and various phenomena of fauna and flora, including the predictions: "(*a*) If the sound of the drum beat does not carry far, there will be rain. (*b*) If metal or stone surfaces get moist, there will be rain."

Such weather forecasts are, of course, far removed from modern meteorological methods, but careful scrutiny reveals links between the two. The techniques of weather forecasting can be divided into two categories: those derived from the treatises of ancient China, influenced by the method of forecasting rain derived from the *yin-yang* theory;[29] and those obtained from experience. The ancients were aware of the close connections existing between the shapes of clouds and the weather; they tried to unravel cause-effect relationships between wind direction and the weather; and they were aware that variations in atmospheric moisture expressed themselves in the sound of a drum or the surface appearance of metal or stone objects.

Korean weather forecasting mostly reflects accumulated empirical knowledge. Kang Hi-maeng, in his book, points out the inconsistency of the method of forecasting rain in terms of *yin-yang* theory. The fact that the south wind brings heavy rain and the north wind does not, he insists, is incompatible with the method. He emphasized the role of experience and observation.

Other Meteorological Observations

OBSERVATION OF WIND VELOCITY AND DIRECTION

The second most important meteorological phenomenon for agriculture is wind velocity and direction after a rain. Kang Hi-maeng 姜希孟 (1424–1483) in his famous *Kŭmyang chamnok* 衿陽雜錄 (Miscellaneous records of farming in the Kŭmch'ŏn District, Kyŏnggi Province) lists wind damage as the most serious bane of agricultural crops after flood damage.

The land of Korea borders the sea on the east and south and is flat in the west. Rugged mountains in the north cover the east and run down to the south. The northern and eastern parts are predominantly mountainous and the west and south are plains. The wind from the sea is warm and is easily converted into cloud and rain, which enables plants to grow. The wind

[29] *Meiji zen Nihon butsuri kagaku shi* (A history of Japanese physics and chemistry before the Meiji era; Tokyo, 1964), pp. 446–451.

blowing over the mountain is cold and thus harmful to plants. During the farming season, the people in Yŏngdong 嶺東 area want the east wind, while those in Hosŏ 湖西, Kyŏnggi 京畿, and Honam 湖南 prefer the west wind to that from the east because the latter comes over the mountain.

In the worst cases, according to him, watercourses in fields and rice paddies are dried up and plants are burned out. Even in a less severe case, the rice shoots wither so early that the grain cannot develop. Katayama maintains that Kang Hi-maeng's theory about the damage caused by the wind, if simple in content, is based on his understanding of the Föhn phenomenon.[30] His wind theory is one of the first systematic approaches to the relation of wind and farm crops to be made in Korea.

Thoroughly familiar with the influence of wind on farm crops, the meteorologists of the Yi Dynasty took special care to observe wind direction closely. An anemoscope, called *p'ung-gijuk* 風旗竹, was set up for that purpose. There is no clear indication when the observation of wind direction was actually started, but the wind gauge was erected in the era of King Sejong to determine wind direction from the flow of a streamer.[31] The stone stand was made in 1770 and installed in the Ch'angdŏk and Kyŏnghŭi palaces. These wind gauge stands are still extant (see Fig. 2.2).

Wind direction was expressed in one of 24 directions. The wind velocity is presumed to have been classified, as in the case of the volume of rainfall, into eight degrees. For instance, a wind strong enough to uproot a tree was referred to as *taep'ung* 大風 (great wind) and one violent enough to strip the roof tiles was called *p'okp'ung* 暴風 (violent wind). There are specific mentions of these two classes of wind.

There are 24 records of observations of wind velocity from the era of Silla, four from the Koguryŏ era and four from the time of Paekche. In the Koryŏ Dynasty, there is mention of a "great wind" in Sŏgyŏng (the present P'yŏngyang) which in

30Katayama Ryūzō, "Konyō zatsuroku no kenkyū" (A study of the *Kŭmyang chamnok*), *Chōsen gakuhō*, 1958, *13*: 163–178.

31Wada, *Chōsen kodai kansoku kiroku chōsa hōkoku*, pp. 12–13; Hong, *Chosŏn kwahaksa*, p. 158.

Fig. 2.2. The granite stand of a Korean anemoscope. This kind of anemoscope was first made in the fifteenth century and the example here probably in the seventeenth or eighteenth century. Ch'anggyŏng Palace.

May 932 blew away roof tiles and caused the collapse of government offices. Again in August 1391 there was a great wind in Chŏlla Yanggwang Province that uprooted trees. Altogether, there are 135 records of wind during the 359 years of the Koryŏ Dynasty. There is, curiously enough, a sharp decline in the number of recorded observations in the Yi Dynasty. There are only 21 references in the later period, beginning with a record of 1412 stating that a great wind uprooted trees, and ending with the record of a great wind in June 1739 that blew away roof tiles as it raged on its path from Ch'ungch'ŏng Province to the western borders (P'yŏngan Province).[32] While the records of the Koryŏ era took note of the day and month, those of the Yi Dynasty often lacked dates.

EARTHQUAKES AND "WHITE RAINBOWS"

While most meteorological phenomena were treated as commonplace, earthquakes and "white rainbows" were regarded as extraordinary.[33] Earthquakes frightened people and caused damage to property and injury or loss of life to humans. The white rainbow, however, was regarded as an especially evil omen. The solar manifestation was apparently a type of halo; the lunar was perhaps a cloud formation that resembled haloes and parhelia.

Earthquakes were divided into two classes. Until the days of Koryŏ a strong earthquake was called *taejin* 大震 (great earthquake) but in the Yi era it was referred to as *kangjin* 强震 (strong earthquake).[34] An earthquake was minutely observed and recorded, both as to its time and degree of strength, and as to its epicenter and the area affected. In the Yi Dynasty, every earthquake was reported immediately without regard to the time of day or night.[35]

Earthquakes were recorded on a total of 1,661 days from the Age of the Three Kingdoms down to the end of the Yi Dynasty, a span of around two thousand years. The actual number should have been far greater, however. Earthquakes of great severity, that is, those that caused property damage or loss of human life, as well as those accompanying volcanic action and those causing earth fissures or ejections of magma or hot springs, are recorded a total of 12 times in the Age of the Three Kingdoms and Unified Silla, 4 times in the Koryŏ era (11 including those referred

[32] *Munhŏn pigo*, ch. 9. pp. 1b–7a.
[33] *Sŏun'gwanji*, vol. 1, ch. 1.
[34] Ibid.
[35] Ibid.

to as "great earthquakes"), and 42 times during the Yi Dynasty era. This means an average interval of 40 to 80 years in the span of approximately twenty centuries.[36]

Records of both solar and lunar "white rainbows" noted their color, shape, hour, the names of any of the Five Planets (Mars, Mercury, Jupiter, Venus, and Saturn) in their vicinity and whether the rainbows were "double" or not.[37] Solar haloes and white rainbows are recorded six times in the age of the Three Kingdoms and Unified Silla, 99 times in the Koryŏ era and 250 times in the Yi Dynasty era for a total of 355.[38]

The *Myŏngjong sillok* 明宗實錄 refers to the halo around the sun on February 13, 1551.[39] It gives a detailed description of the color of the inner part of the halo and of the change of color as time went on. Of lunar haloes and white rainbows, 23 are recorded in the Koryŏ era and 46 in the Yi Dynasty era.

TEMPERATURE AND BAROMETRIC PRESSURE

Temperature was perceived until the sixteenth century in terms of everyday experience. In case of an especially warm winter when there was no snow or ice, for instance, a public ritual was held to pray for cold weather. Or, when an untimely cold prevented the flowering or budding of plants in the spring, when the summer weather was cold like autumn or the autumn weather was frigid like winter, or when the winter weather was warm like spring, they were referred to as *hanoni* 寒溫異 (unusual cold or warmth).[40]

There are extant records of atmospheric or barometric measurements. In the *Chŏngjo sillok* mention is made in 1798 of an order relaxing the ban on the use of mercury for preparation of Western-style barometric instruments.[41] This is the first indication that a mercury barometer was already in

[36]*Munhŏn pigo*, ch. 10, pp. 6b–13b.
[37]*Sŏun'gwanji*, vol. 1, ch. 1.
[38]*Munhŏn pigo*, ch. 7, pp. 4b–17a.
[39]*Myŏngjong sillok*, ch. 11, pp. 19ab, 20b, 21ab, 22ab, 23ab–24a.
[40]*Munhŏn pigo*, ch. 10, pp. 5b–6b.
[41]*Chŏngjo sillok*, ch. 52, p. 10b.

use in that age. However, no further description is available as to the manner in which the barometer was made or used, nor is there any record of barometric readings obtained from the use of such a device. As for other meteorological observations, there are both records and descriptions (or definitions) of such phenomena as aurorae, clouds, lightning, rainbows, thunderclaps, sleet, fog, frost, and snow.[42] Observations were made as to the shape, color, size, time of appearance, and movement of clouds, and there are more than 500 records of extraordinary phenomena involving clouds from the time of the Three Kingdoms to the end of Yi Dynasty.[43] Observations and records were also made of rainbows, thunder, and lightning, including both descriptions of the phenomena and time of occurrence.

AURORAE AND OTHER ATMOSPHERIC OBSERVATIONS

The record in the *Chungjong sillok* 中宗實錄 concerning the aurora observed in Kyŏngju, southeast Korea, in June 1519, is of particular interest.[44]

> That night a natural calamity took place in Kyŏngju, Kyŏngsang Province. In the evening the moon was very bright. No sooner did a slight cloud become visible in the west than light appeared through the cloud. It looked like lightning, but it gave off no heat. Sometimes it moved slowly in the sky in the shape of a flying arrow and sometimes passed away very fast like a shooting star. It looked like a leaping red snake or sparkling flame. Sometimes it was bent like a crooked bow and sometimes it was split like a pair of scissors. It showed immense variety in shape. . . . It began to move from the west to the northeast and disappeared shortly after midnight.

This is perhaps one of the most detailed records of aurorae in that time. In the *Munhŏn pigo,* there are over 200 records on the observation of aurorae from 35 B.C. to the nineteenth century. They were described variously as blue or red clouds, vapor, snakelike red vapor, waterfall-like white vapor, white

[42] *Sŏun'gwanji*, vol. 1, ch. 1.
[43] *Munhŏn pigo*, ch. 8, pp. 1a–15b.
[44] Ibid., ch. 8, p. 11a.

vapor like the march of spears and swords, red fire, sunshine at night, snaky arrows, white arcs, blue-violet cloud-like pavilions, and so on.[45]

In the case of hail, there are recorded measurements of its size, ranging from that of a red mung-like bean through that of a soya bean to that of a bird's egg. Similar observations and records were made of fog, frost, and squalls.[46]

Fog was observed from early days. The *Samguk sagi* records fog in Koguryŏ in April, 34 B.C. "The fog was so dense that nobody could distinguish one color from another for a week." Another record: "In February, A.D. 22, a sudden heavy fog prevented one from recognizing men or things in front of him for ten days."[47]

Observations of snow were made in a manner similar to those of rain, but only the hour of snowfall and the depth to which it fell seem to have been recorded. There are records of such extraordinary phenomena as scarlet snow, rain with sleet, and heavy snows as late as April or June. In recording these observations particular care seems to have been taken to distinguish whether they occurred before frost and snow or after. Records of lightning and thunder were kept only when they were so severe as to cause damage on the ground.

In the parts dealing with the west coast and coastal islands of the *Sejong sillok chiriji* 世宗實錄地理志 (Treatise on Geography of the Veritable Records of the King Sejong Era), there is mention of tidal rise and fall. From the mention of "at each time of ebb tide," it is obvious that measurement was made of the time of flood and ebb. Mentions of tsunami and tidal waves appear in the *Samguk sagi* and other records. The *Samguk sagi* records: "There were sea battles on the East Sea in September 699 and the sound was heard even in the capital" (Kyŏngju, the capital is about twelve miles away from the Sea of Japan). In June 915, "the water of Champ'o clashed with that of the East Sea resulting in a wave 20 *chang* (over 40 m) high which

[45]Ibid., ch. 6, pp. 31a–33a, ch. 8, pp. 1a–15b.
[46]*Sŏun'gwanji*, vol. 1, ch. 1.
[47]*Samguk sagi*, ch. 13; *Munhŏn pigo*, ch. 9, p. 21a.

lasted for three days."[48] The tidal theory was first advanced by Han Paek-kyŏm 韓百謙 (1552–1615) and again appears in a fuller form in Yi Ik's 李瀷 (1579–1624) *Sŏngho sasŏl* 星湖僿說 (Sŏngho's detailed discourses).[49] It attained perfection in Yi Kyu-kyŏng's 李圭景 treatise on sea tides in the early nineteenth century,[50] based on bibliographical studies and observation.

In the Yi Dynasty era, meteorological observations were made and recorded according to regulations that appear in the *P'ungun'gi* 風雲記 (Records of meteorological observation, 1754–1904) and the *Sŏun'gwanji*. The daily recordings were made in two shifts for the day and night, with two men assigned to the day shift and one man for each of the five *kyŏng* into which the night was divided. The total force of seven men recorded such meteorological phenomena as clear or cloudy skies, rain, snow, frost, and fog, as well as other extraordinary occurrences referred to earlier.[51]

[48]*Samguk sagi*, ch. 8; *Munhŏn pigo*, ch. 10, p. 17b.
[49]*Sŏngho sasŏl yusŏn*(Sŏngho's detailed discourses, classified), ch. 1.
[50]*Oju yŏnmun changjŏn san'go* (Collected works of Oju = Yi Kyu-kyŏng), vol. 2, ch. 53, pp. 720–726.
[51]*Sŏun'gwanji*, vol. 1, ch. 1.

3

Physics and Physical Technology

Elementary Physics and Its Application

WEIGHT AND MEASUREMENT

Weight and measurement as instruments of science in Korea made their appearance during the time of Lo-lang, when the civilization of Han China was first introduced. Records of length, volume, and weight are contained earliest in *Samguk sagi* 三國史記 (History of the Three Kingdoms), regarding Paekche, A.D. 500.[1]

The record refers to the Imnyugak 臨流閣 (Waterside Pavilion), which was located east of the royal palace, and claims that the structure was "five *chang* 丈 high." Another reference, in the *Samguk yusa* 三國遺事 (Memorabilia of the Three Kingdoms) speaks of the huge scale of the nine-story pagoda of Hwangyong Temple 皇龍寺, built in A.D. 643. The pagoda, according to the reference, stood "42 *ch'ŏk* 尺 above the iron foundation and 183 *ch'ŏk* below it."[2] Judging from the fact that a method of weight and measurement had been introduced in Japan from Korea between A.D. 588 and 592, it seems fair to presume that metrology as such had been systematized by that time. It is known that plans for many tombs and buildings of Koguryŏ were actually drafted mathematically by the use of Han units from the time of Lo-lang.[3]

No more data are available to further our knowledge of the weights and measures of the Three Kingdoms period, but it may be presumed that the system was fairly similar to that used in T'ang China. How it changed up to the time of Koryŏ is not easy to comprehend, but records in the *Koryŏsa* 高麗史

[1]*Samguk sagi* (History of the Three Kingdoms), ch. 26.

[2]*Samguk yusa* (Memorabilia of the Three Kingdoms), twelfth year of King Sŏndŏk.

[3]Yoneda Miyoji, *Chōsen jōdai kenchiku no kenkyū* (Studies in ancient Korean architecture; Tokyo, 1944), pp. 174, 200 ff.

are good enough to convince us that the system of weights and measures had been grasped and understood.

Reference to length appears in the *Koryŏsa* "Sikhwaji" 食貨志 (Treatise on commodities and economics of the history of the Koryŏ Dynasty) as follows:[4] "In the twenty-third year of King Munjong (1069), the principles of land survey were determined. One *kyŏl* 結 of farmland is 33 *po* 步 [literally, "paces"] square; 2 *kyŏl*, is 47 *po* square; 3 *kyŏl*, 57 *po* and 3 *p'un* square . . . 10 *kyŏl* being 104 *po* and 3 *p'un* square."

Reference to volume appears in the "Hyŏngpŏpji" 刑法志 (Treatise on the penal law) of the *Koryŏsa*, stipulating the size of the official container (*kwan'gok* 官斛), in the seventh year of King Munjong (1053). A rice container (*migok* 米斛) was a cube 1 *ch'ŏk*, 2 *ch'on* on a side. A container for taxation grains (*p'aejogok* 稗租斛) was 1 *ch'ŏk*, 4 *ch'on*, 5 *p'un* on a side. A container for soy sauce (*maljanggok* 末醬斛) measured 1 *ch'ŏk*, 3 *ch'on*, and 9 *p'un*. Finally, a container for beans and peas (*tugok* 豆斛) measured 1 *ch'ŏk* and nine *p'un*.[5]

Early references to weight also appear in connection with various commodities that Mongol emissaries requested of Korean royalty in August 1221, during the eighth year of King Kojong 高宗. Among the items that appear are 10,000 *kŭn* of cotton seed, 5 *kŭn* of gromwell, 50 *kŭn* each of safflower, indigo-plant sprout, and vermilion; 10 *kŭn* each of orpiment and paulownia seed oil.[6] And again in the "Sikhwaji" of the *Koryŏsa*, in February of the first year of King Ch'ungsŏn 忠宣, there is the following reference: "[The price of] 1 *kŭn* of silver equals 64 *sŏk* of salt; 1 *yang* of silver equals 4 *sŏk* of salt."

In the time of Koryŏ, the *p'obaekch'ŏk* 布帛尺 (cloth measure) was used for measuring cloth, and the *chuch'ŏk* 周尺 was used to measure farmland. In determining volume, the *migok* was used for rice; the *maljanggok* for soy sauce; the *p'aejogok* for other grains; and the *tugok* for corn and beans. In weight, both *yang* and *kŭn* (16 *yang* equaled 1 *kŭn*) were used, but these

[4]*Koryŏsa* (History of the Koryŏ Dynasty), ch. 78, p. 3b.
[5]Ibid., ch. 84, p. 19a.
[6]Ibid., ch. 79, p. 6b.

weights and measures were firmly regulated by the P'yŏng-duryang Togam 平斗量都監 (Supervisorate of Weights and Measures).[7] In February of 1040, official standards for weights and measures were established, and six years after, the court decided to verify various official and private volume-measuring containers twice a year, between spring and autumn. But the social and economic confusion that prevailed in the latter days of Koryŏ corrupted the use of standards. This can be explained partly by the loss of the government's standard measures. According to documents of the Yi Dynasty, various standards at the early time of King T'aejo 太祖 had been so corrupted that even the length of the *chuch'ŏk* could not be officially determined. But it seems that standard measurements were not entirely nonexistent, for it was said, "Everybody at first depended upon rubbings of Ssu-ma Kung's 司馬公 stone inscription as preserved in each household, but because the original was worn out, many errors were registered." But the temporary confusion was settled in 1393 (second year of King T'aejo), when Hŏ Chu 許稠 borrowed a *chuch'ŏk* from Chin Li 陳理, the son of Chin U-ryang 陳友諒, and compared it with another *chuch'ŏk* made of ivory belonging to Chin Kang 金剛. The new standard measure that resulted was used for roads, archery ranges, astronomical instruments, and graduated scales (*kyup'yo*). It was equal to 0.6 *ch'ŏk* of the *hwangjongch'ŏk* 黃鐘尺.[8] For all that, a standard measure was still not available in 1393, so that in April 1431 (thirteenth year of King Sejong), the Board of Works (Kongjo 工曹) instructed all local officials to submit bamboo measures that would in turn be standardized at provincial or capital offices called Kyŏngsisŏ 京市署 (Marketing Control Office).[9] Later on, measures were further standardized by production of the *hwangjonggwan* 黃鐘管 (literally, yellow bell pipe), the first of a series of 12 standard pitchpipes, which also defined early Chinese volume measures. On the basis of the length of the *hwangjonggwan,* it was possible to

[7]Ibid., ch. 77, p. 26a.
[8]*Sejong sillok*, ch. 77, p. 11ab.
[9]Ibid., ch. 52, p. 5b.

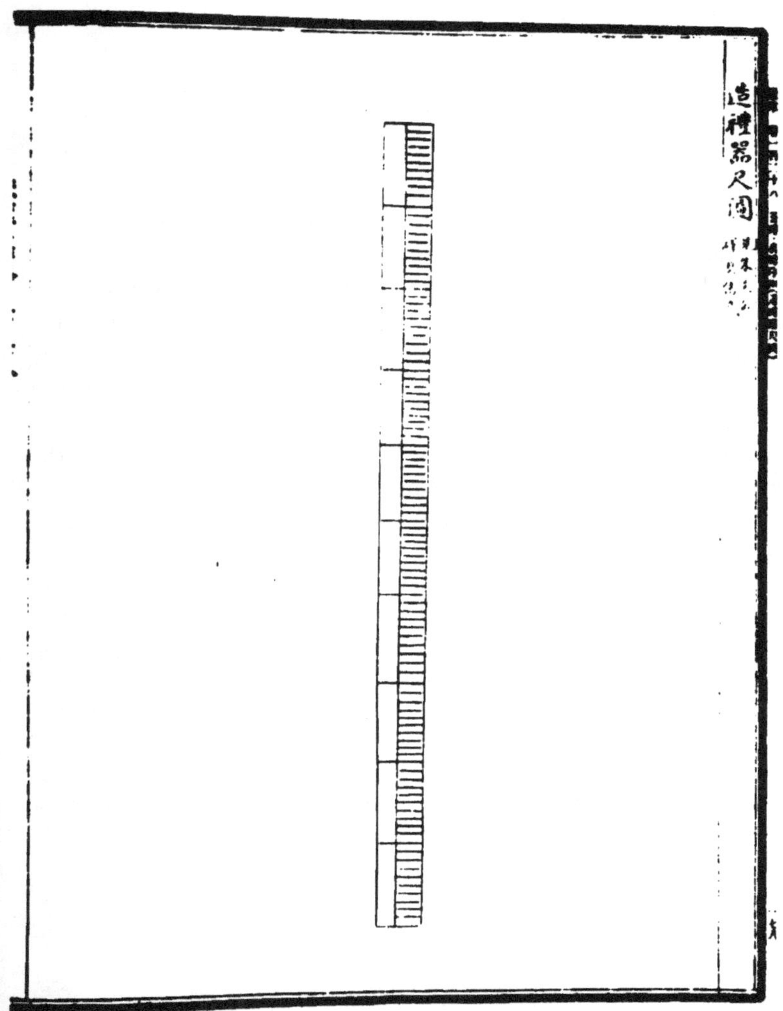

Fig. 3.1. A *choryegich'ŏk*, an official measure for making ritual wares, from a page of the *Sejong sillok* (fifteenth century). This is the only reliable surviving example of the measures of the early Yi Dynasty period.

produce *yŏngjoch'ŏk* 營造尺 (measures for architectural carpenters), in September 1446 (twenty-eighth year of King Sejong). With this, various measures, such as the *hwangjongch'ŏk*, *yegich'ŏk* 禮器尺, *chuch'ŏk*, and *p'obaekch'ŏk* were cast in bronze for standardization.[10] The length of the *hwangjonggwan* is unknown, but the *choryegich'ŏk* 造禮器尺, which was illustrated in the *Sejong sillok* (see Fig.3.1), is 28.9 cm. When this figure is used as a basis, the length of the *hwangjongch'ŏk* becomes 35.11 cm and that of the *chuch'ŏk* 21.27 cm.[11] These new measures served to produce a standard system to determine the volume of various containers. The *hwangjonggwan* was used by filling the pitchpipe with water to determine its capacity. In other words, the system of length, weight, and volume was established during the reign of King Sejong, on the basic unit of the *hwangjonggwan*, as had been done long before in China.[12]

The following is the account of the system of weights and measures as given by the *Kyŏngguk taejŏn* 經國大典 (National code):[13]

The Kongjo (Board of Works) will be responsible for establishment of weights and measures in various offices and throughout the kingdom. (Weights and measures of various counties will be determined by submitting standard measures to the provincial offices so that they may be stamped with the official seal

[10]Ibid., ch. 113, p. 36a. Of the many kinds of measures, the most popular were the *chuch'ŏk* and the *yŏngjoch'ŏk*. The *chuch'ŏk* was used mainly for determining the length of roads and for calibrating astronomical instruments and the *yŏngjoch'ŏk* for measuring volume and for weapons, shipbuilding, and architecture, especially building castles.

[11]*Sejong sillok*, ch. 128, p. 5a.

[12]According to *Sejong sillok*, ch. 113, volumes were determined by the following table:

Container	Size (in *ch'ŏk*)	Volume (in *ch'on*)
Large *sŏk* (20 *tu*)	2 × 1.12 × 1.75	3920
Small *sŏk* (15 *tu*)	2 × 1 × 1.47	2940
Tu	0.7 × 0.7 × 0.4	196
Sŭng	0.49 × 0.2 × 0.2	19.4
Hop	0.2 × 0.09 × 0.14	1.96

[13]*Kyŏngguk taejŏn* (National code), ch. 6; *Taejŏn hoet'ong* (Collection of national codes), ch. 6.

and sanctioned by the governor.) Private measures will be examined once a year at equinox by Py'ŏngsisŏ 平市署 (Marketing Control Office). The system of measurement will be determined thus: ten *li* 釐 equal 1 *p'un* 分; 10 *p'un* equal 1 *ch'on* 寸; 10 *ch'on* equal 1 *ch'ŏk;* and 10 *ch'ŏk* equal 1 *chang.* Comparing the *chuch'ŏk* with the *hwangjongch'ŏk,* 1 *ch'ŏk* of *hwangjongch'ŏk* is equal to 6 *ch'on* and 6 *li* of *chuch'ŏk*. One *ch'ŏk* of *hwangjongch'ŏk* is equal to 8 *ch'on,* 9 *p'un,* and 9 *li* in *yŏngjoch'ŏk.* As for the *choryegich'ŏk,* 8 *ch'on,* 2 *p'un,* and 3 *li* of it are equal to 1 *ch'ŏk* of *hwangjongch'ŏk;* and 1 *ch'ŏk,* 3 *ch'ŏn,* and 8 *li* of the cloth measure (*p'obaekch'ŏk*) are equal to 1 *ch'ok* in *hwangjongch'ŏk*. As for volume, 10 *chak* 勺 are worth 1 *hop* 合, and 10 *hop* one *toe* 되 (or *sŭng* 升), ten *toe* one *tu* 斗. According to this system, 15 *tu* are designated as *sogok-p'yŏngsŏk* 小斛平石 and 20 *tu* as *taegok-chŏnsŏk* 大斛全石. As for the system of weights, the water which fills the *hwangjonggwan* is divided into 88 *p'un,* each *p'un* weighing 10 *li,* 10 *p'un* weighing 1 *chŏn* 錢, 10 *chŏn* weighing 1 *yang* 兩, and finally 16 *yang* weighing 1 *kŭn* 斤. In addition, the *taech'ing* 大稱 (big scale) weighs 100 *kŭn, chungch'ing* 中稱 (medium scale) 30 *kŭn* or sometimes 7 *kŭn,* and *soch'ing* 小稱 (small scale) 3 or sometimes 1 *kŭn.*[14]

Fig. 3.2. A Korean official grain container, Yi Dynasty. Photo from *Chŏsen no Kōgei.*

[14]For this double system of weights in China, see Nathan Sivin, *Chinese Alchemy: Preliminary Studies* (Cambridge, Mass., 1968), pp. 252–254.

Years ago, an examination of the *hwangjongch'ŏk* preserved in Tŏksu Palace in Seoul revealed that it was 34.10 cm long. On the basis of this figure, the *chuch'ŏk* measure was 20.66 cm and the *yŏngjoch'ŏk* 30.65 cm. The *Kyŏngguk taejŏn* also gives the same sizes for various volume measures as those set down in the *Sejong sillok*. Thus, the system of weights and measures was firmly established during the reign of King Sejong, but a Japanese invasion destroyed the system because of the subsequent loss of metallic standard measures.

As the chaotic economic situation following the invasions of the sixteenth and seventeenth centuries receded little by little, the readjustment of weights and measures became the most important task. In February 1715 (forty-first year of King Sukchong), the King ordered the Hojo 戶曹 (Board of Taxation) to recast bronze measures to determine the volume of grains (Fig.3.2) and send them to the provinces.[15] Following the example of China, this volume measure was made with a wide bottom and narrow top so that falsifications could not be perpetrated. In February 1740, *tugok* 斗斛 (rice-containing measures) were reshaped in the pattern of the system that prevailed during the reign of King Sejong, and in April of the same year the system of weights and measures was completely reestablished on the basis of the cloth measures, or *p'obaekch'ŏk*, which had been used during the reign of King Sejong and were preserved at the Samch'ŏk provincial office.[16] The system was further legalized in 1745 through the publication of the *Soktaejŏn* 續大典 (Supplement to the national code) and strongly enforced.

THEORY OF OPTICS

The oldest record on optical phenomena refers to a bead called *hwaju* 火珠 (fire pearl), which allegedly was found inside the nine-story pagoda of Hwangyong Temple. The record is contained in the *Tonggyŏng chapki* 東京雜記 (Miscellaneous records of the Eastern Capital, Kyŏngju). According to this

[15] *Sukchong sillok*, supplement, ch. 56, p. 1a.
[16] *Yŏngjo sillok*, ch. 51, p. 8a and p. 17a.

reference, the bead was in the shape of a *go* (chess) stone; it shone clear as crystal, and one was able to see things more clearly through it. It was said that black cloth caught fire when focused under it in the sun.[17] Thus we understand the bead to have been a spherical or double-convex lens, leading us to believe that by the seventh century the people of Silla already knew about lenses.

It is therefore presumed that this "fire pearl," a sort of burning lens, was brought into Silla early in the seventh century by way of China or directly from India. In China, the first historical mention of *huo-chu* 火珠 (fire pearl) is made in the Old History of the T'ang Dynasty (*Chiu T'ang shu* 舊唐書, completed 945): ". . . fire pearl big as a hen's egg, round, white, and spotless, its rays shining several feet; it resembles crystal. If at noon you put in the sun, it will ignite mugwort placed around it."[18]

According to the study of Laufer, burning lenses were transmitted to India, not from Hellas, but from the Hellenistic Orient of the Roman Empire, in a period ranging between the fourth and the sixth century, to be passed on to China in the beginning of the seventh century.[19]

Convex mirrors and concave mirrors were other types of optical and burning instruments. The remains of convex mirrors have been found among relics dating back to the sixth century B.C.,[20] while those of the concave mirror have been found in ruins of the second century B.C. These are believed to have been made in Korea, but one can easily surmise that behind these mirrors lay the influence of Chinese culture and that of other countries. On the other hand, it is at

[17] *Tonggyŏng chapki* (Miscellaneous records of the Eastern Capital, Kyŏngju), ch. 2.

[18] *Chiu T'ang shu* (Old history of the T'ang Dynasty), ch. 197, p. 1b; cf. *T'ang shu* (History of the T'ang Dynasty), ch. 222, p. 1b, and Berthold Laufer, "Optical Lenses," *T'oung Pao*, 1915, *16*: 208–209.

[19] Ibid., p. 228.

[20] Chŏng Paek-un and To Yu-ho, *Najin Ch'odo wŏnsi yujŏk palgul pogo* (Report on the excavation of the primitive ruins in Ch'odo Island, Najin, North Korea; P'yŏngyang, 1956), p. 45. Japanese summary in *Chōsen kenkyū nempō*, 1959, *1*: 76–90.

least possible that the artisans involved in the production of bronze trinkets originally found them by chance as they polished the bent surfaces of metal. It took the Koreans quite a long time before they began to call the concave mirror by the Chinese names for burning mirrors, *yangsu* 陽燧 (*yang-sui*) or *kŭmsu* 金燧 (*chin-sui*).

One may note the symbolic form of a golden concave burning mirror cup (*kŭmsu, chin-sui*) in belts with pendants that have been found in the tombs of the fifth to sixth century in Silla.[21] These belts were made of seventeen long ornaments, of which fourteen consisted of golden oval hemispheres, resembling the *chin-sui* that were attached to the belts described in the *Li chi* 禮記 or the *T'ang shu* 唐書.

Many concave mirrors were produced in Koryŏ. According to Sudzuki, these concave mirrors were not for ladies' makeup, but were used either for some mystical purpose or for starting fires.[22] The *Tongŭi pogam* 東醫寶鑑 (Precious mirror of Eastern medicine), a renowned Korean medical encyclopedia of the sixteenth century, mentions the concave mirror as follows:[23] "When a *yangsu* is placed toward the sun it creates fire. According to Hŏ Chin, a *yangsu* is made of gold. When a golden cup is polished and mugwort absorbs the reflected sunbeams at noon it catches fire. These sunbeams are the true fire of the sun. This can be noted when mugwort begins to burn after absorbing sunbeams through a crystal ball or concave bronze mirror."

Other optical theories on lenses are found in Yi Kyukyŏng's 李圭景 section on spectacles in *Oju sŏjong pangmul kobyŏn* 五洲書種博物考辨 (Oju's book on the investigation of phenomena) published in the early nineteenth century. He called all kinds of lenses *an'gyŏng* 眼鏡 (the common word for spectacle lenses) and claimed that pure crystal, amethyst, and smoky quartz were inferior to the glass used in Dutch telescopes

[21]Osamu Sudzuki, "A Concave Mirror of Koryŏ Dynasty and its Earlier Phases," *Chōsen gakuhō*, 1959, *14*: pp. 650–651.
[22]Ibid., p. 663.
[23]*Tongŭi pogam* (Precious mirror of Eastern medicine), "Volume of Miscellaneous Diseases," ch. 9, Various Prescriptions.

and Western microscopes. He also explained the principles of concave and convex lenses, and how through their use came the production of *noan'gyŏng* 老眼鏡 (literally, old men's spectacles, for presbyopia), *changan'gyŏng* 壯眼鏡 (literally, spectacles for men in the prime of life), and *kŭnan'gyŏng* 近眼鏡 (literally, spectacles for myopia). Yi Kyu-kyŏng also explained the structure and manufacturing method of the magnifying glass, which he called *ch'ungan'gyŏng* 虫眼鏡 (literally, insect-eye glass) and of telescopes and microscopes. He said telescopes could be used not only for observation of the heavens but also for military purposes. Yi said that by using smoky quartz one could observe sunspots, the occurrence of which had been recorded throughout history.[24] Thus we know how lenses were used during the days of Silla and that they were made by carving crystals. Whether or not lenses were also made of silica glass is not clear.

Other references to optical phenomena include expositions on mirrors and lamps, but particularly interesting to us is the explanation of the "magic mirror." The magic mirror was so called because light reflected from its curved surface could throw grotesque or decorative patterns on the wall. It was for this reason that in ancient China and Japan they were carefully preserved. Scientific explanations of the magic mirror were attempted from the latter part of the nineteenth century, but no decisive exposition has yet been made. In recent years, certain Western scholars have shown that the design was chased from the back (which was later filled in with alloy) and the raised dots polished off the face, so that the bumps of strained metal sprang back to form a series of practically invisible tiny concave mirrors.[25] Yi himself explains the phenomenon as follows:

They say that when you look into the mirror in sunlight, you can see the engraved dragon on the back, but not so when

[24]*Oju sŏjong pangmul kobyŏn* (Oju's book on the investigation of phenomena), pp. 1141–1143.

[25]Cyril Stanley Smith, "Note on a Japanese Magic Mirror," *Archives of The Chinese Art Society of America*, 1963, *17*: 29–31.

looking into it in a dark room. The bronze mirror, when possessed for a long time, wears greenish specks at the back surface. Without knowing how these specks are formed, people value them as treasures. The following explains how this is so. You inscribe the shape of a dragon or flower on the back of a mirror made of pure bronze, and pour melted bronze, alloyed with twice the amount of tin normally used in mirrors, on the back surface. The front surface then is well rubbed with lead until it is smoothly coated. When looked into against a backdrop of sunlight, this mirror produces the image of a dragon inside. The natural historian Shen Kua 沈括 considered this phenomenon very enigmatic, but the Yuan scholar Wu Yen 吾衍 finally succeeded in explaining its cause. The Chinese scholar Fang I-chih 方以智 offered the following simple explanation. With the passage of time, copper assumes a greenish color, and an even greener color with the passage of more time; mercury and cinnabar (*chusa*, Chinese *chu sha* 朱砂) are then formed. Admixture of tin causes a dark greenish color, which, with the passage of time, becomes even darker. *Honggi* 汞氣 (*hung ch'i,* mercury pneuma) causes cinnabar and mercury to be formed. This process is caused by either *chigi* 地氣 (yin pneuma) or *yŏmjogi* 鹽錯氣 (literally, pneumata of salt and vinegar); and [it] is different from the transformational process of tin.[26]

In other words, Yi, quoting the theories of Chinese scholars, explained that the reflection of an engraving on the face of the mirror was caused by an inlay of different substances on the back of the mirror which reflect the sunlight. Yi's theory is certainly more interesting than earlier expositions, if no better. His description of the process is almost accurate.

THERMAL PHENOMENA AND THEIR APPLICATION

Our study of thermal phenomena may begin with an examination of Korea's room-heating system. Korea, because of its extremely cold winters, has elicited many studies on the application and preservation of heat in early times. Academic research into this field has yielded the discovery that the peculiar system of *ondol* 溫突 floor heating was in use even in the Stone Age. Early abodes of Stone Age men found near

[26] *Oju sŏjong pangmul kobyŏn*, pp. 1115–1116.

Haenggwan-myŏn 行管面, Chongsŏng-gun 鍾城郡, north Hamgyŏng Province, yielded to archaeologists large earthen jars that seem to have served as fireplaces for primitive men, and it is believed that they even had more specialized stoves like the Russian and Manchurian *pechka* to keep themselves warm. And at the Stone Age mound discovered at Songp'yŏng-dong 松坪洞, Unggi-ŭp 雄基邑, we find the remains of an *ondol* floor-**heating system.**[27]

It appears that the Koreans had a comparatively well-developed *ondol* system, using two or more tunnels, around the time of the birth of Christ. The system originated from an early one that featured a single square tunnel set against the wall. It is believed that the present-day hot floor system consisting of stone ducts under the whole floor dates back to the twelfth century, or mid-Koryŏ period.

Preservation of heat has diversified and developed since then. Utilizing the principle of heat dissemination and reflection, the *ondol* system best fitted the peculiar climate of the country. Since the whole surface of the room floor reflects the heat from the fireplace, which is used twice a day for cooking, the flame passing through multiple flues was effective for room heating as well. Room temperature could be maintained at the level of 13 to 16 degrees centigrade. The efficiency of the Korean radiant-heating system, in principle, is very similar to those of North America and Britain.[28] But because of the walls, paper windows, and paper doors, a brazier was necessary as an additional source of warmth to counteract the cold wind leaking through thin partitions. Another heating device borrowed from China was a heat-preserving bottle that was partly filled with burning charcoal. The round brass container had a mouth that could be closed after the charcoal was inserted, and ventilation holes to keep it burning.

[27]*Han'guksa* (A history of Korea; Seoul, 1959), vol. 1, pp. 11–12; Chŏng Ch'an-yŏng, "Urinara kutulŭi yuraewa palchŏn" (Origin and development of the Korean heated floor), *Kogo minsok* (Archeology and ethnology), 1966, *4*: 15–24.

[28]W. Viessman, "Ondol—Radiant Heat in Korea," *Transactions of the Royal Asiatic Society, Korea Branch*, 1948, *31*: 12–13.

ICE STORAGE

Ice was used for food storage even though it was impossible to manufacture it in ancient Korea. According to historical records, natural ice was collected in winter and stored for all-season use during the reign of King Yure 儒禮王 of the Silla Dynasty around the end of the third century and also during the reign of King Chijŭng 智證王 around the beginning of the sixth century.[29] The ice-storage house was built by digging into the foot of a mountain, and the ice chamber was con-

Fig. 3.3. Stone ice-storage house in Kyŏngju. Many scholars believe this kind of storage might date from the Silla Dynasty. This is the best example of the Yi Dynasty icehouses that are found in Kyŏngsang Province. It is a subterranean rectangular chamber with a tunnel-like ceiling supported by five equally spaced parallel arches. The heights of the arches are gradually diminished from the entrance toward the back wall so that water from the melted ice can flow out through a hole at the other end of the floor. Three vents with capstones are arranged along a longitudinal axis on the ceiling. From *Korean Arts*, ed. Ministry of Education (Seoul, 1960), vol. 3, pl. 105. Kyŏngju, first half of the eighteenth century.

[29]*Samguk sagi*, ch. 4, p. 2b.

structed of wood. But such a structure was ineffective, for the ice melted rapidly and the chamber had to be repaired almost every year.

Two such ice-storage houses existed in Seoul in 1396. *Tong-guk yŏji sŭngnam* 東國輿地勝覽 (Geographical conspectus of the Eastern Kingdom) records that one of them, mainly in use to store ice to preserve ritual offerings of food, existed at Tumop'o 豆毛浦, now Oksu-dong 玉水洞, Seoul, near the mouth of the Han river. Another one, called Sŏbinggo 西氷庫 (West Ice Storage), was located at the foot of Tunjisan 屯智山 mountain. Ice kept there was used mainly to preserve food for the palace or for consumption by court officials. The storage house accommodated eight ice-preserving chambers. A separate ice-storage house was established inside the royal court, and it was called Naebinggo 內氷庫 (Palace Ice Storage). In November 1420, the structure of the ice-storage houses was further improved for more effective preservation, and they were built with stone. An example of this new storage system, shown in

Fig. 3.4. Ice-storage house in Kyŏngju (see Fig. 3.3).

Figs. 3.3 and 3.4, is extant at Kyŏngju, the capital of the Silla kingdom.[30]

Ice was collected in Seoul from the Han River when the water froze four *ch'on* thick; staff from various royal offices were mobilized to saw the ice and transport it to the storage houses. Ice was strictly allotted to court officials or royal kinsmen according to regulations stipulated in the *Kyŏngguk taejŏn* 經國大典 (National code): only officials or soldiers of the highest grades, or retired top-level officials above the age of 70, were eligible to receive an ice supply. Ice was never lacking in the palace between May and September. Judging from the fact that ice was stored in massive quantities, it is possible that iceboxes were used at the time, but no evidence now exists to corroborate this opinion. A cold-storage boat was also built in 1789.

MAGNETISM

Although it is not clear exactly when magnetic attraction was first discovered in Korea, judging from the date of manufacture of the *sikjŏm ch'ŏnjiban* 式占天地盤 or diviner's board found in a Lo-lang tomb (reflecting Chinese settlement), the time is generally presumed to be the first century A.D. This assumption was strengthened by Wang Chen-to 王振鐸 through his research on the development of the magnet in China published in recent years. His study began with an interpretation of the writings of Wang Ch'ung 王充 (first century A.D.) and led to the conclusion that various tools for magnetic use had been in existence several centuries earlier.[31] A magnet was made in the shape of a spoon, signifying the Big Dipper. This was spun on a diviner's board, called a *shih-p'an* 式盤 and a fortune was told.

A precursor of the lodestone spoon was the *sikjŏm ch'ŏnjiban* of the type discovered in a Lo-lang tomb (before 100 B.C.).

[30] *Tongguk yŏji sŭngnam* (Geographical conspectus of the Eastern Kingdom), ch. 2, p. 30b.

[31] Wang Chen-to, "Ssu-nan chih-nan-chen yü lo-ching-p'an" (Discovery and application of magnetic phenomena in China), I, II, III; *Chung-kuo*

This instrument consisted of one round and one square plate signifying heaven and earth respectively, for heaven was considered to be round and the earth square. Over the square plate were various signs denoting the eight trigrams (*pa kua* 八卦) of the Book of Changes, the ten celestial stems (*shih kan* 十干), and the twenty-eight lunar mansion constellations; the round plate was put over the square plate and made to revolve when used to tell fortunes. The round plate developed into a spoon and evolved into the magnetic needle in Korea by the fourth or fifth century.[32]

The magnetic needle is considered to have developed further into the compass or "south-pointing needle" (*chinamch'im* 指南針), also called *yundo* 輪圖 (literally, "wheel diagram"), by the end of Silla. It was used mainly by geomancers (*chikwan* 地官, literally, land official) in connection with *yin-yang feng-shui* 陰陽風水 (geomancy). The needles were mainly produced at the Bureau of Astronomy. South, which was placed at the top of Far Eastern maps, became the reference direction for the compass, whose declination was known in China by the eighth or ninth century.[33]

Records of declination of magnetic needles are seen in the *Tongŭi pogam* 東醫寶鑑 (sixteenth century):[34]

When one rubs a pointed (iron) needle upon the lodestone, it acquires the property of pointing to the south. The best way is to attach a single fiber of fresh cotton to the middle of the needle with a very tiny quantity of wax [literally, "a piece half the size of a mustard seed"]. When this is hung up in a windless place it always points to the south. Again, if one pierces a small piece transversely with this needle, and floats it on water, it will also point to the south but will always incline toward the compass-point *pyŏng* 丙 [*ping*, i.e., S.15°

k'ao-ku hsueh pao (Chinese journal of archaeology), 1948, *3*: 119 ff.; 1950, *4*:185 ff.; 1951, *5*: 101 ff.

[32]Ibid., 1948, *3*: 119 ff.; W. C. Rufus, "Astronomy in Korea," *Transactions of the Royal Asiatic Society, Korea Branch*, 1936, *26*: pl. 3. Joseph Needham, *Science and Civilisation in China*, vol. 4, pt. 1 (Cambridge, England, 1962), pp. 263–267.

[33]Ibid., pp. 305 ff.

[34]*Tongŭi pogam*, ch. 9, tr. in Needham, *SCC*, vol. 4, p. 251.

E.]. This is because *pyŏng* belongs to the principle of Fire, and the points *kyŏng* 庚 (*keng*) and *sin* 辛 (*hsin*) [in the west], which belong to Metal [the needle being of metal] are controlled by it. Thus its [declination] is quite in accord with the mutual influences of things.—cited from the *Pen-ts'ao yen i* 本草衍義 (Dilatations upon the pharmacopocia, a Chinese source).

However, it is not clear when this discovery was made. By the fifteenth century, compasses were used for practical purposes by travelers and others, free of the monopoly of geomancers. Astronomers used the needle to determine the correct orientation of portable sundials. By the end of the Yi Dynasty, it became popular to attach a needle to almost all portable sundials. Many of the "south-pointing needles" took the shape

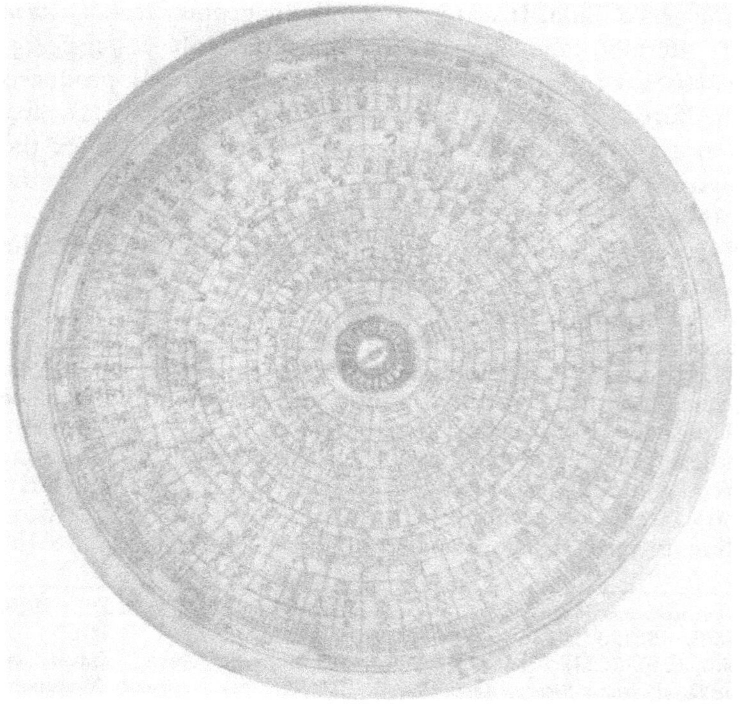

Fig. 3.5. *Yundo* compass, wood. Diameter 30.5 cm. Yi Dynasty, seventeenth to eighteenth century. Koryŏ University Museum.

of geomantic compasses (*yundo*) so that the compass is to this day popularly dubbed *yundo*. An example is shown in Fig.3.5.

The full-sized *yundo* used by geomancy officials consisted of 24 concentric circular disks, according to an official text put out by the Bureau of Astronomy in 1848. According to *Yŏngjo sillok* 英祖實錄 (Veritable records of the King Yŏngjo era), the Korean *yundo* was modeled after a Ch'ing piece imported in November 1742, and was usually made of brass.[35] Other portable forms, with five or seven circles, usually made of wood or ivory, were also used. A smaller compass with 24 compass points was also in wide use; when it was placed over the pole of the 24-circle *yundo* compass, *feng-shui* practitioners could use it for purposes of geomancy; the south-pointing needle, detached, could be used to tell directions. Because it was carried about by geomancy officials, the smaller compass was called *p'aech'ŏl* 佩鐵 (portable iron-piece).

Kim Chŏng-ŭi 金正義, a maker of the traditional Korean compass, gives the following account of compass manufacturing:[36]

The round plate is cut from a plum tree and marked with 24 compass points. Due north is marked *cha* 子 (*tzu*, rat), due south is marked *o* 午 (*wu*, horse), east is marked *myo* 卯 (*mao*, hare), and west is marked *yu* 酉 (*yu*, cock). Each circle of the *yundo* compass is also inscribed if necessary. A piece of iron, cut in the shape of a needle, is then magnetized and mounted on the surface of the compass for direction-pointing.

According to my own research, there were two distinct methods of making the traditional compass in Korea. One was to cut a round hole on the surface of the compass, in the center of which was placed an upright iron needle. A south-pointing needle was then attached to the upright. A bronze cap was pressed on to make an upper pivot. The surface was then covered with a glass cap. Another way was to add an upper needle bearing. These methods are identical with those of China.

[35] *Yŏngjo sillok*, ch. 56, p. 29a.
[36] Ye Yong-hae, *In'gan munhwajae* (Human cultural assets; Seoul, 1963), pp. 406–407.

DYNAMICS

It is no surprise to find that explanations of the theory or principles of fluid dynamics were few in ancient Korea. Discussion of the subject in this chapter is therefore limited to explaining practical tools that were made with dynamic principles in mind, with attention to the siphon, buoyancy by the use of Archimedes' principle, and the problem of kites.

References to siphons (*kalo* 渴烏) first appear in the *Sejong sillok* 世宗實錄 on April 15, 1437.[37] The principle of the siphon was first used to prevent fluctuation in the flow of water due to pressure in the portable water clock. It seems that the principle was well understood even before this, but the official record begins in the fifteenth century.

Evidence that practical knowledge of buoyancy predates this time appears in log canoes that primitive Koreans made. Series of log canoes were made into rafts, and this technique further developed to produce rafts made by binding several earthen jars to a log. Even before this, primitive men had used hollow pumpkins and similar objects to make floats. In the *Oju sŏjong* 五洲書種 there is a diagram of a floating bag made of sheepskin. The sheepskin was the most developed form of float before the appearance of rubber bags. *Oju sŏjong* also relates the existence of animal skin boats in Korea. Cow or horse skin was fastened around bamboo to make them. They needed an auxiliary log when carrying two persons.[38]

The oldest reference to the kite appears in *Samguk sagi* 三國史記 in the chapter dealing with General Kim Yu-sin 金庾信. The reference records that General Kim used flaming kites in A.D. 647, during the reign of Queen Chindŏk 眞德女王 of Silla, in order to scare away enemy troops. The flames showed against the night sky, leading superstitious troops to think the stars were forecasting ill fortune.[39] Kite flying since that time has become one of the most traditional recreations of Korea. Paper kites are designed for efficiency. According to Ch'oe Sang-

[37] *Sejong sillok*, ch. 77, p. 9 ff.
[38] *Oju sŏjong pangmul kobyŏn*, pp. 1052, 1053.
[39] *Samguk sagi*, ch. 41, biography 1.

su 崔常壽, Korean kites were usually rectangular, their sizes varying in accordance with the ages and dexterity of their users.[40] What should be noted here is the fact that bigger kites were necessary in windy regions, and smaller ones in areas less vulnerable to wind. Coastal provinces therefore had bigger kites than the interior provinces. Korean kites were made with a large hole in the middle to help them adjust to the impact of the wind and to stop the kite from turning upside down. Bamboo sticks were attached to their heads. With nothing but paper in their bottoms and with bamboo above, the kites flew easily.

Moving Machines

REVOLVING AXLE AND WHEEL

At the beginning of the Stone Age in Korea, between 2000 and 3000 B.C., primitive men who originated the so-called comb ceramics made their first rotary tool, the quern (*chŏnsŏk* 碾石). With little modification, the quern is still used in rural homes today.

Another simple rotary machine was called *mulle* 물레 in Korean. It was used as a spinning wheel, potter's wheel, and waterwheel. Since remains of these wheels have been discovered dating to the New Bronze Age, they seem to have been used from the sixth to fifth century B.C. It is presumed that men of the New Bronze Age, by utilizing spinning wheels, made rough clothing material from hemp. This theory is further born out by the discovery of clothing material from Yayoi Japan,[41] which was under the influence of Korea's comb-pattern pottery culture.

The primitive spinning wheel developed with the emergence of the Iron Age. Just before or after the birth of Christ, these primitive tribes knew something of sericulture too. There are

[40]Ch'oe Sang-su, *Han'guk chiyŏnŭi yŏn'gu* (A study of Korean kites; Seoul, 1958), with English summary, pp. 6 ff.
[41]Kobayashi Yukio, *Kodai no gijutsu* (Ancient technology in Japan; Tokyo, 1962), pp. 42–55.

historical records, in the form of the *Pien han chuan* 辨韓傳 (Commentary on the Pien Han, Pyŏnhan), to the effect that they knew how to weave wide silk cloth.[42] Lo-lang tombs have produced some silk cloth. By this time, the wooden spinning wheel must have been in use.

By the time the Three Kingdoms were established in Korea, handlooms (*nŭngjikki* 綾織機) developed, and during the latter part of the fourth century some of these products of Silla were shipped to Japan. The art of cloth weaving seems to have been highly developed in Korea.

Another area in which wheels were used is represented by the two-wheeled cart. It seems to have existed at the time of the birth of Christ, judging from remains in Lo-lang tombs. One piece of evidence is a painting of an ox-drawn cart on the wall of the Tomb of the Dancers (*Muyongch'ong* 舞踊塚) of the Koguryŏ Dynasty (see Fig. 3.6).[43] The wheels were large and thin. Another picture of a cart, discovered on the wall of the Tomb of Twin Pillars (*Ssang'yŏngch'ong* 雙楹塚), also from the time of Koguryŏ, shows a thicker wheel, smaller in shape.

In China, iron tires were not used on wheels of vehicles until the fourth century, during the Chin Dynasty. Wheels were made by bending dried wood, with spokes made to protrude half an inch, so that the wheel itself could be protected from constant contact with the ground. But later on, the wheel became smaller, and the spokes no longer protruded. The use of a steel tire on the wooden wheel enabled the manufacturer to make the wheel thinner but larger. The vehicle shown in the Tomb of the Dancers seems to have been an improved model. It used a wooden axle made of plum wood, and animal-fat lubricant to enable it to revolve faster and more smoothly.

A modification of the revolving axle was also applied to the folding fan. Making its first appearance in Koryŏ times, during

[42]Hong I-sŏp, *Chosŏn kwahaksa* (A history of Korean science; Seoul, 1946), pp. 27–28.

[43]Ikeuchi Hiroshi, *T'ung-kou* (The ancient site of Kao-kou-li [Koguryŏ] in Chi-An district, Northeastern China; Tokyo and Hsin-Ching, 1938), vol. 2, plate 12.

Fig. 3.6. Painting of an oxcart. Drawn on the left wall of the main chamber, Tomb of the Dancers. The wheels are distinguished by their large, thin frames. Koguryŏ Dynasty, ca. fifth century (from *T'ung kou*).

its inception it was called *chŏpsŏn* 摺扇, which literally means "folding fan." The Koryŏ fan reached Sung China, where it was called the "new model." During the early part of Ming, the folding fan became popular in China, in a form modeled after the Korean version.[44]

Before the fifth century, two devices in which the principle of the lever was applied to the revolving axle were made in Koguryŏ, and have been used until recent times. One was the lever device used for drawing water from a well and the other was a foot-operated threshing tool.[45] They were simple devices but were significant in that the Korean utilized an important principle of dynamics in designing them. The lever device was also used in hydraulic trip-hammers to replace human power.

POTTER'S WHEEL

The potter's wheel first appeared in the Kimhae area on the southern coast around the time of the birth of Christ. It is believed to have been made under the influence of Han technology via Lo-lang.[46] The potter's wheel was being used by about the fourth century to produce stoneware as well. The skills required for the wheel-building of Japanese Sue pottery were exported from Silla and Paekche. The revolving axle of the potter's wheel in its inception consisted of a horizontal axis that later evolved into a vertical axis. A device was added for attaching a roller at the top of the axle to smooth the surface of ceramics. The axle was revolved by use of a hemp cord wound around the axle and pulled by both hands.[47] The principle of the potter's wheel was also applied to wooden and metallic handicraft manufacturing. One example of this is the numerous *kogok* 曲玉 (*magatama*), or "curved jewels," which have been recovered from Silla tombs. These are of jasper, pink agate,

[44]Ch'oe Nam-sŏn, *Urinara yŏksa* (A Korean history; Seoul, 1952), p. 59.

[45]*Anak chesamhobun palgul pogo* (Report on the excavation of tomb no. 3 in Anak; P'yŏngyang, 1958), ed. by To Yu-ho, p. 19. The lever device is often called the swape.

[46]Kim Wŏn-yong, *Han'guk misulsa* (A history of art in Korea; Seoul, 1968), p. 29.

[47]Kobayashi, *Kodai no gijutsu*, pp. 38–39.

crystal, jade, and other stones, but the most numerous were made of jade with a hardness of 7. The potter's wheel was used for grinding and polishing or boring a hole in the comma-shaped jade.

The potter's wheel began to be foot-propelled in the fifth century A.D. Compared with the hand-propelled wheel of China, it was more suitable for large urns and large plates, but because of the small table and slower revolving speed it was not sensitive enough. The foot-propelled wheel developed so that by the end of the fifteenth century and beginning of the sixteenth century it was being exported to Japan. The kickwheel (*ch'uknongno* 蹴轆轤), very popular in the Kyūshū area of Japan, clearly was adopted from Korea during the Hideyoshi invasion.[48] The kickwheel was constructed as follows: A large axial pole was planted three *ch'ŏk* into the ground. A foot plate was fastened just one *ch'ŏk* above ground level. Three *ch'ŏk* above that extended a vertical axis topped by a revolving plate. These two plates were connected to revolve together.

The lathe used in the brassware shops of the Yi Dynasty was also foot operated. It differed from the potter's wheel in that the latter was on a vertical axis, whereas the *sŏnch'a* 旋車 (lathe) was horizontal. The axle was rotated by pedaling. Urns were held at the tip of the wooden axle, and the surface of the object was smoothed with a cutting tool.

WATERWHEELS

Before the invention of the steam engine, the water mill and windmill did much of what later became its work. They were of course low-powered compared with modern types, but were made with the most precision of any machine of their time, with the exception of clock systems. Although it is not clear when the water mill was first used as a power-generating machine, it is thought to have been between the fifth and sixth centuries. What is sure is that it existed in the year A.D. 610. There is a historical record that the Monk Tamching 曇徵 had

[48]See, for instance, Yoshida Mitsukuni, *Yakimono* (Japanese pottery and porcelain; Tokyo, 1965), p. 37.

manufactured a wheel-driven mill in Japan in that year.[49] Although it was not referred to specifically as a water mill, it was identified variously as *suyŏn* 水碾 (waterwheel-driven roller mill), *sŏgyŏn* 石碾 (stone roller mill), *konyŏn* 輥碾 (roller mill), *soyŏn* 小碾 (hand roller), *suae* 水磑 (waterwheel-driven stone gristmill), *p'ungae* 風磑 (wind gristmill), or simply as *ae* 磑 (gristmill). In this instance, the two words, *yŏn-ae* 碾磑 (gristmill and roller mill) combine to render the meaning of "mill" in English. The *suyŏn* and *suae* that Monk Tamching had reputedly made meant "waterwheel-driven roller mill" in the case of the former, and "waterwheel-driven stone gristmill" in the case of the latter. It would be reasonable to presume that the water mill he had invented at the time would have been similar to those depicted on Koguryŏ tomb murals.

Water mills therefore went by various names, depending on the nature of their uses. When used in connection with grinding, they were called *suyŏn, suae, suma* 水磨, and *sunong* 水礱 (waterwheel-driven grinding mill). When used for milling, they took the name of *kidae* 機碓 (trip-hammer, mechanical waterwheel). The structures of these machines, one of which is shown in Fig. 3.7, are so well known that they do not require separate explanations.

Besides being used for husking or grinding grains, water mills were also used to pump water and for other irrigational purposes. It is extremely difficult to trace their development because there is only one reference to the use of a water mill in the *Koryŏsa*.[50] According to the exhortation quoted there, the Chinese were using waterwheels in times of drought, whereas Korean farmers depended upon reservoirs and embankments. The court was therefore to encourage provincial farmers to adopt the new system. In other words, the water mill was not used generally by the population until the end of the Koryŏ Dynasty.

[49]*Nihon shoki* (Chronicle of Japan [to 697]), ch. 22; Yi Kwang-nin, *Yijo surisa yŏn'gu* (A study of irrigation under the Yi Dynasty; Seoul, 1961), pp. 11–12.
[50]Yi Kwang-nin, *Yijo surisa yŏn'gu*, pp. 87–88.

Fig. 3.7. Waterwheel-driven trip-hammer. This type of water mill was the most popular in Korea. From *Imwŏn simyukchi.*

Without many variations, the waterwheel, whether used as a mill or quern, has been handed down to posterity. Many of them could still be seen during the 1940s in their ancient forms and some are in evidence even at the present in rural areas. These are Yi Dynasty models.

The Treatise on Geography of the *Sejong sillok* states that there was a large watermill at Segŏmjŏng 洗劍亭, on the outskirts of Seoul. These wheel-driven mills seem to have been operated by the Sant'aeksa 山澤司, one of three official organizations of the Board of Works, for the purpose of husking or polishing government grain. Some of the mills of the Yi Dynasty have been described as wheel-driven, but they were in fact operated by gear trains. The Yi Dynasty saw popularization of horizontal axles, rather than vertical axles.

There existed also various kinds of rotary water pumps. Representative were the *yonggolch'a* 龍骨車 ("dragon backbone" machine, or square-pallet chain pump) and *t'ongch'a* 筒車 (noria), and it was after the sixteenth century that such Western types of water pumps as the *yongmich'a* 龍尾車 (Archimedes screw) and *okhyŏngch'a* 玉衡車 (double-cylinder force pump) were imported. Beginning in the latter days of Koryŏ, water pumps were generally called *yonggolch'a* or *bŏnch'a* 翻車 (square-pallet chain pump). Invented by the Chinese during the Han Dynasty, around A.D. 170, this mechanism was handed down from generation to generation without the slightest modification until the beginning of the Yi Dynasty. The *yonggolch'a* consisted of a wooden channel in which an endless chain of square wooden plates revolved to transport water upward from the river or reservoir to the place where water was needed. Various resources were used to generate power, including hands, feet, animals, and water, but at the beginning of the Yi Dynasty the *bŏnch'a* (foot-driven pump) was the most widely used machine.

Pak Sŏ-saeng 朴瑞生, a Korean emissary who visited Japan in December 1429, submitted a report to the royal palace which stimulated production of a new kind of water pump.[51] When conventional water-driven pumps were used against rapid currents, the wheel ran too quickly; and foot power had to be applied when the current was too slow. This new machine was so improved over the conventional foot pedal pumps that King Sejong ordered dissemination of the new machine to provincial areas. The new wheel was called *chagyŏk such'a* 自激水車 (barrel wheel) or, in English, peripheral pot pump. Water

[51] *Sejong sillok*, ch. 46, p. 14ab.

was pumped by means of pots attached to the wheel. From that time, the pot pump was called *waesuch'a* 倭水車 (Japanese waterpump), because of its Japanese origin, and the conventional pump was called *tangsuch'a* 唐水車 for its Chinese origin. Despite much effort of the government to propagate the new pot pump, it failed to gain adoption, largely because of the poverty of the farmers, which made new machines unattainable, and because of climatic conditions under which people had to rely on rainwater much of the time. When drought attacked, the pot wheel's use was limited. Although primitive in form and requiring much energy, well buckets were easier to handle and more useful. Labor they had in plenty, but money was scarce among the farmers, and waterwheels were expensive in Korea because of a shortage of the wood suitable for them. The most important reason that the pot pump did not gain acceptance, as Yi Kwang-nin 李光麟 and Kim Yong-sŏp 金容燮 pointed out,[52] was the fact that Korea's natural conditions, such as the quality of soil and rainfall, brought about the development of irrigation ponds and dikes. These facilities were sufficient for growing rice, which dominated Korean agriculture, and their use made the manufacture of waterwheeels unnecessary.

Whenever drought attacked the country, efforts continued even after the Sejong reign to popularize water pumps, but to no avail. The problem produced official discussions of administrative measures to be taken, but not much more was done. Between the reign of King Sŏngjong and 1496, efforts were exerted to popularize a hand-operated water pump imported from China by Ch'oe Pu 崔溥. In 1502, Chŏn Ik-kyŏng 全益慶 reputedly manufactured very delicate and effective water pumps. The *Yijo sillok* contains other references which show that in 1546 efforts were again directed to popularizing water pumps imported from China.[53] But, as mentioned earlier,

[52]Kim Yong-sŏp, "Chosŏn hugiŭi sudojak kisul—iyŏnggwa surimunje" (The techniques of paddy rice cultivation in the late Yi Dynasty), *Asea yŏn'gu*, 1965, *8*.2: 293.

[53]Jeon Sang-woon, *Han'guk kwahak kisulsa* (A history of science and technology in Korea; Seoul, 1966), p. 152.

social, economic, and climatic conditions in Korea prevented the Yi Dynasty from gaining acceptance for the more convenient pumps; and absence of government policy discouraged the development of any simple substitute. Under such conditions, it would have been unreasonable to expect a speedy development of water pumps, so much so that discussion of inherent technological limitations becomes unnecessary.

But sporadic efforts continued up to the middle of the Yi Dynasty to fight drought by using improved forms of water pumps. King Hyojong also endeavored to promote a new type of pump that he had seen in Shenyang (Mukden), where he had been exiled. But he too failed. After that followed the efforts of physicist Yi Min-ch'ŏl, during the reign of King Sukchong, who produced a very effective waterwheel-operated pump in 1683. He dispatched several models to each province but failed to produce satisfactory results.[54]

The effort to popularize new forms of waterwheels, however, was again taken up by the time King Yŏngjo assumed the throne. Behind this revival of the discussion lay the introduction of the *yongmich'a*, imported from China. Neither the efforts of the government nor private encouragement succeeded in persuading the farmers. The most popular model was the so-called *tapch'a* 踏車, or foot-pedal type. It was not only simple to make and thus cheap, but also more effective than well buckets. The foot pedal types are still extant in southern provinces, and they are much in use in salt mines.

PULLEY AND CRANE

It is true that it was Chŏng Yag-yong 丁若鏞 (1762–1836) who first used cranes, when he oversaw the construction of Suwŏn Castle in 1784.[55] However, the models Chŏng used were based on pulleys employed at the time of the Three Kingdoms to fly Buddhist flags on staffs. Such iron pulleys were

[54]Yi Kwang-nin, *Yijo surisa yŏn'gu*, p. 95; *Sukchong sillok*, ch. 8, pp. 1ab, 8b.
[55]*Hwasŏng sŏngyŏk ŭigwe* (Records and machines of the Emergency Capital Construction Service), introduction, pp. 1a–4b, 74b–76b; Hong, *Chosŏn kwahaksa*, p. 229.

generally called *nongno* 轆轤 (pulley wheel). Chŏng himself described his pulleys as *nongno*, in his *Hwasŏng sŏngyŏk ŭigwe* 華城城役儀軌 (Record and machines of the Emergency Capital Construction Service), published in 1800, as fixed pulleys.

Chŏng Yag-yong made a crane creatively employing the principles of pulley and windlass from the *Ch'i ch'i t'u shuo* 奇器圖說 (Illustrated explanations of wonderful machines), a seventeenth-century Chinese Jesuit treatise. The structure and parts of the crane are shown in Figs. 3.8 and 3.9. The thing to be lifted is connected to eight movable pulleys, four of which are above and the rest below, which in turn are con-

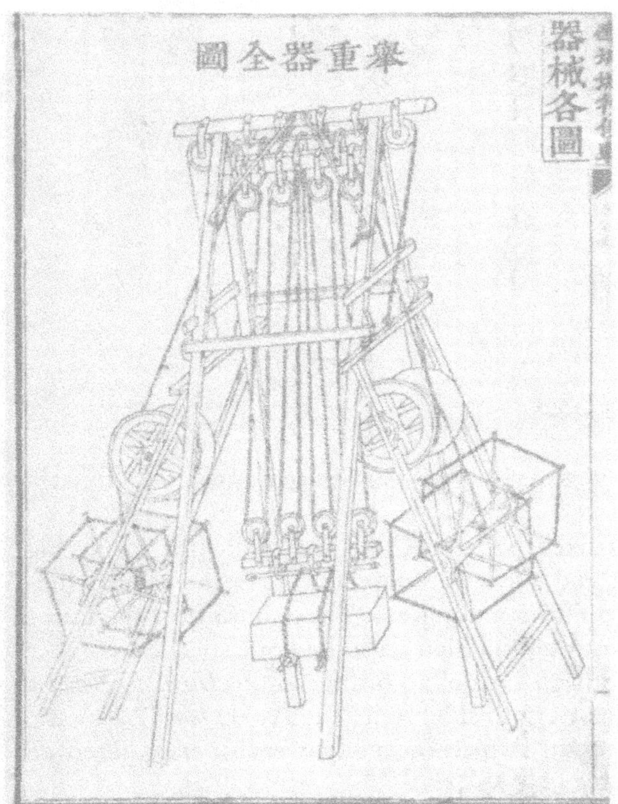

Fig. 3.8. Diagram of crane made by Chŏng Yag-yong in 1791. From *Hwasŏng sŏngyŏk ŭigwe*.

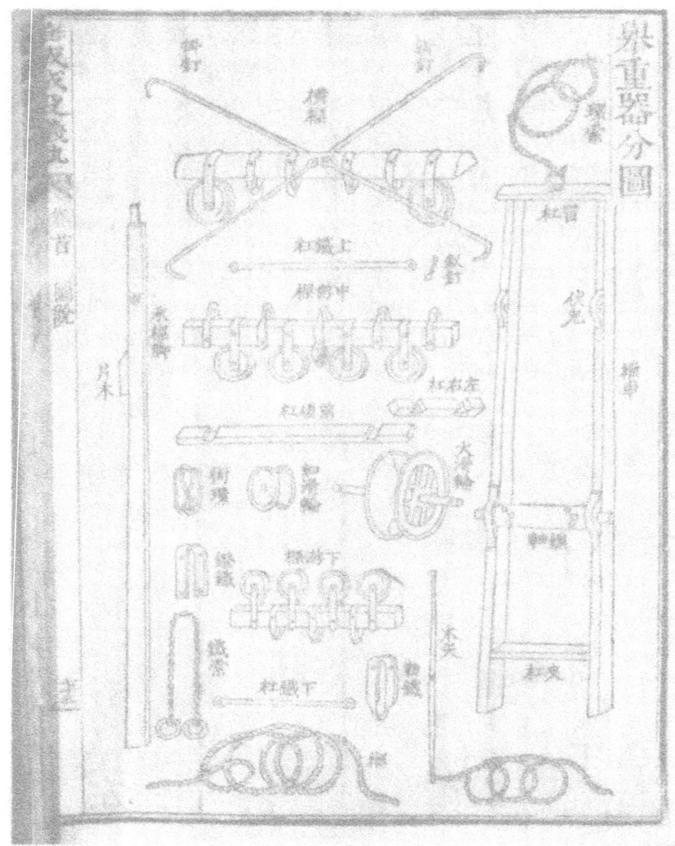

Fig. 3.9. Diagrams of parts of Chŏng Yag-yong's crane. From *Hwasŏng sŏngyŏk ŭigwe*.

nected to eight fixed pulleys. These are again tied to the big pulleys at right and left, and wound on the frame of the windlass. In order to lift something, one has only to turn the frame of the windlass at right and left with the same speed.

In building Suwŏn Castle, according to the *Hwasŏng sŏngyŏk ŭigwe*, 30 laborers at both sides could lift 12,000 *kŭn* (7.2 metric tons) with this crane, which means that each person lifted an average of 400 *kŭn* (240 kg).[56]

[56]*Hwasŏng sŏngyŏk ŭigwe*, introduction, pp. 74b–80a; Needham, *SCC*, vol. 4, p. 220 n.

AUTOMATIC MECHANICAL TOYS OF SILLA

During the period 763–764, Silla mechanics contrived a set of automatic mechanical toys, called the Manbulsan 萬佛山 (literally, Mountain of Ten Thousand Buddhas), which utilized the force of the wind.

Samguk yusa describes the structure of the Manbulsan as follows:[57]

When the king was informed that Emperor T'ai-Tsung of T'ang was an ardent believer in Buddhism, he wanted to send a gift to the Emperor and ordered his mechanics to make one. It was supposed to be a miniature mountain, about 2 meters high, on a multicolored carpet decorated with a carved wooden base and beautiful jewels. The mountain was composed of exquisite rocks and stones, and there were dolls of various countries which were singing, dancing, and playing musical instruments in the valleys and caves. One could hardly distinguish the miniature from a real mountain where breezes blew, bees and butterflies flew, and swallows and sparrows danced. Many Buddha images the size of 1 *ch'on* to 8–9 *p'un* were also placed on the mountain. The size of the heads ranged from that of Indian millet to that of green beans, but they had distinct ears, eyes, mouths, and noses. The mountain was such that one could only see vaguely what it looked like. It was called Manbulsan. Furthermore, gold and precious stones were used for various kinds of flowers, trees, streams, hermitages, towers, and pavilions. Although small in scale, they appeared lifelike. There were more than a thousand dolls in the image of monks at the front, and underneath three golden bells were hanging in a pavilion with whale-shaped belfries. When wind blew, one saw all the monks bow to the ground and chant sutras, since the bells contained automatic devices for this. It was really difficult to describe what Manbul looked like. Emperor T'ai-Tsung, on receiving that automatic mechanical toy, used all kinds of superlatives in praising it. He commented that the craftmenship was not that of a man, but that of a genius endowed by Heaven. In fact, all who looked at it were amazed at its delicacy.

It is interesting that Silla, which was under the profound influence of T'ang culture and technology, produced an elaborate mechanical toy that amazed even the T'ang emperor.

[57] *Samguk yusa*, ch. 3, Manbulsan.

Evidence that Silla mechanics were capable of producing such automata can be found in the artificial channel for floating wine cups at royal garden-parties in P'osŏkchŏng (Pavilion), which was built in the seventh or eighth century. It seems certain that this, as Needham has pointed out,[58] was made in imitation of a similar channel in China. If we assume that it was made to carry a kind of mechanical toy, it would be reasonable to think that the Manbulsan was influenced by China or by Islamic mechanical toys via China.

Mechanical Clocks

INTRODUCTION AND MANUFACTURE OF THE STRIKING CLOCK

It is difficult to say exactly when the striking clock was first introduced and manufactured in Korea. *Injo sillok* 仁祖實錄 (Veritable records of the King Injo era) and other sources claim it was first introduced by Chŏng Tu-wŏn 鄭斗源. Chŏng, returning from Ming China in July 1631, brought with him a striking clock together with Chinese translations of Western science books and a telescope.[59]

My own investigation, however, suggests that it was first introduced in Korea between 1568 and 1607, during the reign of King Sŏnjo 宣祖, and it had come from Ming China.[60] Two factors support my opinion: one is a record telling us that Korea sent a striking clock to Tokugawa in Japan in 1607; and another is a presumption that it might have been used by Ming troops during that sixteenth-century Japanese invasion of Korea. That Korea had sent a striking clock to Japan soon after the restoration following the Japanese invasions cannot be disputed. The Yi Dynasty therefore must have owned it before that. Since few Koreans knew how to use it, the striking clock must have been only a curiosity.

58 Needham, *SCC*, vol. 4, pp. 160–161.

59 *Injo sillok*, ch. 25, p. 5b; *Kukcho pogam* (Precious mirror of the Yi Dynasty), ch. 35.

60 Takabayashi Hyōgo, *Tokei no hanashi* (Story of timekeepers; Tokyo, 1925), pp. 47–48; Yamaguchi Ryūji, *Nihon no tokei* (Japanese clocks; Tokyo, 1944), pp. 17–18.

Kim Yuk 金堉, in his *Chamgok p'iltam* 潛谷筆談 (Collected jottings of Chamgok = Kim Yuk), said: "Chŏng Tu-wŏn brought a striking clock back from Peking; and we knew not the delicate method of handling it or how it told time."[61] It is clear, therefore, that the use of the striking clock was not known until that time. In 1636, Kim Yuk himself went to Peking and saw one, but he, too, professed ignorance of its mechanism.[62] According to *Chamgok p'iltam*, "a man named Yu Hŭng-bal 劉興發 who hailed from Miryang made a long study of the striking clock brought in by a Japanese between 1650 and 1659, and succeeded in learning its principles. He explained that the alarm struck every hour (i.e. every two Western hours) in the following manner: the alarm struck nine times at *cha* 子 (*tzu*, rat) and *o* 午 (*wu*, horse) hour; eight times at *ch'uk* 丑 (*ch'ou*, ox) and *mi* 未 (*wei*, sheep) hour; seven times at *in* 寅 (*yin*, tiger) and *sin* 申 (*ch'en*, dragon) hour; six times at *myo* 卯 (*mao*, hare) and *yu* 酉 (*yu*, cock) hour; five times at *sul* 戌 (*hsu*, dog) and *chin* 辰 (*shen*, monkey) hour; four times at *sa* 巳 (*ssu*, snake) and *hae* 亥 (*hai*, boar) hour; and once in the middle of each hour "[63] If the record is correct, Yu undoubtedly was the first person to understand the working of the clock in Korea.

The clearest record of clock manufacturing appears in *Hyŏnjong sillok* 顯宗實錄 (Veritable records of the King Hyŏnjong era), of October 1669, which claims that professor of astronomy Song I-yŏng 宋以穎 first manufactured one.[64] But what he made was a striking astronomical clock, of which more later.

In April 1715 Hŏ Wŏn 許遠, an official of the Bureau of Astronomy, had a reproduction of a Chinese striking clock made and installed at the Bureau of Astronomy.[65] In the year 1723, the Bureau of Astronomy also had a clock (*munsinjong* 問辰鐘)

[61] Yi Nŭng-hwa, *Chosŏn kitokkyo kŭp oegyosa* (History of the Christian missions and foreign relations in Korea; Seoul, 1928), p. 4.
[62] Ibid.; Hong I-sŏp, *Chosŏn kwahaksa*, pp. 260–261.
[63] Ibid.
[64] *Hyŏnjong sillok*, ch. 17, p. 35ab.
[65] *Sukchong sillok*, ch. 56, pp. 3ab; Jeon, *Han'guk kwahak kisulsa*, p. 106.

made on a Chinese model, "for it was of very delicate structure, very sensitive to day and night and shadow and rain, so that it was easy to know the position of the sun."[66] In 1759, during the reign of King Yŏngjo, a man named Na Kyŏng-jŏk 羅景績 manufactured an astronomical clock (*hujong* 候鐘), the history of which is very fragmentary.[67] The fact that clock manufacturing remained stagnant, despite the possession of technical knowledge, reflects the inclination of Yi Dynasty people to rely on waterclocks and sundials, a preference that reflected socioeconomic conditions of the time as well as lack of a policy of promoting technological development.

Nam Pyŏng-ch'ŏl 南秉哲 claims that a *chamyŏngjong* 自鳴鐘 (automatic striking clock), *hŏmsiŭi* 驗時儀 (clock) and other mechanical clocks were manufactured between the time of King Chŏngjo 正祖 and King Ch'ŏlchong 哲宗. His book, *Ŭigi chipsŏl* 儀器輯說 (Collected writings on astronomical instruments) presents detailed explanations of their structures and principles.[68] Judging from another of his statements, that in France there were some 2,000 clockmakers at the time, and that annual production of mechanical clocks numbered 12,000, Koreans knew that clock manufacturing had become a large enterprise in the West. Meanwhile, in China a new specialist called *shih-chi-shih* 時計師, or clocksmith, had appeared, and the Japanese had developed their own "Japanese clock" (*wadokei* 和時計) which told time in reverse order as described earlier. In Japan, such clocks circulated fairly widely. Mechanical clocks of the Yi Dynasty (Fig. 3.10) thus had two main origins: China and Japan. But since both of them had originated in the West, they were more or less alike.

Because of the poor guild system and government policy of the Yi Dynasty, it failed to lay the foundation for modern me-

[66]*Munhŏn pigo* (Comprehensive study of civilization), ch. 3, p. 6a.
[67]*Tamhŏnsŏ* (Collected writings of Tamhŏn = Hong Tae-yong), *Oejip*, ch. 3; Jeon Sang-woon, "Yissi Chosŏnŭi sige chejak sogo" (A study of timekeepers in the Yi Dynasty). pp. 107–108.
[68]*Ŭigi chipsŏl* (Collected writings on astronomical instruments), vol. 2, p. 1 ff.

Fig. 3.10. Striking clock. Iron and bronze, 34.0 cm × 12.5 cm. Ca. seventeenth century. Seoul National University Museum.

chanical industries, beginning with the production of clocks. I believe that Korea's present-day underdevelopment in the field of machine manufacture actually began then.

STRUCTURE OF THE STRIKING CLOCK

Since the structure and principles of the striking clock have been explained in connection with astronomical clocks, I shall limit myself in this section to examination of the escapement of the striking clock and its dial. The escapement of Western mechanical clocks at first was the verge and foliot. After Galileo

Fig. 3.11. Mechanism of Song I-yŏng's armillary clock. This automatically striking clock is based on the principle of Huyghens' pendulum and the Chinese-Arabic tradition of clockmaking. Yi Dynasty, 1664–1669. Koryŏ University Museum.

discovered the principle of the pendulum, Huyghens invented the pendulum clock between 1656 and 1657. During the Yi Dynasty, pendulum clocks with verge escapements were made. The astronomical clock of Song I-yŏng (see Figs. 3.11, 3.12) was of this type. In China, the verge and foliot was called *t'ien-heng* 天衡 or *t'ien-p'ing* 天枰; in Japan, it was called *hōtenpu* 棒天符.

The face of the striking clock was divided into a 12-hour day, with an *o* (horse) situated at the top and a *cha* (rat) at the bottom. The bell struck 4 to 9 times to indicate the hour. Although the striking clock was first made of iron and was contained in an iron box, this was later replaced by bronze and brass. Still later, brass was also used for the works.

Fig. 3.12. Song-I-yŏng's armillary clock (see Fig. 3.11).

Printing Technology

SOME PROBLEMS IN THE HISTORY OF PRINTING TECHNOLOGY IN KOREA

It has not yet been determined exactly when wood-block printing was invented in Korea, but many seem to agree that it started around the early eighth century. A printed scroll of the *darani* (Sanskrit *dharani*, charm or prayer) scripture, found in the Sŏkkat'ap pagoda of Pulguk Temple 佛國寺 in October 1966, is known to be the oldest of this kind in the world, printed between 704 and 751 (see Fig.3.13). This means either that wood-block printing was quickly transmitted from China

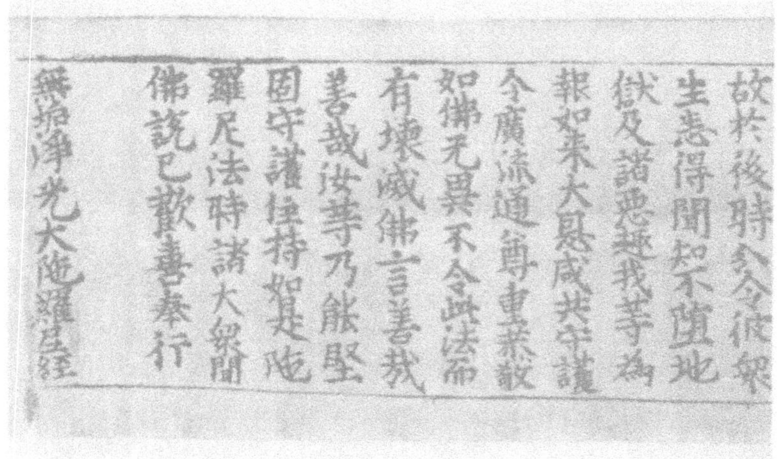
故於後時令彼衆
生悉得聞知不墮地
獄及諸惡趣我等爲
報如來大慈咸共守護
令廣流通尊重恭敬
如佛無異不令此法而
有壞滅佛言善哉
善哉汝等乃能堅
固守護住持如是陁
羅尼法時諸大衆聞
佛說已歡喜奉行
無垢淨光大陁羅尼經

Fig. 3.13. A printed scroll of the *darani* scripture, 704–751.

to Korea, or that Silla developed it first.[69] Although various inventions that served as the basis of printing were made in China, it is at least possible that the synthesis that led to the invention of wood-block printing occurred in Korea.

It had been only thirty years since Silla unified the peninsula. It had just completed consolidating the ruling structure of the kingdom by introducing the Chinese culture of the T'ang period, including Buddhism, which was meant to be a state religion. Pulguksa was thus built out of a fervent desire to secure Buddha's protection for the kingdom. Silla must have been in great need of making many copies of the *darani* scripture, of which a newly translated version had been introduced from China. Turning to the technology of Silla, provincial governments used copper seals that were made and

[69]Goodrich estimates that the Chinese invented wood-block printing between 712 and 756. T. F. Carter and L. C. Goodrich, *The Invention of Printing in China and its Spread Westward* (New York, 1955), pp. 37–41; 44–45. On the other hand, Son Po-gi holds that Korea was the first country in the world to develop wood-block printing, on the grounds that the scroll in Sŏkkat'ap was printed about 706. See Son Po-gi, "Han'guk insoe kisulsa" (A history of Korean printing technology), *Kwahak kisulsa*, in *Han'guk munhwasa taege* series, vol. 3 (Science and technology; Seoul, 1968), pp. 974–975.

distributed during the reign of King Munmu 文武王 (661–680),[70] and ceramists produced a number of beautiful tiles using wooden molds. Therefore, conditions for inventing wood-block printing were ripe in Silla, as they had been for many centuries in China.[71] One can proceed further to believe that the printing technique of Silla was transmitted to the Japanese and helped them to print the *darani* scripture in the 770s.

By the beginning of the eleventh century, during the reign of King Hyŏnjong 顯宗, the wood-block printing system had improved to such a degree that a decision was made to engrave 6,000 wood blocks of Buddhist scriptures.[72] This served as the occasion for the art of wood-block printing in Koryŏ to reach the level it had attained in Sung China.[73] The chief motivation that led to this printing of Buddhist scriptures was religious: to invoke the power of Buddhism to repel the Khitan invaders, who had already occupied the capital of Songak 松岳 (present-day Kaesŏng) castle. But the collapse of the Koryŏ Dynasty and new internal problems prevented development of Korea's printing technology, which did not have the impact that its Western counterpart would later have. In addition to these, the prejudice of Koreans themselves against Korea's culture and civilization, produced between 1900 and 1945

[70]*Samguk sagi*, ch. 7, King Munmu.

[71]"In 847, Silla mobilized all able wood-block printing technicians, and had them print the voluminous Buddhist scriptures called the Eighty Thousand Scriptures on Kŏje Island. They were published with gorgeous design." This passage in the *Chosŏn sach'al saryo* (Historical documents of the Korean temples; Seoul, 1911), vol. 1, pp. 496–499, which has been regarded as unacceptable at face value, deserves reexamination. The project referred to is not believed to have been as huge as The Eighty Thousand Scriptures, but was probably a collection of wood-block sutras. Pak Si-hyŏng wrote in his article "Chosŏnesŏ kŭmsok hwalchaŭi palmyŏnggwa kŭ sayong" (On the invention of metallic type and its employment in Korea), *Yŏksa kwahak* (Historical science), 1959, *5*: 33–41, that it is hard to believe what is written in the *Chosŏn sach'al saryo.*

[72]See for instance, Son, "Han'guk insoe kisulsa," pp. 978, 979 n.

[73]Chang Hsiu-min, *Chung-kuo yin-shua-shu ti fa-ming chi ch'i ying-hsiang* (The invention of printing in China and its influence; Peking, 1958), p. 131.

by foreign rule and national turbulence, did no justice to one of our most brilliant cultural achievements.

WOOD-BLOCK PRINTING OF KORYŎ

Printing of books other than Buddhist scriptures began during the reigns of King Chŏngjong 靖宗 (1036–1046) and King Munjong 文宗 (1047–1083). Printing was also done in the provinces. By the time of King Sukchong's 肅宗 reign, between 1096 and 1105, numerous books printed from Koryŏ blocks were stored at the Pisŏgak 秘書閣 (Secretariat Library).[74]

The 10,000 blocks of the Tripitaka originally engraved during the eleventh century were burned up during the Mongol invasion, and the government was removed to Kanghwa Island. When hopes of restoring the Tripitaka were given up, King Hyŏnjong, to implore the benevolence of Buddha, again ordered the engaving of 80,000 blocks, a process that consumed sixteen years, beginning in 1236.[75] Now enshrined at Haein Temple 海印寺 (see Fig. 3.14), these constitute the so-called Eighty Thousand Scriptures, the largest ancient relic of block printing in the world.

The art of wood-block printing in the Koryŏ Dynasty thus developed from government projects on two different occasions. The printing blocks are mainly divided into types from three different sources: Sung types, Yuan types, and Koryŏ types. The clearest and most beautiful were those of Sung origin: representative of these is the Tripitaka of Koryŏ. The rest of the types were not as delicate as those of Sung design, nor were the letters as even in size. Wood-block printing in Koryŏ therefore did not progress linearly, but was much affected by who paid the expenses of engraving, and whether the project was official or private.

I shall now examine wood-block manufacturing, citing the example of the Eighty Thousand Scriptures. The blocks used

[74] *Koryŏsa*, ch. 7, p. 38b; *Hsuan-ho feng shih Kao-li t'u ching* (Illustrated record of an embassy to Korea in the Hsuan-ho reign-period, 1124), ch. 6, Palace Offices.

[75] See, for instance, Kim Sang-gi, *Koryŏ sidae sa* (A history of the Koryŏ period, Seoul, 1961), pp. 917–918.

Fig. 3.14. A block of the Eighty Thousand Scriptures. The blocks, which are the oldest in existence, were made from magnolia wood to dimensions of 24 cm × 65 cm × 4 cm. Koryŏ Dynasty, 1236–1252. Haein Temple.

were made from magnolia wood, soaked and boiled in saltwater to extract gum, and dried for years in the shade. The wood came from Cheju 濟州, Wando 莞島, Kŏjedo 巨濟島, Ullŭngdo 鬱陵島, and other islands. This process, used to this day in the manufacture of wooden furniture, prevented the blocks from developing cracks and warps. Each block was 24 cm × 65 cm × 4 cm in size, capped by bronze on its corners and held straight by two wooden sticks. Its surface was painted with lacquer. In other words, much effort was directed to preparation for long storage. Each block weighed between 2.4 and 3.75 kg. The vertical length of the printing surface was 22 cm. Horizontal lines were drawn across the space only at top and bottom, with no vertical lines. Each surface of the block contained 23 lines of characters, each line consisting of 14 letters, each 1.5 cm square. Characters were engraved on both sides of the blocks. These blocks were stored in two separate warehouses at Haein, each warehouse 60 m × 10 m. Ventilation was regulated inside the warehouse for ideal preservation of the blocks. Other details of the manufacturing process are given by Mun Ki-hyŏn 文琦鉉, the engraver who is quoted in *In'gan munhwajae* 人間文化財 (Human cultural assets) as follows:[76]

> The ideal material for blocks is pear-tree wood, which is sensitive to the knife because of its soft quality and has a smooth surface. The next best material is *kŏjesu* 거제수 (trautvetteria), because it is hard and does not break open. Persimmon wood is also welcome for its tough quality. The wood is good for use during autumn and winter, when damp wood can be avoided. Damp wood corrupts easily. The cut tree is soaked in sewage for more than two months so that the bark comes off. To repel insects, the wood is boiled in saltwater. The surface is then cut smoothly. Paper on which the text has been written is then glued to the surface with the characters facing the block. If thick paper is used the excess layers are stripped off, leaving the reversed characters clearly visible on the surface. It is the job of engravers to carve out each letter with a knife. When a good calligrapher is not available, pages of the book to be reproduced may be torn and

[76]Ye, *In'gan munhwajae*, pp. 305–306.

glued to the surface. The man sits before a table when carving. Normally, each engraver engraves 10 lines, each line composed of 21 characters. An expert may engrave 40 lines on both surfaces in one day. Character size on the block surface varies between text and annotations.

Paper was then pressed against the inked surface to print. Each volume did not exceed 100 pages. In Japan and China, the fascicule was sewn at four places. In the case of Korea, the book was bound in five places, and the string used was thicker than that used in the other two countries. The cover paper was also thicker than that of Japan and China; five or six pages of white paper were glued together, then soaked in honey, pressed hard, and then ornamented. Before placing the cover on a *nŭnghwa* 菱華 (rhombic designed) plate for pressing, the cover is dyed in an infusion of yellow pagoda-tree petals or simply left white. In the case of a dyed cover, red string is used for binding; when the cover is not dyed, blue string. The cover was often also dyed blue, a color favored for Buddhist literature.

TECHNICAL BACKGROUND OF TYPECASTING

The wood-block printing system evolved cast metal blocks, and for cheaper printing, movable type finally emerged. But the use of metallic type involves several important conditions, including special ink, clean but hard paper that can sustain the pressure of metallic type, techniques for casting, and finally, bronze trays. These conditions were met by the early thirteenth century.

The paper manufacturing techniques that Korea derived from China had so developed by the late eleventh century, during the reign of King Munjong, that paper manufactured in Korea was exported to China. Among items officially traded between Sung and Koryŏ and offerings sent by the latter to the former, were included, according to records, 2,000 sheets of white paper and 400 ink sticks. Later on, Yuan demanded that Koryŏ export paper for Buddhist scripture printing, often as much as 100,000 sheets at a time.[77]

[77]Kim, *Koryŏ sidae sa*, pp. 917–918.

A special type of ink must have been used for printing, in all probability a glue-processed ink of some kind. Ordinary ink would have sufficed for wood-block printing, but not for metallic type. In 1314, well after the invention of metallic type, Wang Chen 王禎 began to experiment with its use but was unsuccessful. He printed his book on agriculture with wooden type. Sarton thinks that the main reason for doing so was probably lack of oily ink.[78] The special ink for printing, made mainly of pine resin smoke, and *yuyŏnmuk* 油煙墨 ink sticks made of lamp-black, were exported to Sung China. Judging from the fact that much perfumed oil was processed in Korea for export to neighboring China, I believe that the people of Koryŏ indeed knew how to use special oil.

In 1126 and 1170, there were big fires in the Koryŏ kingdom in which tens of thousands of books were destroyed.[79] At that time China was suffering from the ceaseless war between Sung and the Chin Tartars. The war made it extremely difficult for Koryŏ to import books from Sung. Consequently, Koryŏ had no choice but to replace the books with its wood-block printing. However, wood blocks had the disadvantage of costing too much money and time for small editions. In addition, Koryŏ was short of the wood suitable for wood blocks or wooden type, while it abounded in bronze. The artisans of Koryŏ had inherited Silla's metalworking and bronze casting techniques.[80] They were also experienced in minting bronze coins of good quality and in casting letters on bronze bells by the blast-furnace (*ku-chu* 鼓鑄) method that they learned from Sung in 1101.[81] It was natural that the Koryŏ dynasty should have seen the shift from wooden to metal type; and by sometime around 1234, Koryŏ artisans had printed with cast type twenty-eight copies of *Sangjŏng yemun* 詳定禮文 (Detailed and

[78]George Sarton, *Introduction to the History of Science*, vol. 3, pt. 2 (Baltimore, 1948), p. 1563.
[79]*Koryŏsa*, ch. 127, pp. 18–19a, ch. 15, p. 11a, and ch. 16, p. 44b.
[80]See Chapter 4, pp. 239, 248.
[81]*Koryŏsa*, ch. 79, p. 11b.

authentic code of ritual) on Kanghwa Island, where the Koryŏ government took refuge to resist the Mongol invasion.[82]

TYPECASTING AND TYPESETTING TECHNIQUES

Before the emergence of the matrix, the only method of casting metallic type was casting into sand molds. All other methods had failed. There is one clear reference to early casting methods in *Yongjae ch'onghwa* 慵齋叢話 (Collected essays of Yongjae), written by the Yi Dynasty scholar Sŏng Hyŏn 成俔 (1439–1504).[83] The character was carved out on a wood blank. Then the wooden type was pressed against a matrix of sea sand. Metallic bronze was poured into the indentation to form a piece of type. The rough type was then polished for final shaping. Early type is said to have been hollow inside, with very sharp ends. The end was sharpened, it is believed, to fasten it against the typesetting plate, and as for the hollow inside, it is believed it was meant to save copper. In spite of difficulties in the manufacturing of type ordered in February 1403, several hundred thousand pieces were produced in several months.[84] King T'aejong overruled all dissent from his advisors against the endeavour, and the type that resulted was called Kemi 癸未 type.

This first Yi Dynasty bronze type, however, possessed several defects. For one thing, because the type was cast in sand, its surface was rough and uneven, often with one corner broken. Because of the uneven surface, it did not print well. One reason that this happened was that the craftsmen did not get a clean impression from the sand molds. Another problem was that the characters were not uniform in size, and thus did not fit well into the line. This is attributable to uneven engraving of the wooden characters.

[82]Pow-key Sohn (Son Po-gi), "Early Korean Printing", *Journal of the Americal Oriental Society*, 1959, *79*. 2: 101–103.
[83]*Yongje ch'onghwa* (Collected essays of Yongjae = Sŏng Hyŏn), ch. 1; Hong, *Chosŏn kwahaksa*, p. 213.
[84]*T'aejong sillok*, ch. 5, p. 7a; *Yangch'onjip* (Collected works of Yangch'on = Kwŏn Kŭn), ch. 22.

The printing technique for Kemi type was substantially identical to that developed by Pi Sheng of China in the eleventh century for movable type. This system involved the following process: the type was set in a bronze tray filled with beeswax. Then the surface of the type was pressed even with a plate, inked, and pressed against white paper. Because of the many stages involved in printing, this system was not efficient since it was possible to print only several sheets a day.[85] Because the setting was done with beeswax, the type often moved under impact of the upper plate, and whenever this happened, the printer had to stop the press and rearrange the type. To produce a clean sheet, the type had to be cleaned every time, and this also disarranged the type.

A printed version of *Sipch'ilsach'an kogŭmt'ongyo* 十七史纂古今通要 (*Shih-chi shih tsuan ku chin t'ung yao*, Critical review of the seventeen Chinese dynastic histories), printed in 1412 with Kemi type, shows unclear letters, which may be attributable not so much to bad inking as to the complexity of the process.

When efficiency is taken into consideration, Kemi type is greatly inferior to Koryŏ wood blocks. However, it was economical in that the type could be used over a long period to print limited editions of different books. King T'aejong thus insisted on using this type, and herein lies his great contribution. The farsighted conviction of the king was recorded by Kwŏn Kŭn 權近, in his *Chujabal* 鑄字跋 (Colophon on metallic type): "His Majesty, in the spring of the first year of Yung-lo 永樂 (1402), told his ministers that in order to rule, and rule well, it was necessary to grasp the principles of governing man through books. Separated by the sea, books from China come rarely; wood blocks break easily; and it has been impossible to print all the books of the Empire. From now on, type for individual characters shall be made of copper, and whenever a new book is acquired, it shall be printed and disseminated widely."[86] This was a new motive for the development of printing technology.

[85]Kim, "Yissi Chosŏn chuja insoe sosa—chujasorŭl chungsimuro," pp. 124–125.
[86]*Yangch'onjip*, ch. 22.

From Kwŏn Kŭn's account, we realize that the king's decision was also based on a desire to be financially independent. He did not attempt to collect printing expenses from his subjects, but paid them out of the treasury, supplemented by contributions from others interested in the project.

IMPROVEMENT OF TYPE AND PRINTING WITH MOVABLE TYPE

The conviction of King T'aejong was handed down to King Sejong, another wise monarch who promoted science and technology in the early Yi Dynasty. With his accession to the

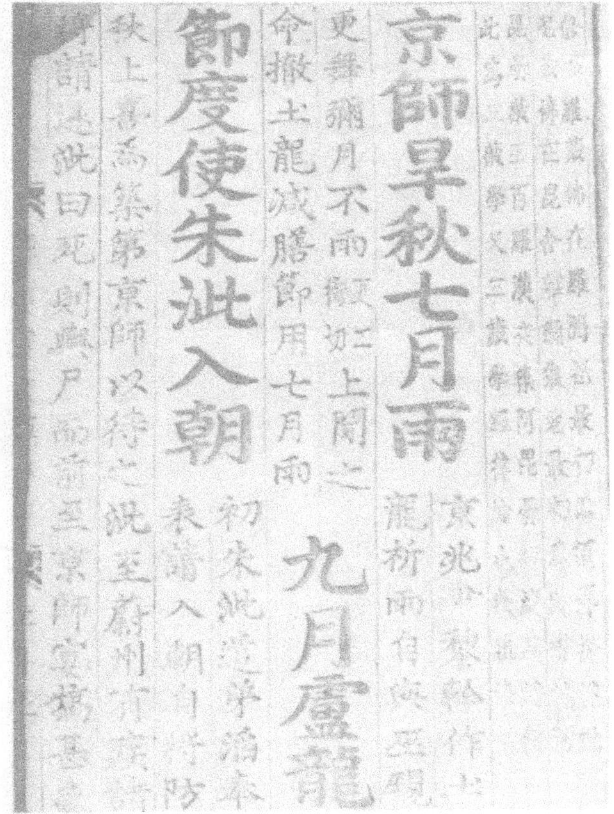

京師旱秋七月雨

命撤土龍減膳節用七月雨

九月盧龍

節度使朱泚入朝

秋上書爲築第京師以待之

Fig. 3.15. Copy printed from the Kyŏngja bronze type of 1420, measuring 1.2 × 1.1cm. Because its characters were designed to conform to a module, it made typesetting more efficient.

throne, printing made spectacular progress. In November 1420, two years after he was enthroned, the King ordered invention of a new printing press and bronze type to solve problems presented by the Kemi type.

The record for March 24, 1421 in *Sejong sillok* 世宗實錄 (Veritable records of the King Sejong era) gives the following account: "The King personally ordered Yi Ch'ŏn 李蕆, the Minister of Works, and Nam Kŭp 南汲 to improve the typesetting process to prevent misalignment. The experts succeeded in inventing a new technique that accomplished this aim even

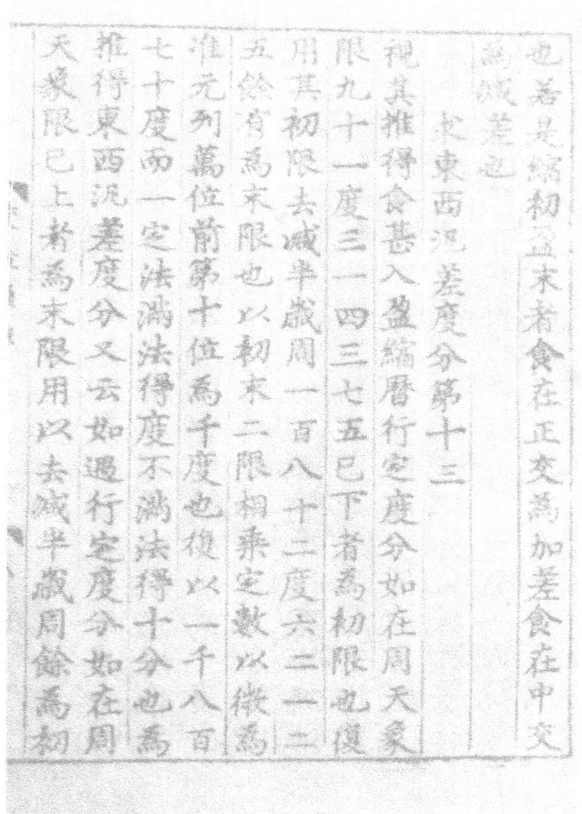

也若是縮初盈末者食在正交為加差食在中交
為減差也
求東西汎差度分第十三
視其推得會甚入盈縮曆行定度分如在周天象
限九十一度三一四三七五已下者為初限也復
用其初限去減半歲周一百八十二度六二一二
五餘者為末限也以初末二限相乘定數以微為
准元列萬位前第十位為千度也復以一千八百
七十度而一定法滿法得度不滿法得十分也為
推得東西汎差度分又云如遇行定度分如在周
天象限已上者為末限用以去減半歲周餘為初

Fig. 3.16. Copy printed from the Kabin bronze type of 1434

though the type was still bedded in beeswax. The King congratulated the artisans by offering them much wine and meat."[87]

Under the supervision of Kim Ik-chŏng 金益精 and Chŏng Ch'o 鄭招 a new type called Kyŏngja 庚子 was developed (see Fig. 3.15). It was identical to Kemi in size. Based on a modular arrangement of exactly 21 letters in a single line, it made typesetting much easier. As Pyŏn Kye-ryang 卞季良 pointed out in his *Chujabal* 鑄字跋, the shape of the type was so precise and the system so efficient that it was possible for one man to set type for some twenty sheets a day.[88] The finished pages were much cleaner than wood-block prints. It took twenty years to develop Kyŏngja type, but in the fourteen years between Kyŏngja and Kabin 甲寅 types there was even greater progress. By 1434, it was possible to print forty sheets a day.[89]

Fig. 3.17. Bronze form tray, traditional Korean model, Yi Dynasty, eighteenth century. The text is a page of the *Kukcho pogam* (Precious mirror of the Yi Dynasty) set in the Chŏngyu (Kabin) type of 1777. Koryŏ University Museum.

[87] *Sejong sillok*, ch. 11, pp. 15b–16a, ch. 18, p. 10b.
[88] Ibid., ch. 18, p. 10b.
[89] Ibid., ch. 65, pp. 3b–4a.

Fig. 3.18. Japanese bronze form tray, Tokugawa period, early seventeenth century. This example seems to copy the Korean design shown in Fig. 3.17. Japanese bronze type printing technology made its start in 1593 with Korean bronze type that was brought into Japan during the Hideyoshi invasion. Photo from Nakayama Kushirō, *Sekai insatsu tsūshi*.

Kyŏngja type was invented to solve technical problems, but Kabin type (Figs. 3.16 and 3.17) was a result of efforts to produce a better-looking letter. The project to produce new type, which began in 1434, with Yi Ch'ŏn leading such distinguished scientists as Kim Ton 金墩, Kim Pin 金鑌, Chang Yŏng-sil 蔣英實, Yi Se-hyŏng 李世衡, Chŏng Ch'ŏk 鄭陟, and Yi Sun-ji 李純之, succeeded finally in producing some 200,000 pieces of type.[90] They set a precedent by producing both small and large type, so that a choice could be made depending on need.

With the emergence of Kabin type, the Yi Dynasty system of printing with movable type had reached success. It became a traditional printing technique in Korea. At about the same time, Gutenberg succeeded in mechanization of printing by using a press, but no such machinery was used by the Yi Dynasty officials. Although it is true that absence of mechaniza-

[90]Ibid.

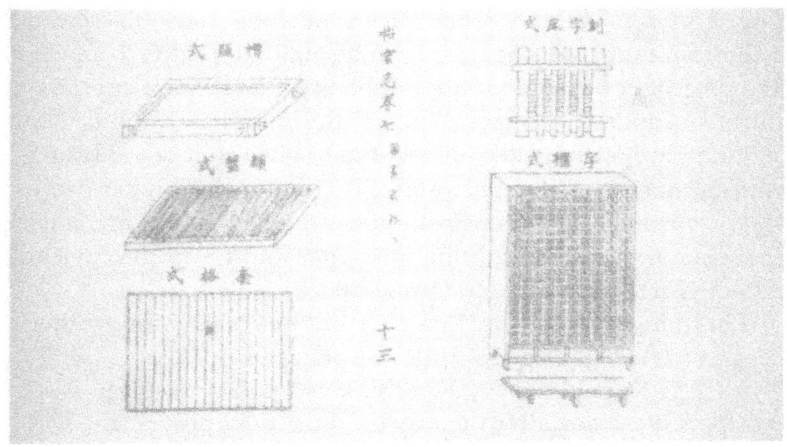

Fig. 3.19. From the *Wu-ying-tien chu-chen-pan ch'eng-shih* (Imperial Printing Office manual for movable type) of 1776. Top: "beds" for cutting type; lower right: type cases; lower left: type trays. Sketch from the *Imwŏn simyukchi*.

tion limited the progress of printing technology, the achievement of the Yi Dynasty should not be minimized, for the most important problems involved in early printing techniques included making of type, trays, and ink (Fig. 3.19).

The Kabin type was the longest used type of the Yi Dynasty, and it should be specially noted that it contributed greatly to the publication of various astronomical books, which was one of the scientific highlights of Sejong's reign.

Thus in about two centuries, from the late Koryŏ to the early Yi Dynasty, the Koreans made remarkable advances in printing techniques. The Koryŏ Dynasty developed the mold necessary for typecasting, and the Yi Dynasty improved typesetting techniques.

As mentioned above, the type of 1403 was made with a sharp tail and hollow body so that the pieces could be easily put into beeswax at the time of setting. However, this method of typesetting, an imitation of the Chinese one, proved to be inefficient: the type moved easily and required large amounts of wax, which took too much time to solidify. In 1421, the method was improved by remodeling the type and standardizing its size. Lines between pieces of type were narrowed to the minimum so that even with a small amount of beeswax the type might be fixed without difficulty. There were repeated efforts to make a copper alloy with improved fluidity for molding hard, clean, and durable type. The result of analyzing some bronze type demonstrates that these advantages were largely attained. The bronze type made in 1455 included some zinc and lead in addition to copper and tin in the typical ratio of Korean bronze, 75:25 and 80:20.

After that, and during first half of the Yi Dynasty, other types, such as Pyŏngjin 丙辰, Kyŏngo 庚午, Ŭrhae 乙亥, Ŭryu 乙酉, Kapchin 甲辰, Kyech'uk 癸丑, and Pyŏngja 丙子 were produced one after the other. Marked improvement was made in the economy of the printing method. Type became gradually smaller, as the state of the art allowed legibility to be maintained. Books became cheaper, and problems of binding and storing were lessened. Printing technology had so progressed by the time of King Chungjong's 中宗 reign that it was possible to publish the *Munwŏn yŏnghwa* 文苑英華, an enormous collection of Chinese literature, in only a thousand volumes by using Pyŏngja type.

The two centuries of progress in printing technology, however, were reversed with the Japanese invasion of the sixteenth century. Most of the bronze type possessed by the dynasty was either lost or looted by the Japanese. But printing with movable type survived, and book publishing was continued with movable wood type. This was similar to what had happened during King T'aejong's reign when the government relied on wood type. In February 1668, printing with bronze type was revived and Sulsin 戊申 type, modeled on Kabin type, was produced. A total of 100,000 pieces were manufactured: 61,000 pieces of a large size and 46,600 smaller pieces. This supply of type contributed to the survival of the printing system in the second half of the Yi Dynasty.[91]

[91]See Kim Wŏn-yong, "Yijo hugiŭi chuja insoe" (On cast-type printing in the latter period of the Yi Dynasty), *Hyangt'o Sŏul*, 1959, *7*: 7 ff.

Table 3.1. Chronological Listing of Korean Metallic Fonts

Year	Name of Type	Material	Size in Centimeters	Type Extant?	Printed Examples Extant?
1234	Koryŏ	bronze	?	no	no
1403	Kemi-ja	bronze	1.3 × 1.4	no	yes
1420	Kyŏngja-ja	bronze	1.2 × 1.1	no	yes
1434	Kabin-ja	bronze	(L) 1.4 × 1.4	no	yes
1436	Pyŏngjin-ja	lead	(L) 2.0 × 2.5	no	yes
1450	Kyŏngo-ja	bronze	1.7 × 1.7	no	yes
1455	Ŭrhae-ja	bronze	(L) 1.6 × 2.0 (M) 1.2 × 1.5 (S) 1.0 × 0.7	no	yes
1457	Chŏngch'uk-ja	bronze	(L) 1.9 × 2.0	no	yes
1465	Ŭryu-ja	bronze	1.0 × 1.0	no	yes
1484	Kapchin-ja	bronze	1.0 × 1.0	no	yes
1493	Kyech'uk-ja	bronze	(L) 1.7 × 2.0 (S) 1.3 × 1.3	no	yes
1516–1519	Pyŏngja-ja	bronze	1.2 × 1.2	no	yes
1573–1580	Chaeju Kabin-ja	iron	1.4 × 1.4	no	yes
1618	Muo-ja	bronze	1.5 × 1.7	no	yes
1668	Samju Kabin-ja	bronze	1.4 × 1.6	yes (?)	yes
1677	Hyŏnjong sillok-ja	bronze	1.3 × 1.5	yes	yes
1682	Han'gu-ja	bronze	(L) 1.0 × 1.3 (S) 1.0 × 0.6	yes	yes
1721–1724	Munjip-ja	iron	1.1 × 1.1	yes	yes
1772	Imjin-ja (Saju Kabin-ja)	bronze	(L) 1.5 × 1.8	yes	yes
1777	Chŏngyu-ja (Oju Kabin-ja)	bronze	(M) 1.5 × 1.3	yes	yes
1782	Chaeju Han'gu-ja	bronze	(L) 1.0 × 1.4	yes	yes
1795	Chŏngni-ja	bronze	(L) 1.1 × 1.4 (S) 1.1 × 0.7	yes yes	yes yes
1815	Ch'wijin-ja	bronze	(L) 1.1 × 1.5 (S) 1.1 × 0.7	yes	yes
1816–1821	Chŏnsa-ja	bronze	(L) 1.0 × 1.5 (S) 1.0 × 0.7	yes	yes
1858	Chaeju Chŏngni-ja	bronze	(L) 1.1 × 0.7	yes	yes
1858	Samju Han'gu-ja	bronze	(L) 1.0 × 1.3	yes	yes

From the end of the seventeenth century, distinguished families had their own private types made.[92] But it was the government that controlled the casting and printing,[93] so that civilian contribution in this field was very limited.

Printing was done by the Chujaso 鑄字所 (Office of Type-founding) and the Kyosŏgwan 校書館 (Office of Bibliography). Casting and printing were done by division and specialization of labor[94] into functions such as letter cutter, caster, sorter, collator, and typesetter.

Modern Firearms

FIREARMS IN KOREA

Historically, Korea has always cultivated constructive over destructive technology. The development of armaments was limited to the minimum needed for deterrence or defense and self-protection, until long and repetitive enemy depradations necessitated the study of warfare. It was not cultivated as a political means as in China and Europe, but was rigidly controlled by the government, and ambitious research into its development was discouraged. Perhaps related sciences advanced more slowly than they would have if they had been included under the umbrella of military research. The failure

[92]Kim Tu-jong, "Yissi Chosŏnŭi hugi hwalchaŭi kaejuwa Chamgok Kim Yuk sŏnsaeng samdaeŭi konghŏn" (Innovations in typecasting during the late Yi Dynasty by Kim Yuk and his immediate descendants), *Paek Nak-jun paksa hwan'gap kinyŏm kukhak nonch'ong* (Commemorative papers for the sixtieth birthday of Dr. Paek Nak-jun; Seoul, 1955), pp. 142–170.

[93]Goodrich and Sarton have pointed out that a Korean invented the type mold. See Goodrich, *The Invention of Printing in China*, and Sarton, *History of Science*, vol. 3, pp. 734–735, 1562–1563.

[94]Judson Daland, "The Evolution of Modern Printing and the Discovery of Movable Metal Type by the Chinese and the Koreans in the Fourteenth Century," *Journal of the Franklin Institute*, 1931, *212*: 208–234; *Chōsen bunkashi* (Cultural history of Korea; Tokyo, 1966), vol. 2, p. 7; Sohn, "Early Korean Printing," p. 99. Chemical analysis of the Ŭrhae-ja type (cast in 1455) by Daland: Cu 79.45%, Sn 13.2%, Zn 2.3%, Fe 1.88%, Pb 1.66%, Mn 0.48%, Spectroanalysis of eighteenth-century type: Cu 84%, Sn 7%, Pb 7%, Zn 1%, Fe 0.1%, Bi 0.001%, Si 0.01%, Ni 0.05%, Ag 0.01%–0.05%, P 0.05%.

later to take an active interest in the manufacture of modern weapons should be attributed partly to Korean national characteristics, but is also explained by Korea's peaceful tributary relationship with China, its closest geographical neighbor, and by the dependent state of mind that this relation caused.

The backward military technology of Korea, which greatly conditioned the foreign relations of the Yi Dynasty, must be studied by anyone who wants to understand fully the development of the modern state.

INTRODUCTION OF EXPLOSIVES AND FIREARMS

Although the exact time is hard to determine it is presumed that explosives and firearms were first introduced to Korea during the first half of the fourteenth century. The military section of the *Koryŏsa* refers to arrows being shot through a tube in 1356, which suggests that firearms came from Yuan China.[95] After discovering the long range of the tube-launched arrow, the Koryŏ people tried hard to learn how they could be mass-produced and how explosives could be made. The need to repel Japanese invaders further heightened their desire for these new weapons. Because the Koreans did not know how to produce explosives, in November 1373 the government sent to Ming China an emissary specifically instructed to import explosives.[96] This measure was taken following a decision actively to seek out the Japanese on the surrounding seas and capture their prowling ships. Although the Ming court at first rejected this request, they nevertheless were persuaded to share explosive materials the next year when they, too, were faced with increased Japanese maneuvers off Chinese coasts. On confidential instruction of the emperor,

[95]Hŏ Sŏn-do, "Yŏmal sŏnch'o hwagiŭi chŏllaewa paltal" (The introduction and development of firearms in Korea, 1356–1474, pt. 1), *Yŏksa hakpo*, 1964, *24*: pp. 6–13; Arima Seiho, *Kahō no kigen to sono denryū* (The origin of firearms and their early transmission; Tokyo, 1962), pp. 225–230; Hong, *Chosŏn kwahaksa*, p. 121.

[96]*Koryŏsa*, ch. 44, pp. 19a, 29b.

the Koryŏ court was given 50 *kŭn* (ca. 30 kg) of saltpeter and 100,000 *kŭn* of sulfur.[97] Other necessary materials were also included. Since 50 *kŭn* of saltpeter was an insignificant amount, the whole deal meant little more than that Koryŏ received sulfur, which is but one material for producing explosive gunpowder.

Convinced that firearms were the most effective weapon against the Japanese, Ch'oe Mu-sŏn 崔茂宣, following a long and involved study, succeeded ca. 1375 in learning from a Chinese named Li Yuan 李元 the technique of extracting saltpeter. One theory has it that Ch'oe had traveled all the way to Yuan China to learn the secret, and this does not seem to be an impossibility.[98] One way or another, he learned how to extract saltpeter, the major material for gunpowder, from the soil. By October 1377 Koryŏ formally established a government organization, the Hwat'ong Togam 火㷁都監 (Office of Explosives Handling), to oversee production of explosives.[99]

Soon after the establishment of this organization, production of explosives increased rapidly, and no fewer than twenty different firearms were made. By April 1378, Koryŏ not only had units armed with firearms,[100] but had assigned them to duty on battleships as a deterrent against the Japanese (perhaps despite their inaccuracy). Koryŏ records speak proudly of the Chinp'o 鎭浦 battle of 1380 and the Chindo 珍島 Island battle of three years later.[101]

Early firearms, however, were not designed to hurl percussive or explosive projectiles. Mainly, they shot flaming arrows to set fire to their targets.[102] Iron balls were used as missiles only toward the end of Koryŏ, because in the early days, iron could not cover the distance of a rocket, about 200 *po* 步 (paces).

[97]Hŏ, "Yŏmal sŏnch'o hwagiŭi chŏllaewa paltal," pp. 12–13.
[98]*Tongguk yŏji sŭngnam*, ch. 2, pp. 24ab; Hŏ, "Yŏmal sŏnch'o hwagiŭi chŏllaewa paltal," pp. 13–16.
[99]*Tongguk yŏji sŭngnam*, ch. 2, pp. 24ab; *Koryŏsa*, ch. 133, p. 31b.
[100]*Koryŏsa*, ch. 81, Treatise on Military Affairs, section 1.
[101]Ibid., ch. 114, Biography of Na Se; ch. 126, Biography of Pyŏn An-yŏl.
[102]Hŏ, "Yŏmal sŏnch'o hwagiŭi chŏllaewa paltal," p. 21.

As for kinds of firearms, the *Koryŏsa* and *T'aejo sillok* list eighteen.[103]

Ch'ong 銃 and *t'ong* 筒 were in the main small firearms; *t'ong* 烔 were medium-sized firearms; *p'o* 炮, 砲 (cannon) heavy firearms; the "great commander," "second commander," and "third commander" were large launchers using huge rockets. The largest of these, of course, was the "great commander," the arrows of which, according to *Yungwŏn p'ilbi* 戎垣必備 (Manual for a military commander), were 11 *ch'ŏk*, 9 *ch'on* (ca. 2.5 m) long.[104]

FIREARMS OF THE EARLY YI DYNASTY

With the fall of Koryŏ and the rise of the Yi Dynasty, the use of explosives, mostly because of the negative policy of Yi Sŏng-gye 李成桂, the founder of the new monarchy, was largely limited to making firecrackers for holiday amusement, and firearms were left to rust. What traditions of manufacture there were, were feebly maintained by the personal efforts of Ch'oe Mu-sŏn and his son.

Unlike his father, however, King T'aejong took an active interest in the field, employing Ch'oe Hae-san 崔海山 in 1401 to continue researches into new weapons.[105] Behind his active interest lay renewed harassment of the Korean coasts by the Japanese and the necessity to repel the Manchus. In 1404, the court activated the Kun'gigam 軍器監 (Bureau of Weaponry) and increased the number of soldiers in charge of

[103] *T'aejo sillok*, ch. 7, p. 8b. The Koryŏsa lists the following firearms: *ch'ongt'ong* (gun tube firearm), *hwat'ong* (fire barrel), *hwachŏn* (fire arrow), *hwap'o* (fire cannon). *T'aejo sillok* adds the following: *taechanggun* (great commander), *ichanggun* (second commander), *samchanggun* (third commander), *sŏkp'o* (stone cannon), *sinp'o* (signal cannon), *ch'ŏllyŏngjŏn* (iron arrow feather), *p'ilyŏngjŏn* (leather arrow feather), *chillyŏp'o* (caltrop cannon), *ch'ŏlt'anja* (iron pellet), *chaksan* (mountain drill), *oryŏngjŏn* (five-dragon arrow), *yuhwa* (drift fire), *chuhwa* (running fire), *ch'okch'ŏnhwa* (heaven-touching fire).

[104] *Yungwŏn p'ilbi* (Manual for a military commander), page on "great commander arrow."

[105] *T'aejong sillok*, ch. 1, p. 20b.

explosives.[106] By 1407 the arsenal had doubled, with 33 people aiding in the manufacture of explosives at the office. In October 1409, *kun'gisogam* 軍器少監 (Section Chief of the Bureau of Weaponry) Yi To 李韜 and *kun'gigamsŭng* 軍器監丞 (Supervisor of the Bureau) Ch'oe Hae-san completed construction of a "fire vehicle" (*hwach'a* 火車) and put it to a successful test. This "fire vehicle" was a kind of armored car that carried scores of *ch'ŏllyŏngjŏn* 鐵翎箭 (iron-fletched rockets launched by gunpowder), and fired them as it moved.[107]

With such marked improvement in firearms and explosives, the king in 1409 constructed an arsenal inside Chamun, and in 1417 the Hwayak Kamjoch'ŏng 火藥監造廳 (Directorate of Gunpowder Manufacturing) was established. Between 1410 and 1415, manufacture of firearms and explosives so developed that new tests were held almost annually, and conventional firearms were transformed from their defensive and naval origins to destructive land use. By that time, some 10,000 firearms had been manufactured and distributed to 100 provincial castles and installed on 160 battleships at various ports.[108]

It took much experimentation before Korea could produce truly destructive weapons. The most decisive stop in such development lay in the successful production of stone bullets launched with true explosive gunpowder. Evolution from rockets to bullet-firing firearms was completed when Ch'oe Hae-san manufactured 20 gun barrels (*wan'gu* 碗口) of large, medium, and small sizes made from Chinese models found on a wrecked Ming ship. Tests of these weapons produced excellent results: the stone bullets covered a range of 150 paces (*po*).[109]

The *wan'gu* (bowl) mortar, an innovation on the *chan'gu* 盞口 (bowl) mortar retrieved from a Ming ship in or before 1413, was the first effective siege weapon. It was named for its very short bowl-shaped tube. It was capable of firing large blocks

[106] Ibid., ch. 14; Hŏ, "Yŏmal sŏnch'o hwagiŭi chŏllaewa paltal," p. 25.
[107] *T'aejong sillok,* ch. 18, p. 36b.
[108] *Tongguk yŏji sŭngnam,* ch. 2, p. 25a.
[109] Ibid., ch. 2, p. 25b.

of stone, but with much scatter and limited range. *Yungwŏn p'ilbi* describes an especially large *wan'gu* as follows:

> It is the best firearm for attacking castles or walls because it is loaded with *pichinch'ŏlloe* 飛震天雷 (literally, heaven-shaking thunder shells) and *tansŏk* 團石 (stone balls). The gun is made of 1,100 *kŭn* (ca. 0.66 tn) of cast bronze; the barrel is 4 *ch'ŏk,* 3 *ch'on* (ca. 90 cm) long. The interior diameter of the barrel is 1 *ch'ŏk,* 5 *ch'on* (ca. 31.4 cm). The tube is 2 *ch'on* (ca. 4.3 cm) thick; the external diameter of the tube is 2 *ch'ŏk,* 4 *ch'on* (ca. 50.4 cm). From the top of the tube to the bottom it is 2 *ch'ŏk,* 6 *ch'on* (ca. 55 cm). The bottom diameter is 1 *ch'ŏk,* 6 *ch'on* (ca. 3.4 cm), and from the bottom to the touchhole it is 8 *ch'on,* 7 *p'un* (ca. 18.3 cm). The amount of gunpowder needed is 70 *yang* (ca. 2.6 kg), the range of the heaven-shaking thunder shell is 350 *po* (ca. 480 m); that of the stone balls is 400 *po* (ca. 560 m).[110]

Although the original *wan'gu* firearms must have been primitive, they were greatly improved with the invention of the "heaven-shaking thunder shell." By 1416, the production of 10,000 firearms had led to a corresponding increase in the troops manning them. The next year, the number of firearms further increased to 13,500. Keeping pace with this supply, the store of explosives, which had amounted to 6 *kŭn* 4 *yang* at the beginning of the dynasty, increased by a thousand times to 6,980 *kŭn* in 1417.[111] The firing range had also lengthened to between 300 and 500 *po,*[112] but it was far from satisfactory, and the cannon lacked mobility because of their weight. The Chinese "machine crossbow," which shot many arrows, remained unknown, or at least unused.

Casting of so many cannon later increased the demand for copper, causing a serious shortage. In addition to expropriating bells from Buddhist temples to meet this demand, officials instructed provincial people to collect copper, but they never seem to have considered prospecting for new copper mines (of which they probably knew) or smelting the ores. They could not foresee that this reluctance on their part to institute

110 *Yungwŏn p'ilbi, wan'gu* section. See Arima, *Kahō no kigen,* pp. 297–305.
111 Hŏ, "Yŏmal sŏnch'o hwagiŭi chŏllaewa paltal," p. 31.
112 *Sejong sillok,* ch. 107, pp. 19b–20b.

new projects would hinder further development of metallurgy in the later part of the dynasty.

As pioneering of the northwest frontier region became very active during the reign of King Sejong, demand for firearms and explosives also increased. From 1423, therefore, production of explosives had to be increased on a nationwide scale.[113] As against the private local production of 3,000 *kŭn* (ca. 1.8 tn), the actual need for explosives reached as high as 8,000 *kŭn* (ca. 6.4 tn) a year. The central government had to meet the shortage by producing 5,000 *kŭn* (ca. 3 tn) itself. Production of sulfur in the provinces was rigidly controlled by official supervisors. The provinces of P'yŏngan, Hwanghae, and Kangwŏn provided a steady supply. In the provinces of Kyŏngsang, Chŏlla, and Ch'ungch'ŏng, however, officials from the capital continued to oversee and control its production, which was limited by preference to areas far from the seaside.[114] This was probably because of the government's concern that the production process might be leaked to Japanese pirates. Explosives were also produced inland. In fact, the suspicion of the Korean officials was so deep that some even proposed, in the face of dire shortage, to limit production to the three northern provinces. These officials were willing to do without the ordinary supply from the southern provinces.[115] Firing practice was also reduced to the minimum because of the explosives shortage. Use of firecrackers inside the palace also had to be restrained. By 1433, the shortage and the need to conserve the supply became so acute that the amount of explosives to be used for firecrackers was reduced from 1,000 *kŭn* (ca. 0.6 tn) to a bare 30 *kŭn* (ca. 18 kg).[116]

As manufacturing techniques improved, there followed efforts to produce weapons different from Chinese models. The *wan'gu* improved and the names of various firearms changed to differentiate them from Chinese models. A new

[113]Ibid., ch. 19, p. 3a.
[114]Ibid., ch. 52, p. 32a.
[115]Ibid., ch. 34, p. 16b–17a.
[116]Ibid., ch. 59, p. 5b.

weapon by the name of *palhwa* 發火 ("firer") emerged. Other new arms included the *sinp'o* 信砲 (signaling firearm) and the *sohwap'o* 小火砲 (small portable firearm). Continual improvements in their quality finally led to the invention of iron balls (*ch'ŏlt'anja* 鐵彈子). By 1425, Chŏlla province alone produced and offered 1,578 Chinese-style small iron balls (*tangsoch'ŏlt'anja* 唐小鐵彈子), 616 smaller balls, and 1,104 iron balls for the *ch'ŏnja*-type cannon (*ch'ŏnja ch'ongt'ong* 天字銃筒).[117] Largest of the Yi Dynasty firearms, it was a heavy weapon comparable to the "great commander" cannon of Ming. The fourfold classification of various types of heavy firearms was thus clearly set out by the time King Sejong was ruling the dynasty —in order of size, *ch'ŏnja, chija* 地字, *hyŏnja* 玄字, and *hwangja* 黄字, named for the first four characters of a Chinese classic. Small portable types of firearms went by the name of *sŭngja* 勝字 (victory type). Specifications of these weapons will be explained in the next section.

In November 1425, an experimental firing practice was held near Chŏnch'ŏn River, Seoul, in the presence of the king and prince, to compare the efficiency of bullets and rockets. Iron shells were fired at 300 armor-clad targets from a distance of 150 *po* (ca. 210 m); six famed archers followed.[118] The practice ended in a decisive victory for the shell-firing team, resulting in training and dispatching of artillerymen to each provincial and county seat in June 1430.[119] By January 1433, artillerymen were not only divided into "gunners" and "loaders," but there was an "aimer" to adjust firing as well.[120] Artillery was subsequently drawn by horses, and moved as occasion demanded. Mounted artillerymen would dismount when ordered, aim the cannon, and fire at the target. The cannon had finally become an effective weapon.

During the reign of King Sejong, cannon were installed on the coast to deter Japanese invaders. Another small mobile

[117]Ibid., ch. 27, p. 11b.
[118]Ibid., ch. 30, p. 18a.
[119]Ibid., ch. 48, pp. 30b–31a.
[120]Ibid., ch. 59, p. 5b.

cannon was developed by July 1426, when 120 of them were dispatched to the northwest frontier.[121] Along with these weapons went firing instructors; soldiers who displayed talent were picked regardless of their background. By February 1432, cannon were installed on fixed emplacements in the northwest, along with signal firearms.[122]

In that year also, the first step was taken toward the invention of a multiple rocket launcher powered by gunpowder (*ilbal tajŏnp'o* 一發多箭砲) when Korean technicians successfully developed a firearm which could launch two rockets (*ibal chŏnp'o* 二發箭砲) simultaneously. By September of the next year, they could shoot four at a time with a firing range of 200 *po* (ca. 280 m), as was shown in the presence of King Sejong.[123] Although the bow firing four rockets was still not considered successful, the two-rocket weapon was used for the conquest of Jürchen barbarians to the north, and it proved a success. Arima Seiho 有馬成甫 commented in his book, "Although the Koreans were not the only people who launched rockets through a tube, credit for the perfection of this weapon must be given to the Korean people." He added that it was introduced to Japan as a rocket-launching bow.[124] In October 1441 the *hwach'o* 火鞘 (literally, "fire scabbard") was invented.[125] Four to five *ch'ŏk* long, it consisted of a length of bamboo filled with *kŭmch'ok* 金鏃 (literally, "metal arrowheads") and *sojuhwa* 小走火 (literally, "small running fire"), a deflagrating mixture combined with small projectiles to make an antipersonnel grenade. It was made to be suspended from a cord and thrown at the approaching enemy when one became wounded.

With qualitative and quantitative improvement of various firearms, the dynasty found itself increasingly pressed to open copper mines and to cast iron in order to meet new and increased needs. The government also tried to import Chinese tech-

[121]Ibid., ch. 33, p. 2a.
[122]Ibid., ch. 55, p. 12b.
[123]Ibid., ch. 61, pp. 51a, 53a.
[124]Arima, *Kahō no kigen,* p. 293.
[125]*Sejong sillok,* ch. 94, p. 1a.

niques of casting cannon with pig iron. Efforts were exerted to reproduce a mortar taken from the Japanese near Tsushima island, but the attempt bore no fruit because the Koreans did not know how to cast pig iron. So copper still had to be imported from Japan, as supplies mined in Korea remained insufficient. Only in 1444 did the great metallurgist, Yi Ch'ŏn, succeed in acquiring the Chinese technique of iron smelting from the Jürchen of the northern frontier.[126]

PUBLICATION OF THE *CH'ONGT'ONG TŬNGNOK*: PERFECTION OF KOREAN CANNON

From the autumn of 1444, there were administrative measures to raise the salaries of experts working on cannon so that they would not be distracted by economic worries.[127] At the same time King Sejong ordered the building of a cannon foundry under the charge of the Kun'gigam 軍器監 (Bureau of Weapons) near the palace. Often, the King personally intervened by undertaking researches himself into increasing the firing range. By the beginning of 1445, repeated test-firings of model weapons paved the way for the emergence of arms capable of firing many rockets simultaneously. A revolutionary development of the time was the *ch'ŏnja*-type firearm, capable of firing 1,000 *po* with a very small amount of explosive.[128] Many old weapons became obsolete.[129]

From June 1445, the King not only ordered a countrywide collection of bronze, but also encouraged the study of copper smelting,[130] so that by August of the same year, supervisors were sent from Seoul to the provinces to oversee the casting of cannon.[131] By September 1448, the court was able to publish the *Ch'ongt'ong tŭngnok* 銃筒謄錄 (Complete records on firearms), a work that explained the method of casting cannon and manufacturing explosives introduced three years earlier

[126]Ibid., ch. 106, p. 18b.
[127]Ibid., ch. 107, pp. 19b–20b.
[128]Ibid.
[129]Ibid.
[130]Ibid., ch. 108, p. 18b.
[131]Ibid., ch. 108, pp. 22ab.

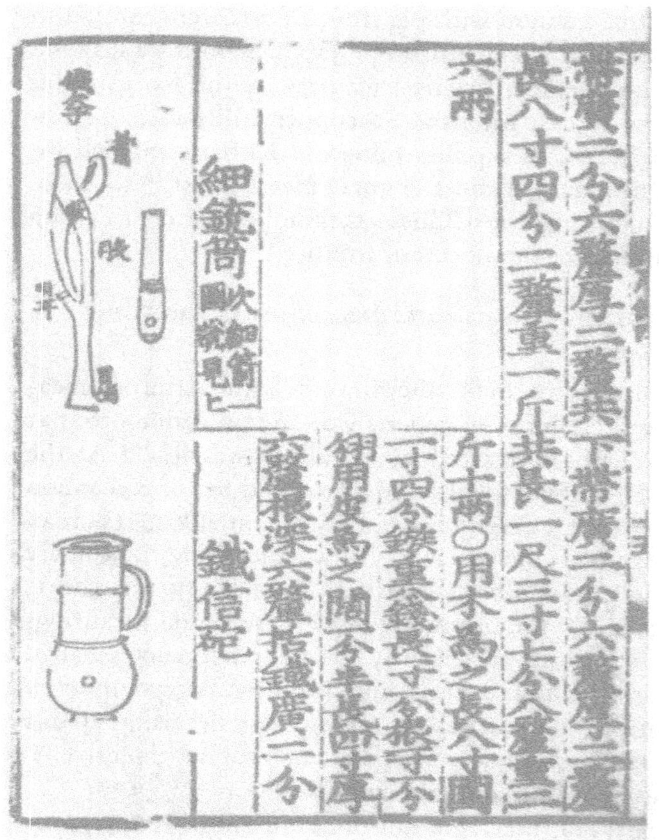
帶廣二分六釐厚二釐共下帶廣二分六釐厚二釐
長八寸四分二釐重一斤其長一尺三寸七分八釐重三
六兩
斤十兩○用木爲之長八寸圍
一寸四分鏃重八錢長一寸六分莖長二寸分
翎用皮爲之闊一分半長四寸厚
窮釐根深六釐括鐵廣二分

細銃筒 次細筒圖說見上

鐵信砲

Fig. 3.20. Fifteenth-century Korean firearms, a mortar cannon (*ch'ongt'ong wan'gu*) and commander firearm (*changgun hwat'ong*) made in accordance with specifications in the *Ch'ongt'ong tŭngnok* (Complete records on firearms) published in 1448 and no longer extant. Drawing from the *Kukcho ore sŏrye* (Introductory remarks on national rituals).

Table 3.2. Names and Ranges for Firearms Listed in the *Sejong sillok*

Firearm	Range For One Rocket	For Four Rockets Fired Simultaneously	Before Improvements
Ch'ŏnja	ca. 1,600 m (ca. 1,300 *po*)	ca. 1,200 m	400–500 *po*
Chija	ca. 1,000 m (ca. 800 *po*)	ca. 800 m	500 *po*
Hwangja	ca. 900 m (ca. 800 *po*)	ca. 600 m	500 *po*
Kaja 架子	ca. 700 m (ca. 600 *po*)	ca. 450 m	200–300 *po*
Sehwap'o 細火砲	ca. 600 m (ca. 500 *po*)		200 *po*

and listed the new weapons and their specifications with illustrations (see Figs. 3.20–3.22).[132]

The publication of this book was notable as a turning point in the art of casting Yi Dynasty cannon. It meant nothing less than that the Yi Dynasty had stopped copying Chinese models and had created distinctively Korean styles. The firearm production of the later Yi Dynasty conformed strictly to specifications contained in the *Ch'ongt'ong tŭngnok*. The contents of this publication never became available because of its status as a dynastic secret. Part of it, however, was included among the illustrated description of military hardware in *Kukcho ore sŏrye* 國朝五禮序例 (Introductory remarks on national rituals) published in 1474.[133]

The contents of the *Ch'ongt'ong tŭngnok* were probably published with amplifications in *Yungwŏn p'ilbi* 戎垣必備, compiled by the Hullyŏntogam 訓鍊都監 (Bureau of Military Training). This conclusion follows from the hypothesis that firearms extant today, dating back to the first half of the Yi Dynasty, between the reigns of King Chungjong and King Myŏngjong, were based on specifications contained in *Yungwŏn p'ilbi*.

[132]Ibid., ch. 121, p. 44b.

[133]*Kukcho ore sŏrye* (Introductory remarks on national rituals) ch. 4, pp. 8a–24a.

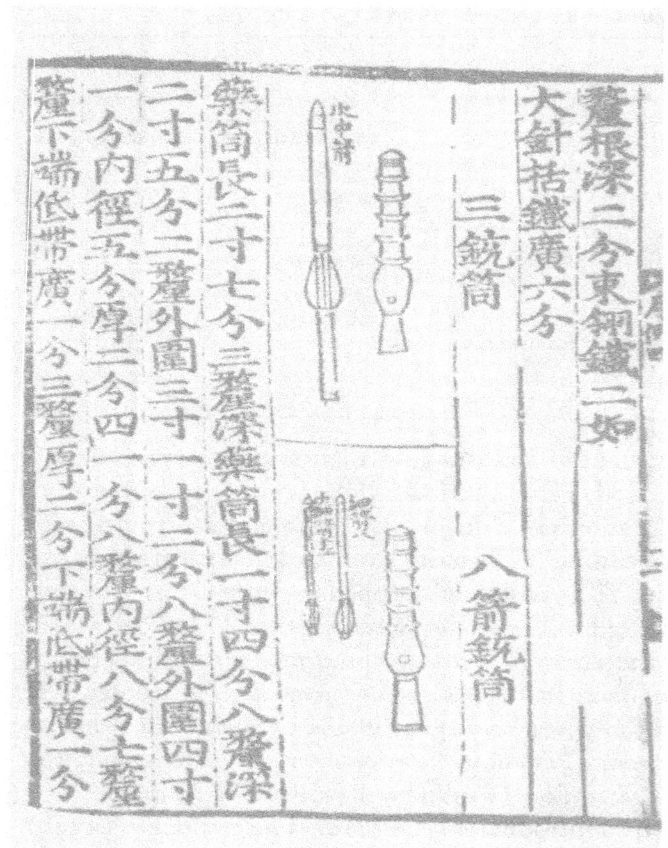

釐根深二分東鋼鐵二如
大針括鐵廣六分
三銃筒
八箭銃筒
藥筒長二寸七分三釐深藥筒長一寸四分八釐深
二寸五分二釐外圍三寸一寸二分八釐外圍四寸
一分內徑五分厚二分四一分八釐內徑八分七釐
釐下端低帶廣一分三釐厚二分下端低帶廣一分

Fig. 3.21. The "three-arrows firearm" (*samjŏn-ch'ongt'ong*) and the "eight-arrows firearm" (*p'aljŏn ch'ongt'ong*), shown in a fifteenth-century source (see Fig. 3.20).

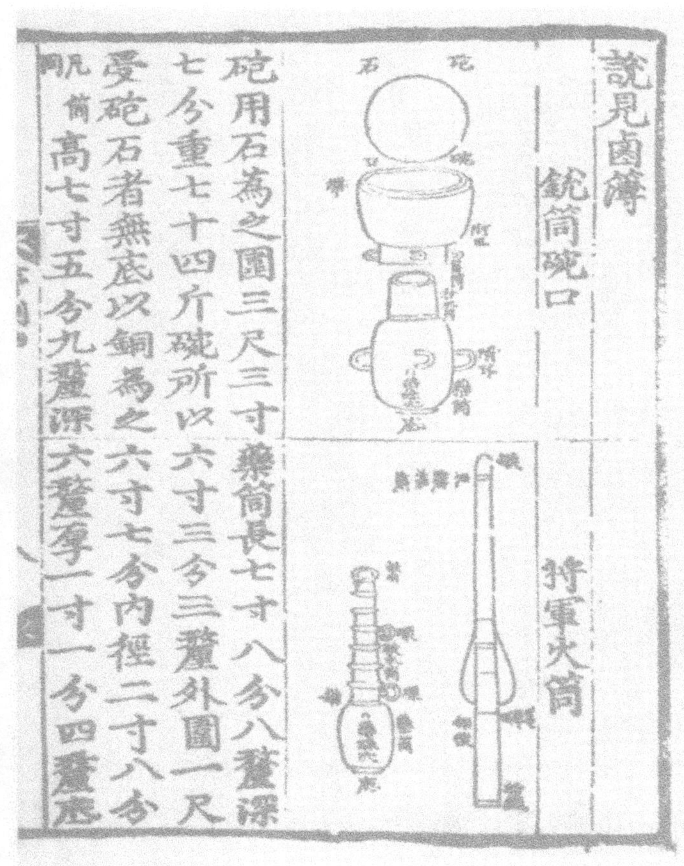
說見圖簿

銃筒碗口

將軍火筒

砲用石爲之圍三尺三寸藥筒長七寸八分八釐深
七分重七十四斤碗所以六寸三分三釐外圍一尺
受砲石者無底以銅爲之六寸七分内徑二寸八分
凡筒高七寸五分九釐深六釐厚一寸一分四釐底

3.22. Small arm (*sech'ongt'ong*) and iron signaling (*ch'ŏl-sinp'o*) firearms shown in a fifteenth-century source (see Fig. 3.20).

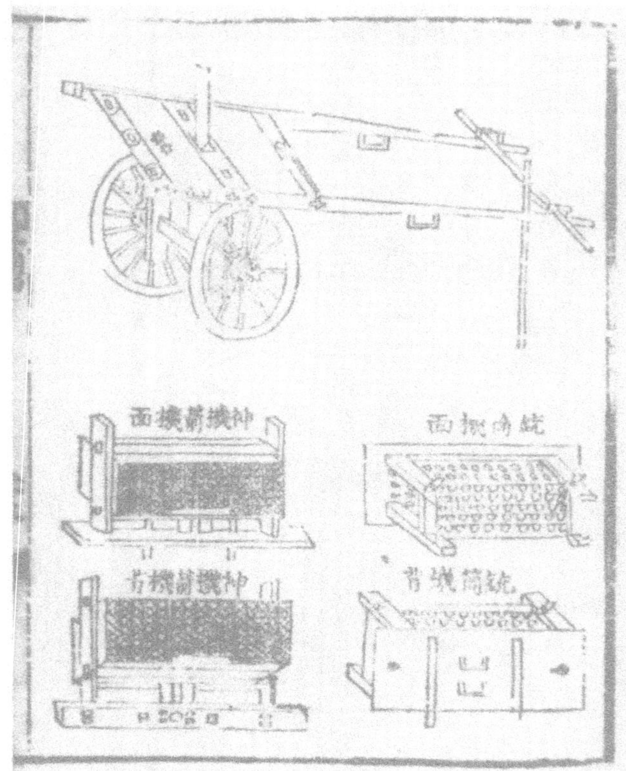

Fig. 3.23. Armored artillery vehicle (*hwach'a*), reportedly invented by King Munjong in 1451. From *Kukcho ore sŏrye*.

Chapter 107 of the *Sejong sillok* gives names for a variety of firearms and their performances as shown in Table 3.2.[134]

King Munjong 文宗 inherited his father's interest in development of cannon. He reputedly invented the *hwach'a* 火車 (armored artillery vehicle) shown in Fig. 3.23, although the weapon does not seem to have been used in battle until Pyŏn I-chung 邊以中 employed it during the Hideyoshi invasion. According to the *Munjong sillok* 文宗實錄 (Veritable records of the King Munjong era), King Munjong invented

[134]*Sejong sillok,* ch. 107, pp. 19b–20b.

and tested the *hwach'a* in February 1451. There was a wooden platform on top of this vehicle, on which were installed 100 medium-sized rockets (*sin'gi* 神機) and 50 four-rocket launchers. There was also a device to ignite the rockets. The vehicle could be drawn by two men on level ground; it required another man pushing from behind when going uphill; and two more men had to push the vehicle when it was going up a steep hill.[135] Armor was later added but was quickly abandoned for the sake of mobility. In time of peace, the vehicle, without its platform, was used as a cart. At the time of its appearance, 50 vehicles were stationed in the capital, and 80 at frontier regions. By 1451 hundreds of them were stationed throughout the realm.[136] Another improvement in cannon during the reign of King Munjong was the enlargement of the touchhole from 7 *ri* (ca. 1.5 mm) to 8 *ri* (ca. 1.7 mm) so that ignition would be easier.

PUBLICATION OF THE MANUAL FOR A MILITARY COMMANDER (*YUNGWŎN P'ILBI*): FIREARMS DURING THE REIGN OF KING SŎNJO AND FIREARMS OF THE LATER YI DYNASTY

Manufacturing of firearms in the Yi Dynasty can be considered to have gone as far as it could during the reign of King Munjong, by 1452. It became stagnant from the time of King Sejo's reign,[137] with only one or two inventions added between the time of King Chungjong and King Myŏngjong (sixteenth century). Just before and after 1555, a new series of heavy artillery weapons (see p. 195) were manufactured, and the King ordered the casting of 1,000 such guns; in October, 1566, King Myŏngjong imported 60,000 *kŭn* (ca. 36 tn) of copper and ordered the study of Japanese copper smelting to spur the production of copper for gun casting. During the early part of King Sŏnjo's reign, the emergence of *sŭngja* 勝字 ("victory type") firearms was notable.[138]

[135] *Munjong sillok*, ch. 6, p. 15b.
[136] *Munjong sillok*, ch. 6, p. 22a, ch. 9, pp. 12a, 33a.
[137] Hŏ Sŏn-do, "Yijo chunggi hwagiŭi paltal" (The development of firearms in Korea, 1474–1592, pt. 1), *Yŏksa hakpo*, 1966, *30*:49 ff.
[138] Ibid., pt. 2, *31*, 1966: 92 ff.

A general outline of the development of firearms up to the second half of the Yi Dynasty, following the Japanese invasion, is provided in the introduction of the *Yungwŏn p'ilbi,* but the *sŭngja* firearm is not included in the book. This small portable weapon was used in Korea until the musket (*choch'ong* 鳥銃, literally "bird handgun") was introduced from Japan. Its firing range and accuracy were rather poor. In general, its caliber was between 16 and 18 mm, its length between 550 and 600 mm; there was also the *sajŏn ch'ongt'ong* 四箭銃筒 (four-rocket tube launcher) whose caliber was 23 mm and whose length was between 250 and 260 mm.

All the extant *sŭngja* firearms were manufactured either during or after King Sŏnjo's reign, and the *Sŏnjo sillok* 宣祖實錄 (Veritable records of the King Sŏnjo era) says the firearm was first made in June 1583 by Kim Chi 金墀.[139]

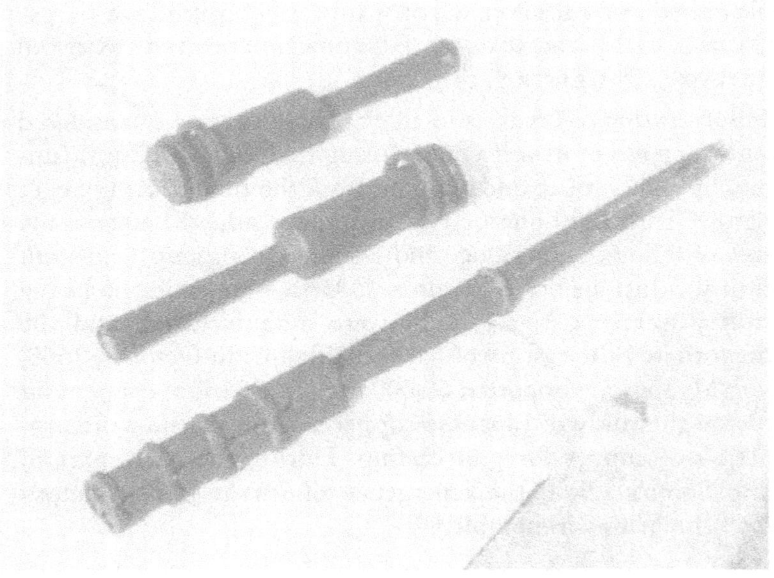

Fig. 3.24. Three-barrelled handgun from the late Yi Dynasty, ca. eighteenth century, caliber 1.3 cm, length 36.8 cm. Korean Military Academy.

[139]*Sŏnjo sillok,* ch. 17, p. 15b.

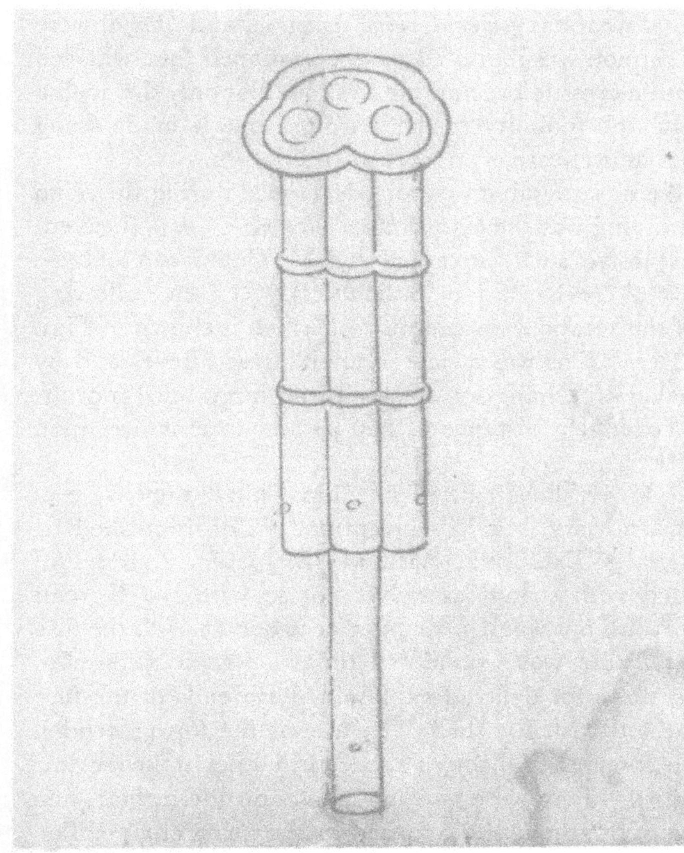

Fig. 3.25. Three-barrelled handgun, sketch from the "Hwagitogam ŭigwe" (Record of the Bureau of Firearms), MS of 1615.

Firearms of the Yi Dynasty thus may be classified as heavy firearms usually identified as *ch'ŏnja* 天字, *chija* 地字, *hyŏnja* 玄字, and *hwangja* 黄字; and three kinds of *wan'gu* 碗口 mortar, extra large, large, and medium-sized. Others (Figs. 3.24 and 3.25) included "Feringi," or Portuguese calivers, the "great commander" cannon, *wiwŏnp'o* 威遠砲, and *hojunp'o* 虎蹲砲. Projectiles included many kinds of rockets, cannonballs (*ch'ŏl-t'anja* 鐵彈子, *tansŏk* 團石, *such'ŏlhwan* 水鐵丸), and "heaven-shaking thunder shells." Tables compare the specifications of these weapons as given in *Yungwŏn p'ilbi*.

The heavy weapons (*ch'ŏnja, chija, hwangja,* and *hyŏnja*) were mobile. A cannon was mounted on a box-shaped four-wheeled vehicle, with a crosstie bracing the cannon. Not only did such a vehicle raise the mobility of the artillery, but it made firing much easier and effectively absorbed recoil.

One of the most notable weapons developed during the reign of King Munjong was the *pigyŏk chinch'ŏlloe* 飛擊震天雷 (heaven-shaking explosive shell) invented by Yi Chang-son 李長孫. *Chingbirok* 懲毖錄 (Record of difficulties) gives the following account of the weapon, an example of which is shown in Figs. 3.26 and 3.27. "This was a new weapon. It was developed by firearm maker Yi Chang-son to shoot in a large bowl mortar. The shell covered a distance of 500 *po* and caught fire upon landing."[140]

Since explosive shells existed in Chin China, they were of course not a novelty, but Yi's invention was distinguished by its fuse system.[141] Explosive devices of Ming China were set off or detonated with string fuses, but not so with the Korean *chinch'ŏlloe.* Within it was a spring device around which the fuse was tied; the fuse was wound ten times for rapid explosion, and fifteen times for delayed explosion. Both ends of the fuse were connected with the shell. The fuse at the top protruded from the enclosure, and through a special channel in the mortar. In other words, there were two touchholes on the mortar, one for the fuse of the shell, and another for the firing charge. The Japanese scholar Arima Seiho made the following comment: "This very delicate device, if it indeed was invented by a Korean, would constitute epoch-making progress in the technique of explosive manufacturing."[142] It is said that this kind of mortar proved itself a tremendous defensive weapon when the Japanese attacked Korea in the sixteenth century.

Thus, firearms in Korea began from copies of various Ming models. Not only did the Yi Dynasty succeed in carrying on the manufacture of traditional weapons, but it went further to

[140]*Chingbirok* (Record of difficulties), ch. 1.
[141]Arima, *Kahō no kigen,* p. 302.
[142]Ibid.

Fig. 3.26. Korean "heaven-shaking explosive shell" (*pigyŏk chinch'ŏlloe*). The photo shows remains, cast iron, 45 cm in diameter. Museum of Korean Military Academy.

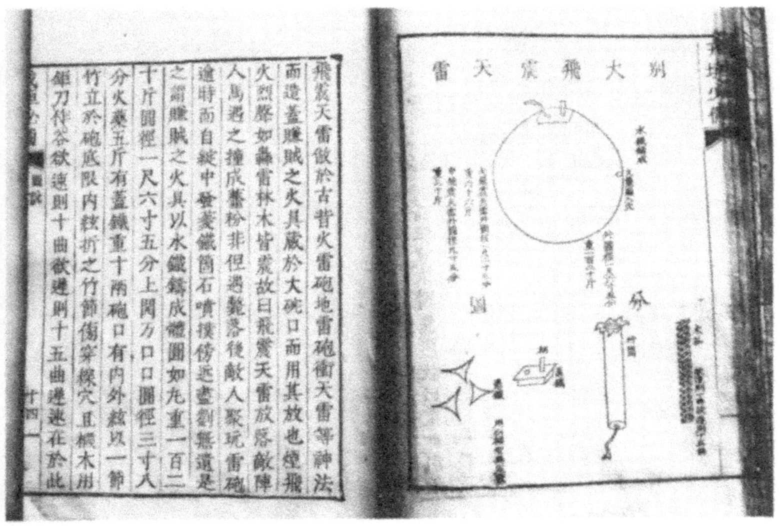

飛震天雷做於古昔火雷砲地雷砲衝天雷等神法
而造蓋賺賊之火具藏於大碗口而用其放也煙飛
火烈聲如轟雷林木皆震故曰飛震天雷放落敵陣
人馬遇之撞成虀粉非但遇難落後敵人聚玩雷砲
適時而自綻中發菱鐵與石噴撲傍近盡劉無遺是
之謂賺賊之火具以水鐵鑄成體圓如丸重一百二
十斤圍徑一尺六寸五分上開方口口圍徑三寸八
分火藥五斤有蓋鐵重十兩砲口有內外紋以一節
竹立於砲底限內紋折之竹節傍穿倮穴且横木用
鉅刀作谷欲速則十曲欲遲則十五曲遲速在於此

Fig. 3.27. Sketch of the "heaven-shaking explosive shell" from the *Yungwŏn p'ilbi*.

Tables 3.3–3.8. Specifications for Weapons in the Yungwŏn p'ilbi

Table 3.3. Firearms

	Specifications				
Type	Caliber (*ch'on*)	Length (*ch'ŏk*)	Weight (*kŭn*)	Gunpowder (*yang*)	Range of Fire (*po*)
Ch'ŏnja	5.6	6.63	1,209	30	1,200
Chija	5.0	5.67	724	20	800
Hyŏnja	3.9	4.01	155	4	2,000 (rocket)
Hwangja	2.2	3.64	130	3	1,100 (rocket)

Note: Metric equivalents are as follows: 1 *ch'on* = ca. 2.3 cm (approximate length of the *chuch'ŏk* in the late Yi Dynasty was 23 cm); 1 *kŭn* = ca. 600 g; 1 *yang* = ca. 37.5 g; 1 *po* = ca. 138 cm.

Table 3.4 Rockets

	Specifications			
Type	Length (*ch'ŏk*)	Weight (*kŭn*)	Firearm	Range (*Po*)
Great commander	11.9	50	*Ch'ŏnja*	900
Commander	9.23	33	*Chija*	2,000
Large	6.37	7	*Hyŏnja*	2,000
P'iryŏng 皮翎	6.3	3.5	*Hwangja*	1,100

Table 3.5. *Wan'gu* (Mortars)

	Specifications			
Size	Caliber (*ch'ŏk*)	Length (*ch'ŏk*)	Weight (*kŭn*)	Gunpowder (*yang*)
Extra large	1.85	4.3	1,100	70
Large	1.31	3.1	528	35
Medium	1.00	2.83	290	35

Table 3.6. "Heaven-shaking Thunder Shell"

	Specifications			
Size	Outer Caliber (*ch'ŏk*)	Weight (*kŭn*)	Mortar Used	Range of Fire (*po*)
Extra large	1.65	120	Extra large	350
Large	1.23	66	Large	400
Medium	0.95	30	Medium	350

Table 3.7. Cannonballs

Used with	Specifications		
	Diameter (*ch'on*)	Weight (*kŭn*)	Material
Ch'ŏnja	3.6	13	Lead-coated cast iron
Chija	2.9	8	Lead-coated cast iron
Hyŏnja	1.7	1.8	Lead
Hwangja	1.3	0.8	Lead

Table 3.8. Stone Balls for *Wan'gu*

Size	Specifications	
	Weight (*kŭn*)	Range of Fire (*po*)
Extra large	120	400
Large	45	500
Medium	35	500

develop its own devices, and they proved to be effective instruments of naval warfare. This is considered the end of developments in firearms that originated in China and were introduced from there. These early weapons were gradually eliminated in favor of more modern ones with the introduction of Portuguese calivers and the Japanese musket. I shall in the following paragraphs discuss these two weapons briefly.

According to *Chingbirok*, the Feringi (weapon of the "Franks," a general term for south Europeans) was first manufactured and test-fired in Korea in 1597.[143] Judging from extant models, these bronze Western guns were mostly cast between the end of the sixteenth century and the middle of the seventeenth. A small one, 0.74 m long, had a caliber of 29 mm; a large one, extending 1.43 m, was 58 mm caliber. This, again, is an example of Koreanization of a European weapon originally introduced through China.

The *choch'ong* (musket, literally, "bird handgun") was first introduced to Korea when Hwang Yun-gil 黃允吉 and several other emissaries visited Japan in 1589.[144] They were given

143 *Chingbirok*, ch. 14.

144 *Chōsenshi* (Chronological history of Korea; Seoul, 1937), pt. 4, vol. 9, p. 354.

several of these muskets by the hereditary lord of Tsushima Island as gifts. The value of the new weapons was unrecognized until the Japanese attacked Korea with them; then all the muskets available were quickly collected for firing practice. It is said that training officer Chŏng Sa-jun 鄭思竣, craftsman An 安, sailor Yi P'il 李必, private slave An Sŏng 安成, and temple slave Ŏnbok 彥福, all under command of Admiral Yi Sun-sin, participated in the making of the first muskets after Japanese models, but the structure of the firing mechanism was slightly different, adapted from that of the native *sŭngja*.

Unlike earlier versions of Yi Dynasty muskets, which were distinguished by bamboolike joints on the barrel, the new model had a long straight barrel of smaller caliber, with sights but without a stock. Admiral Yi sent five samples to the capital, while requesting further efforts in production.[145] The *Chingbirok* refers to it as a "new model of *sŭngja* firearm . . . developed on September 14, in that year of Imjin."[146] The gun model now extant bears an inscription to the effect that it was manufactured in September of the same year.[147] Later weapons must have been close copies of Western handguns. In March 1593, with the war turning to the advantage of the Korean side, local troops were given regular firing exercises, and use of the new weapon was included among the subjects for national examinations. Improvement of methods of making the explosive powder used for this musket was also encouraged.

For all that, it is apparent that techniques for manufacturing muskets at the time were far from satisfactory. In April 1624, several thousand of these guns were imported from Japan, and in July 1656, intructions were given to produce Western type muskets like those taken from Europeans whose ships drifted onto the Korean coast. Probably spurred on by the example of these new weapons, the Yi Dynasty in 1657 requested of Ch'ing China 100 muskets of an improved model.[148]

[145] *Yi Ch'ungmugong chŏnsŏ* (Collected works of Admiral Yi Sun-sin).
[146] *Chingbirok*, ch. 7, pp. 12a–13b.
[147] Hŏ, "Yijo chunggi hwagiŭi paltal," pt. 2, p. 122.
[148] *Hyojong sillok*, ch. 17, pp. 2a, 16b.

Fig. 3.28.Vessel-shaped container, fifth–sixth century. National Museum of Korea, Seoul.

Shipbuilding Technology

VESSELS OF ANCIENT AND MEDIEVAL TIMES

Ancient vessels originated from floating gourds that Stone Age people used for seafaring; and rafts, made of logs fastened together, gradually developed into another form using earthen jars tied to the logs for flotation. The vessel-shaped container now preserved at the National Museum of Korea and dating back to the time of Three Kingdoms (Fig. 3.28) is representative of primitive Korean vessels. By the time canoes appeared, there was an anchor as well, secured with a chain. The steersman used a long steering oar; the hull, sharp at the prow, was wide in the middle for stability.

There is no material evidence to indicate exactly when built-up vessels came into being, but it is recorded that in the year A.D. 32, Koguryŏ sent an emissary by sea to Later Han

China.[149] Written data of the early third century indicates that the navy of the Koguryŏ kingdom was very powerful, and the kingdom of Paekche sent an emissary to China in the year 373 by sea,[150] as did Silla in 381.[151] Sailing ships were thus in widespread use by the end of the fourth century, during the time of Three Kingdoms, when Silla sent shipbuilders to Japan to train Japanese[152] (as did Paekche in 650).[153] Called *kyŏntangsŏn* 遣唐船 (China-bound vessels), the double-masted sailing ships, resembling Chinese junks, each carried 150 passengers.[154] In Silla, the Sŏnbusŏ 船府署 (Office of Shipbuilding), established in A.D. 583, consisted of 14 persons in charge of shipbuilding affairs.[155] The large vessels constructed between Three Kingdoms and Unified Silla were official ships used mainly for the frequent voyages of royal emissaries to T'ang China, but they gradually developed into commercial ships doing business mainly with Japan and China.[156]

Military vessels developed rapidly during the Koryŏ dynasty. Numbering more than 100 at the time, the largest of these is understood to have carried a crew of between 100 and 200 sailors. According to the *Koryŏsa*, one of these ships was so large (96 *ch'ŏk*, about 20 m wide) that one could race a horse on deck.[157] One of the reasons that construction of warships was so animated at the time was the need to defend Korea not only from the Japanese but also from Jürchen pirates. The Treatise on Military Affairs of the Koryŏsa makes a claim that there were no fewer than 75 warships by 1009,[158] most of these presumably of distinctively Koryŏ styles. We mention a distinctive

[149] Ch'oe Sŏng-nam, *Han'guk sugunsa yŏn'gu* (Study of the history of Korean naval forces to 1910; Seoul, 1964), pp. 2–3.

[150] *Samguk sagi*, ch. 24, p. 9a.

[151] Ibid., ch. 3, p. 2b.

[152] *Han'guksa*, vol. 1, p. 614.

[153] Imamura Tomo, *Fune no Chōsen* (Maritime Korea; Seoul, 1930), p. 51.

[154] *Nihon kagaku gijutsushi* (A history of science and technology in Japan; Tokyo, 1962), p. 600.

[155] *Samguk sagi*, ch. 4, p. 12b.

[156] Imamura, *Fune no Chōsen*, p. 51.

[157] Kim, *Koryŏ sidae sa*, pp. 244–245.

[158] *Koryŏsa*, ch. 36, Treatise on military affairs, section 2.

style because of a Japanese record that quotes a Japanese woman's account of warships of peculiar structure engaging in battle with Jürchen pirates.[159] The record said: "When I boarded the ship, having been rescued from the sea, I noticed a wide space different from that of normal vessels. Five or six men were using oars to move the ship, and some twenty men were moving some object [lacuna]. On one side of the ship were seven or eight supernumerary oars not attached to the ship, and the ship's surface had numerous iron spikes for ramming enemy vessels." In other words, this warship was designed to ram an enemy vessel and destroy it. According to *Hsuan-ho feng shih Kao-li t'u ching* 宣和奉使高麗圖經 (Illustrated record of an embassy to Korea in the Hsuan-ho reign-period, 1123), written by Hsu Ching 徐兢, there was yet another model: "The structure of the vessel and oars was especially simple, with a single mast standing in the middle of the vessel; there was no forecastle, and there was a steering oar. I saw ten such patrol craft."[160] Although the Japanese scholar Imamura Tomo 今村鞆 claims that the shipbuilding technology of Koryŏ was inferior,[161] this does not seem to have been true. For the Sung observer, no expert on ships, was making a general comment after seeing numerous patrol craft along the coast. Though it is true that the ships that Hsu Ching had seen were simpler in structure than those of China, he also said, "Official vessels had a cabin with door and portholes, and around the cabin roof were handrails made of wood. The decking of the vessel was of pine board, but the hull itself was of shaped logs fastened with pegs. At the bow was a sail, and another large sail was located amidships. Only one-fifth of a total of twenty canvas sails were unfurled, in deference to the power of the wind. The sail was spread in sections."[162] Judging from these descriptions, Koryŏ vessels were certainly simply designed (Fig. 3.29). But their hardiness is proven by the fact that where-

[159]Kim, *Koryŏ sidae sa*, pp. 244–245.
[160]*Hsuan-ho feng shih Kao-li t'u ching*, ch. 33.
[161]Imamura, *Fune no Chōsen*, pp. 51–52.
[162]*Hsuan-ho feng shih Kao-li t'u ching*, ch. 33.

Fig. 3.29. Koryŏ ship. Bronze mirror of Koryŏ, eleventh to twelfth century.

as all the Yuan vessels, which had been patterned after Sung designs, were destroyed by strong winds during their second invasion of Japan in 1281, not much damage was done to some 900 Koryŏ vessels.[163]

Thus there were two structural styles in Koryŏ ships: One patterned after that of southern Sung, and another a distinctively Korean style derived from Paekche and Silla. Some 900 warships built during the second half of the thirteenth century were of large sizes, capable of carrying 3,000 to 4,000 *sŏk* 石 (ca. 1,380 ft^3) of cargo.[164] It was explained that many Yuan vessels were patterned after those of Koryŏ and not after those of Southern Sung because the Koryŏ version was cheaper and required far less time for building.

Two kinds of vessels were used for coastal transportation in Koryŏ: the *ch'omasŏn* 哨馬船 (utility vessel) capable of loading 1,000 *sŏk* (ca. 320 ft^3) of cargo; and the *p'yŏngjŏsŏn* 平底船

[163]Hong, *Chosŏn kwahaksa*, p. 133.
[164]Ibid.

(cargo barge), carrying 200 *sŏk*.[165] Thus Koryŏ ships were largely divided into war vessels and coastal freighters. Although this had not been true in the Silla era, commercial vessels were still rare. One reason may have been the aggressive commercial activities of Sung ships, which discouraged Koryŏ mercantile activity. Continuation of the trend later was connected to increased Japanese piracy and to the negative trade policy of Ming China. These external factors virtually put a stop to seafaring activities by the end of Koryŏ and the beginning of the Yi Dynasty. Conversely, however, building of warships was much stimulated, and correspondingly less effort was given to developing fishing and passenger vessels.

SHIPBUILDING STYLE OF THE EARLY YI DYNASTY

The shipbuilding technology of the Yi Dynasty is distinguished by its emphasis on military ships (Fig 3.30). This was

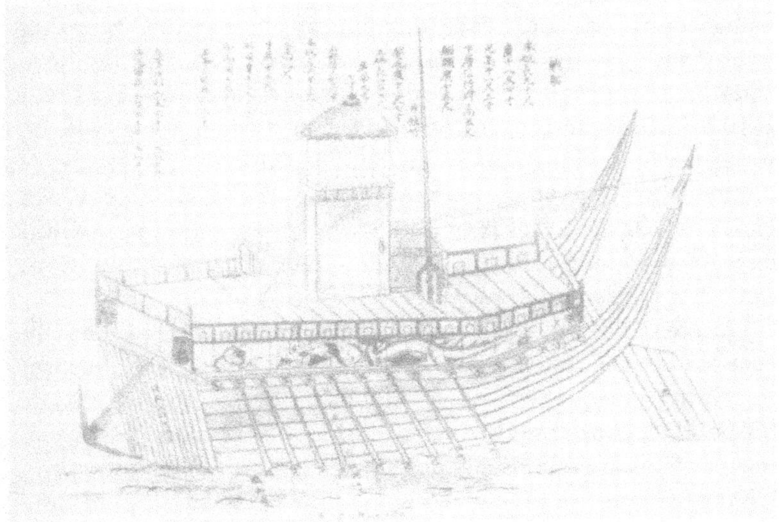

Fig. 3.30. Korean warship of the Yi Dynasty. Description shows a length of 90 *ch'ŏk*, maximum width of 18 *ch'ŏk*, 4 *ch'on*, height of the ship body, 11 *ch'ŏk*, 5 *ch'on*. Sketch from *Kaksŏn tobon* (Illustration of various ships), MS.

[165] *Koryŏsa*, ch. 79, p. 37b; Hong, *Chosŏn kwahaksa*, p. 133.

especially so with emergence of the *kusŏn* 龜船 (*kŏbuksŏn* 거북선) or "turtle ship," improved and developed by Admiral Yi Sun-sin 李舜臣. An examination of the process of building a turtle ship will reveal a general outline of the shipbuilding technology of the Yi Dynasty.

According to the *Kyŏngguk taejŏn* 經國大典 (National code), there were three different kinds of fighting vessels during the early Yi Dynasty: *taemaengsŏn* 大猛船 (big warship), *chungmaengsŏn* 中猛船 (medium warship) and *somaengsŏn* 小猛船 (small warship).[166] The first type carried some eighty sailors aboard, the second sixty, and the third thirty. In addition to these there were reserve vessels called *mugun* 無軍 (literally, "noncombatant"), in big, medium, and small sizes, based on Koryŏ and Southern Sung designs. Yi Dynasty vessels were distinguished by their width (which tended to be excessive) rather than their length (Fig. 3.30). Moreover, because of the thick lumber used, the vessels were heavier than those of neighboring countries, and hence capable of less speed. According to the *Kyŏngguk taejŏn*, "The size of a large vessel will be more than 13 m long and 5.9 m wide; that of a medium vessel will be 10.5 m long and 4.3 m wide; that of a small vessel will be more than 5.9 m long and 2 m wide." In the case of river vessels which were longer and narrower, "the size of a large vessel will be more than 16 m long and 3.2 m wide; that of a medium vessel 14.3 m long and 2.8 m wide; that of a small one more than 12 m long and 2.5 m wide."[167]

It seems that Yi oceangoing ships were built so heavily because, for the purpose of coastal navigation and defense, the solidity of a vessel was of higher priority than its speed. Shipbuilders continued trying to increase the speed of warships, but not of commercial liners or fishing vessels. Between the time of King T'aejo and King T'aejong, the Korean shipbuilders began to adopt Japanese techniques, among them the use of iron nails as an improvement on pegs (by 1430). In 1433, the dynasty built three vessels, one fashioned after those of Ok-

[166] *Kyŏngguk taejŏn,* Military Code, warship section.
[167] Ibid., Works Code, ships and vehicles section.

inawan shipbuilders, using iron nails (total iron requirement: 3,352 *kŭn*, 2 tn), and another constructed by the Sasusaek 司水色 or Chŏnhamsa 典艦司 (Office of Shipbuilding), with the lower half of the hull built with iron nails and pegs, but the upper structure solely with iron nails (iron requirement between 1,800 and 1,900 *kŭn*). The third vessel did not use iron. These three ships were launched on the Han River in the following year for experimentation and a comparison of their functioning. The result of this test was reflected in later improvements in shipbuilding. In 1445, naturalized Japanese were given military rank and asked to build ships. But what they built took longer and was twice as expensive as usual, and the wooden members were so thin as to invite destruction. Yi Dynasty Koreans therefore continued to build ships with raw lumber, mixing wooden pegs with iron nails in order to minimize the weight of the vessels.[168] But the speed of Japanese vessels continued to haunt the Yi Dynasty builders.

Thus the *kapchopŏb* 甲造法 (literally, armoring method), which involved a heavy, seamless, protective wooden shell, as advocated by Yi Ch'ŏn and others during the reign of King Sejong, was abandoned in the latter half of the fifteenth century.[169]

Lack of support by the Yi Dynasty retarded development of its shipbuilding technology. Out of 2,795 official craftsmen involved in 129 different fields, those responsible for building ships at the Office of Shipbuilding were no more than 10, or 0.36 percent of the total,[170] half of the average of 21.7 persons in each of the 71 groups of craftsmen that worked for 30 government offices. Of 6,559 craftsmen in the capital and provinces at the time, the number of shipbuilders constituted 0.15 percent of the total.[171] The number of craftsmen assigned to each sector

[168]Kang Man-gil, "Yijo chosŏnsa" (A history of shipbuilding in the Yi Dynasty), *Kwahak kisulsa*, in the *Han'guk munhwasa taege* series (Seoul, 1968), pp. 915–920.
[169]Ibid., pp. 921–924.
[170]*Kyŏngguk taejŏn*, Works Code, craftsmen section.
[171]See Kang Man-gil, "Chosŏn chŏn'gi kongjang ko" (A study of artisans

reflected not only its level of advancement but political and social demand. Because the latter were low, heavy shipbuilding was neglected by the government.

With the appointment of Sin Suk-chu 申叔舟 to the post of the director of the Office of Shipbuilding in 1465, new priorities were given to military vessels, and a decision was reached to build troop-carrying vessels too.[172] Three types of troop vessels were actually built the following year, all of them blends of Chinese, Japanese, and Ryūkyūan styles. These were called barges (*p'yŏngjŏsŏn* 平底船) because in time of peace the upper structure was removed and they were used for cargo transportation. The hull was regulated to a standard size and water displacement, with the capital authorities sending models to provincial ports. In February 1550, a paddle-wheel vessel

Fig. 3.31. Paddle-wheel boat. Painting "Boat on Rough Sea," ink and color on paper by Yu Suk (1827–1873). Such a vessel was commissioned in 1550 by An Hyŏn, after study of structural information provided in the *Ta-hsueh yen-i pu-i*.

---

in the first half of the Yi Dynasty), *Sahak yŏn'gu* (Journal of the study of history), 1961, *12*: 1 ff.

[172]*Munhŏn pigo*, ch. 120, p. 4b.

was constructed. Foot-powered paddle-wheel vessels had been constructed by 1543 in the West and before the eighth century in China. A Korean version, similar to the one shown in Fig. 3.31, was constructed in February 1550 and launched in September 1553, during the eighth year of King Myŏngjong. It was commissioned by An Hyŏn 安玹, governor of Kyŏngsang Province at the time, after he had studied a description in the Chinese classical commentary *Ta-hsueh yen-i pu-i* 大學衍義補遺. Kim Sun-go 金舜皐 designed and constructed the vessel.[173] When it proved successful, An Hyŏn also ordered it to be used as a cargo vessel. In 1740, a third and fourth wheel were added to the starboard and port of the wheeled vessel at the request of Yi Min-su 李民秀, chief naval commander of Chŏlla province and a descendant of Admiral Yi Sun-sin. With four persons pedaling each of the four wheels, its speed was much increased.[174] For reasons yet undetermined, the wheeled pedal does not seem to have been widely used.

WARSHIPS AND TURTLE SHIPS OF THE YI DYNASTY

Development of warships in Korea reflected the government's changing battle strategy against rampaging pirates. The *kwasŏn* 戈船 (spear vessel), constructed by Koryŏ shipbuilders at the beginning of the eleventh century, was a superior battleship built to ram enemy vessels during engagements with Jürchen marauders. To counter the close-combat tactics of the Japanese pirates, another ramming vessel, a wood-canopied "turtle ship," was launched in February 1413, in the thirteenth year of King T'aejong.[175] Its function is clearly outlined in an appeal that T'ak Sin 卓愼 sent to the King in July 1415:[176] "The turtle ship is capable of ramming the enemy fleet and causing damage to enemy vessels without incurring any itself; it is therefore a decisive tactical instrument. I suggest your Royal Highness build more of such solid and cleverly designed vessels as instruments of victory."

[173] *Myŏngjong sillok,* ch. 10, p. 8b, ch. 15, p. 18a.
[174] *Munhŏn pigo,* ch. 120, pp. 14b–15a.
[175] *T'aejong sillok,* ch. 25, p. 7a.
[176] Ibid., ch. 30, p. 4b.

However, with the gradual decrease in the number of Japanese pirates roaming the Korean coastlines the use of turtle ships correspondingly diminished. As other types of war vessels were built during the reign of King Sejong, they gradually replaced the turtle ship in the table of distribution of war vessels of the eight provinces of Korea.

With rumors of Japanese invasion again circulating during the reign of King Sŏnjo, in February 1591 Yi Sun-sin was appointed naval commander of Chŏlla Province. He placed particular emphasis on constructing and repairing war vessels. Admiral Yi, a superb tactician familiar with Japanese tactics, and his subordinate Na Tae-yong 羅大用, an expert on shipbuilding, came to the conclusion that the turtle ship was the most efficient assault vessel for use against the Japanese.

The turtle ship that existed during the time of King T'aejong seems to have been what is called the *t'ongjeyŏng* 統制營 (flagship) type (see Fig. 3.32) shown in the *Yi Ch'ungmugong chŏnsŏ* 李忠武公全書, the collected works of Admiral Yi. Its specifications were as follows: length: 115 *ch'ŏk* (ca. 34.5 m); maximum width: 15 *ch'ŏk* (ca. 4.5 m); maximum height: 15 *ch'ŏk*; oars: 20; gunports: 74; hatches: 28. The surface of the ship was covered by a canopy. The mast was made to be removed. The bow of the ship took the shape of the head of a turtle, from which bellowed smoke from sulfur or other burning material to frighten the enemy. The cabin was double-decked; on the first deck were, in a 24 *kan* 間 (ca. 144 ft^2) space, an arsenal of five *kan*, and quarters for troops measuring 19 *kan*. On the second deck were cabins for officers and commanders and various offensive firing weapons.[177]

Although the turtle ship developed by Admiral Yi and Na was almost identical to the earlier flagship type, there was a second bow beneath the head of the turtle, shaped like the head of a demon, and the wooden canopy was decorated to resemble a turtle's back. In the bulwarks were twenty gunports, ten on

[177] *Yi Ch'ungmugong chŏnsŏ*, illustration section.

龜船

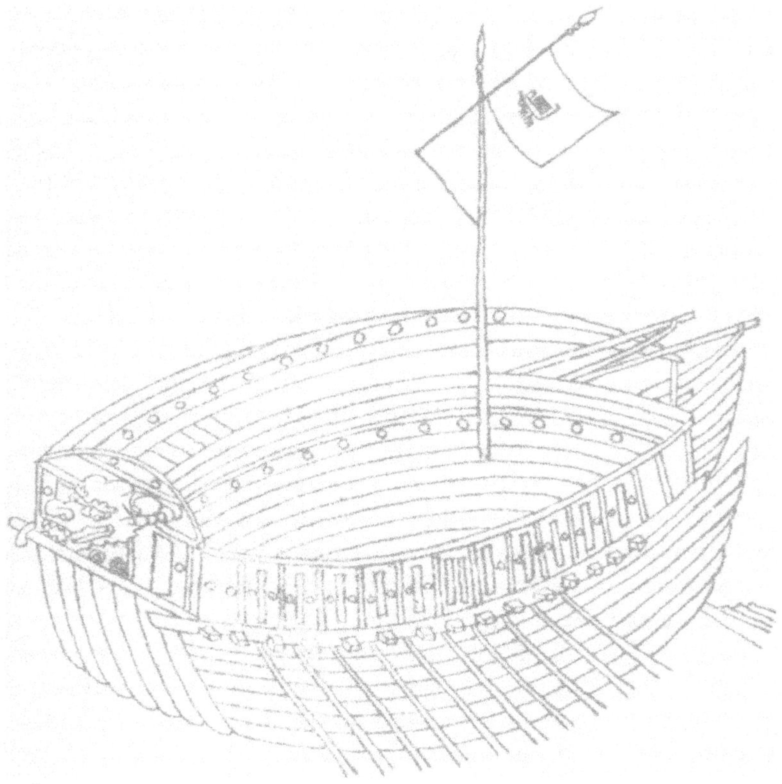

Fig. 3.32. Turtle ship of the *t'ongjeyŏng* type from the *Yi Ch'ungmugong chŏnsŏ* (Collected works of Admiral Yi).

each side, and another six on each side of the canopy, so there was a total of 36 cannon. The ship was driven by 16 oars.[178]

The Admiral suggested to King Sŏnjo in June 1592 the construction of a turtle ship of entirely different design, with cannon firing through the mouth of a dragon head located at the bow. Topside were numerous gimlets to prevent enemy troops from successfully boarding her. The admiral also proposed that this new ship carry an officer in charge of attack, whose duty it was to drive the ship deep into the enemy fleet, and direct the firing of all sorts of weapons, including *ch'ŏnja*, *chija*, and *hwangja*.[179] We are able to learn more about the structure of this later version through a report written by Yi Pun 李芬, a cousin of the admiral. The report said, "Another special war vessel has been developed. About the size of a *p'anoksŏn* 板屋船 (planked-cabin junk), its surface is planked, with narrow passageways through which sailors can move about. The whole surface of the ship, other than these passageways, is covered with blades and spears so that no enemy can walk over it. At the bow is a dragon head from which cannon can be fired; another cannon is installed at the stern. It is also called a turtle ship from its appearance. During battle, the blades and spears are hidden under heaps of seaweed as the ship charges into the enemy fleet."[180]

Several illustrations of this *chwasuyŏng* 左水營 (first commander's flagship) model have survived, the most authentic of which seems to be the drawing in the *Yi ch'ungmugong chŏnsŏ*. A comparison of illustrations of turtle ships shows that the one labeled *t'ongjeyŏng* has a flat planked canopy without blades or gimlets on the surface. The illustration of the *chwasuyŏng* version shows a dragon-head bow and a deck shaped like a turtle shell, and is so described in the text.

The *t'ongjeyŏng* model, the first turtle ship to appear in the reign of King T'aejong, was later developed by Admiral Yi and Navy Sub-area Commander Na into two models, one of which was the so-called Chŏlla *chwasuyŏng* (see Fig. 3.33) that

[178] Ibid.,
[179] Ibid., ch. 2.
[180] Ibid., ch. 9.

全羅左水營龜船

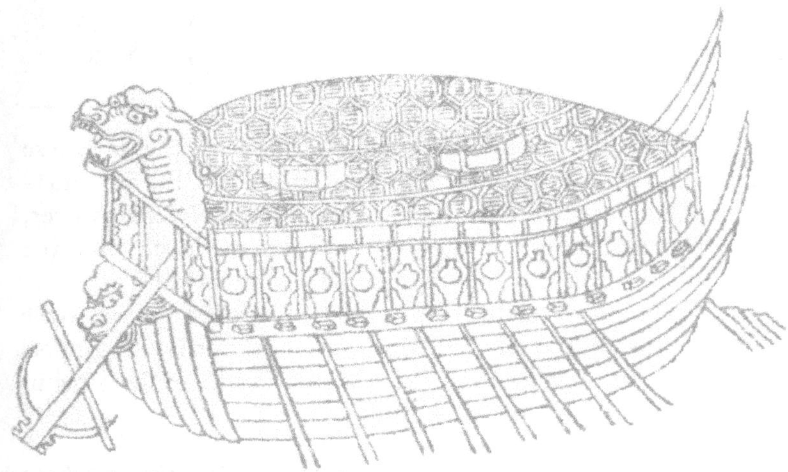

Fig. 3.33. Turtle ship of the Chŏlla *chwasuyŏng* type. From the *Yi Ch'ungmugong chŏnsŏ*.

Yi Sun-sin had his men build after his appointment to the naval commander's base of Chŏlla Province.[181] It has been conjectured that the shell of the turtle ship was clad with hexagonal iron plates held on with nails, like a castle gate protected with iron against fire attack. There is a record in Japan that supports this conjecture. In February 1593, one year after the Hideyoshi invasion, the Japanese government ordered the feudal lords to supply iron plate for use in building warships. It is

[181]Kang, "Yijo chosŏnsa," pp. 940–951; Ch'oe Yŏng-hi, "Kusŏn ko" (A study of the turtle ship), *Sach'ong* (The historical journal of Koryŏ University), 1958, *3*: 2–20; Arima Seiho, *Chōseneki suigunshi* (A history of naval forces during the Hideyoshi invasion; Tokyo, 1942).

Table 3.9. Warships of the Later Yi Dynasty

Kind of Ship	Lengths of Pine Used and Size of Tree: Large	Medium-Sized	Small	Very Small	Number of Men
Chŏnsŏn 戰船	14	89	77	40	198
Kusŏn 龜船	12	76	62	30	175
Haegolsŏn 海鶻船		59	36	12	56
Pangp'aesŏn 防牌船		56	24	15	55
Pyŏngsŏn 兵船		51	9	10	36
Sahusŏn 伺候船		12	4	4	5

believed that this order reflected an attempt of the Japanese to build ironclad ships after being astonished at the invulnerability of Korean vessels in several sea battles in 1592. However, Watanabe maintains on the basis of this evidence that the Japanese built the first ironclad ships in the world.[182]

Another type of turtle ship, which Yi called *pyŏlche kusŏn* 別制龜船 (variant model turtle ship), was equipped with gimlets on the shell to prevent boarding. Others had holes for guns in the shell.

Several other war vessels were constructed after the Japanese invasion of the sixteenth century. One was the *ch'angsŏn* 鎗船 (spear vessel), designed by Na himself in December 1606.[183] Twelve such vessels were constructed and tried out.[184] In 1740, Chŏn Un-sang 田雲祥 designed and tested the *haegolsŏn* 海鶻船 (seabird), a superior war vessel smaller than the turtle ship, lighter and able to navigate even on stormy sea.[185]

Table 3.9 shows various kinds of warships of the later Yi Dynasty, the material of which they were constructed, and the number of men each carried.[186]

[182] Watanabe Yosuke, "Chōsencki to waga zōsen no hattatsu" (The Hideyoshi invasion and the development of shipbuilding in Japan), *Shigaku zasshi,* 1935, *46*.5: 588, 597.
[183] Arima, *Kahō no kigen,* p. 235.
[184] *Sŏnjo sillok,* ch. 206, pp. 16b–17b.
[185] *Yŏngjo sillok,* ch. 52, p. 4b.
[186] Arima, *Kahō no kigen,* p. 240.

PRINCIPLES OF SHIPBUILDING IN THE YI DYNASTY

The little extant material on naval architecture was written in the eighteenth century, the late Yi Dynasty. The literature includes the illustrated descriptions in the *Yi ch'ungmugong chŏnsŏ*, from which we are comparatively well informed about the types and structures of the Yi Dynasty warships in the fifteenth and sixteenth centuries.[187] "Kaksŏn tobon 各船圖本," a collection of simplified plans of some ships, also includes details on the lengths of the main parts. One of the neatest theories in shipbuilding technology is found in the treatise on ships in the *Yŏamjip* 旅庵集 (Complete Works of Yŏam = Sin Kyŏng-jun), written by Sin Kyŏng-jun 申景濬 (1712–1781). In it Sin develops his theory as follows:[188]

The speed of a ship depends not only on the heaviness of its construction but on the shape of the ship. Like a plowshare thrusting into the land, a ship has power when the bow is heavy and the stern is light. A low bow and high stern help the ship go fast as it is pushed by the wind.

His book also tells us that the building of ships from scale diagrams was being introduced into Korea. As shown in Fig. 3.34, a ship was divided by five sections, showing coordinates and measurements for height from the base. Sin's blueprint was drawn so that the plane projection would correspond to the perpendicular projection.

Civil Engineering and Architectural Technology

ARCHITECTURAL TECHNOLOGY OF THE SŎKKURAM GROTTO

The Sŏkkuram 石窟庵 grotto, located inside T'oham-san 吐含山 mountain, Wŏlsŏng-gun 月城郡, north Kyŏngsang Province, is a masterpiece of Silla civil engineering and among the finest works of Buddhist art in the Chinese tradition.[189]

[187]According to Cho Sŏng-uk, the speed of the turtle ship was about 6 knots (*Sintonga*, monthly magazine, January 1970).
[188]*Yŏamjip* (Complete works of Yŏam = Sin Kyŏng-jun), ch. 18, pp. 13ab.
[189]See, for instance, Chewon Kim and Won-yong Kim, *The Arts of Korea: Ceramics, Sculpture, Gold, Bronze and Lacquer* (London, 1966), p. 154. The same book was published in the United states as *Treasures of Korean Art: 2000 Years of Ceramics, Sculpture and Jeweled Arts* (New York, 1966).

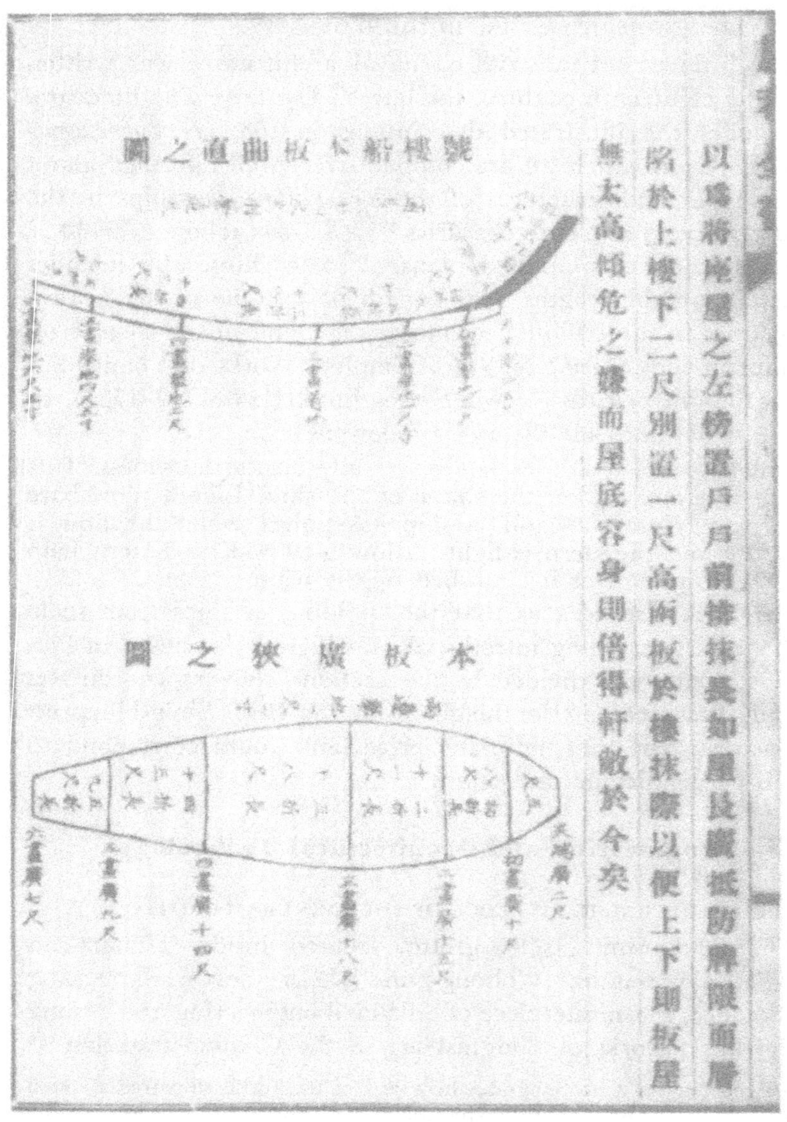

Fig. 3.34. Diagrams of a ship by Sin Kyŏng-jun, eighteenth century. The ship was divided into five parts. Coordinates are given for each line of demarcation. Coordinates for the keel line correspond to height from the base. Top: perpendicular projection; bottom: ground projection. From the *Yŏamjip* (Complete works of Yŏam = Sin Kyŏng-jun).

The grotto is known to have been built by Kim Tae-sŏng 金大成, the prime minister also responsible for reconstruction of Pulguk Temple 佛國寺, in honor of his parents in A.D. 751 (tenth year of King Kyŏngdŏk's 景德王 reign). The geometrical construction of the grotto is based on cosmological and Buddhist symbolism.

To discuss this grotto further, it is necessary to examine the structural technology of old Koguryŏ tombs. These tombs, constructed on a square foundation, were usually built in the shape of a pyramid.[190] The ceiling was truncated, either square or octagonal in shape. Koguryŏ tombs, their ceilings decorated with paintings of the rulers of the four directions or of asterisms, also embody elements of geometric design, as do architectural sites of the Paekche era.[191] The repertory of forms later expanded to include the circle, sphere, hexagon, and octagon.[192] Silla seems to have inherited the use of the circle as a design element from Paekche, where it originated. An evidence for this is the Och'ŭngt'ap 五層塔 pagoda of Paekche. This five-story pagoda was built on the principle of a circle with inscribed octagons and square. That these architectural techniques reached their apex with the construction of the Sŏkkuram grotto is a result of the harmonious use of various geometric figures.[193]

It was the Japanese scholar Yoneda Miyoji 米田美代治 who first attempted, in 1939, a mathematical analysis of the design of Sŏkkuram grotto.[194] Sŏkkuram is of granite, the domed grotto built on a round stone foundation; at the center of the vault is a round stone with lotus designs; on a lotus pedestal located at the center of the grotto is a seated figure of Buddha, 3 meters in height; on the inner wall are reliefs of an

[190]Yoneda, *Chōsen jōdai kenchiku no kenkyū*, pp. 200–207.

[191]Nakamura Kiyoe, "Kokuri jidai no kofun ni tsuite—sono seishō hekiga kōsatsu o chūshin to shite" (On the paintings of various celestial bodies in ancient tombs of the Koguryŏ period), *Kōkogaku ronsō* (Theses on Archeology), 1937, *4*: 380–385.

[192]Yoneda, *Chōsen jōdai kenchiku no kenkyū*, p. 206.

[193]Ibid., pp. 208–219.

[194]Ibid., pp. 3–22. See also *Kōkogaku* (Archeology), 1939, *10*.3.

eleven-faced Avalokiteśvara 十一面觀音 and other Bodhisattvas.[195] Now let us look at the design of the grotto (illustrated in Fig. 3.35) on the basis of Yoneda's research:

The plan of the grotto is a circle with a diameter of 7.15 m (ca. 24 T'ang *ch'ŏk*); the width of the entrance is 12 T'ang *ch'ŏk*, corresponding to the radius of the plan circle and at the same time the side of a hexagon inscribed in the circle. An equilateral triangle circumscribed about the entrance can be projected horizontally to just touch the center of the grotto circle, thus meeting the center of the front of the octagonal pedestal. The spatial composition of the grotto shows the same regularity. The height between the floor and upper wall of stone blocks, as well as that between the tip of Buddha's head and center of the dome, also equals the radius of the plan circle. Another radius is drawn from the tip of Buddha's head to define the dome. Stone layers of the dome are set in a five-layered canopy, topped with a large stone decorated with lotus designs. Each circular layer is formed of ten pieces of stone, which become smaller toward the top. The stones from the third layer upward are placed to prevent water from dripping. Stones carved into special rivet shapes are used in a ring at the base of the dome. The protruding parts of each layer are carved with designs. The lotus designs at the pinnacle of the dome not only constitute the very center, but prevent it from collapsing.

This ceiling construction has much in common with those of old tombs of the Three Kingdoms and Unified Silla.[196] But whereas Koguryŏ tomb ceilings were adorned with sun, moon, and stars, the ceiling of the Silla grotto expresses cosmic themes in a more symbolic way. The 12-*ch'ŏk* radius of the grotto's floor corresponds with the 12 hours of the day, the circle (with its traditional $365\frac{1}{4}$ degrees) symbolizes a year; the dome is heaven and at its center is a lotus-shaped stone representing

[195]Kim and Kim, *The Arts of Korea,* p. 155; *Bukkoku-ji to sekkutsu-an* (Pulguk Temple and Sŏkkuram grotto; Seoul, 1938), pl. 24–65.
[196]Yoneda, *Chōsen jōdai kenchiku no kenkyū*, pp. 3–22.

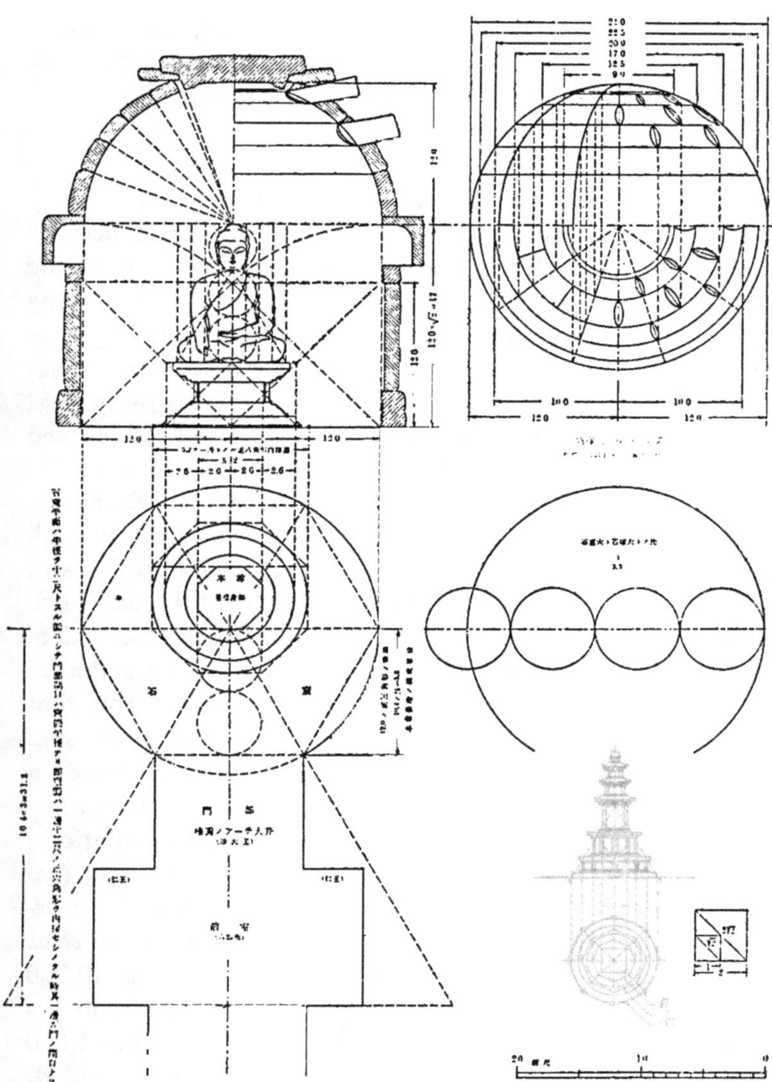

Fig. 3.35. Mathematical analysis of the Sŏkkuram grotto. The diagram shows that the plan of the grotto was based on a unit of 12 T'ang *ch'ŏk*. From Yoneda Miyoji, *Chōsen jōdai kenchiku no kenkyū*.

the sun surrounded by smaller stones symbolizing stars.[197] The grotto structure as a whole seems to be intended to represent the perfect world of the enlightened Buddha.

CONSTRUCTION OF CASTLE WALLS

Koguryŏ people first built castles around the first century B.C. Early structures of course show heavy Chinese influences.[198] Examples include Hwando-sŏng 丸都城 and Changan-sŏng 長安城 of Koguryŏ, Han-sŏng 漢城 of Paekche, Sabi-sŏng 泗沘城 of Puyŏ, and Kŭm-sŏng 金城 of Silla. Walls began as earthen and timber palisades; stone walls led to brick constructions. Forms, too, varied; they included the *chang-sŏng* 長城 (long wall); *ong-sŏng* 甕城 (fortress wall); *kok-sŏng* 曲城 (curved wall); and *na-sŏng* 羅城 (outer wall).[199] Aspects of the castle-building technique of Koguryŏ are revealed in the Kungnae-sŏng, which stands comparatively intact. It is about 6 m high and is equipped with a battlement. The lower part of the wall is of large stones, which decrease in size toward the top. Such a structure ensured sturdiness and stability. This method of using successively smaller stones was continued into the Yi Dynasty and became a unique Korean style, distinct from that of China.[200] The *T'aejo sillok* 太祖實錄 (Veritable records of the King T'aejo era), in a discussion of projects for building walls around the capital city of Seoul in 1396, gives the following account: "Where there were high and rough foundations, stone walls 15 *ch'ŏk* (ca. 3.6 m) high and 19,200 *ch'ŏk* (ca. 4.6 km) long, were built; earthen walls were built over ordinary hills, 24 *ch'ŏk* (ca. 5.8 m) wide at bottom and 18 *ch'ŏk* (ca. 4.3 m) wide at the top, 25 *ch'ŏk* (ca. 6 m) high and 40,300 *ch'ŏk* (ca. 9.6 km) long; over water outlets arch-style culverts were built, with stone walls 16 *ch'ŏk* (ca. 3.8 m) high and 1,050 *ch'ŏk* (ca. 2.4 km) long, constructed at right and left." It added,

[197]Ibid., pp. 149–170.
[198]*Sŏul t'ŭkpyŏlsi sa kojŏk-p'yŏn* (A history and remains of Seoul, the capital city of the Yi Dynasty; Seoul, 1963), p. 504.
[199]Ibid., pp. 504 ff.
[200]*Chōsen bunkashi,* vol. 1, p. 50.

"Redoubled efforts, however, were required in building walls in the Tongdaemun (East Gate) area because the land there was marshy, requiring the use of piles and gravel."[201]

The construction method in use at the time consisted of piling up large stones to form a wall and filling the clefts with smaller ones. This was essentially the method handed down from the Three Kingdoms era. Foundations were laid by planting large stones into the earth. The face of the wall receded at about a 15-degree angle. This kind of structure is still visible in the walls of private homes, mostly in the southern and southwestern provinces.

The techniques of castle-wall building, however, greatly improved from the time of King Sejong's rule. Old earthen walls were completely replaced by new stone walls beginning in 1421. Not only did the form of walls change with the opening of roads 15 *ch'ŏk* wide to enable patrolling, but construction materials, by requiring 106,199 *kŭn* of iron (ca. 63.7 tn) and 9,610 *sŏk* (ca. 17,280 l) of lime for over 16 km of wall, also saw radical transformation.[202] It was possible to build a wall 10 kilometers long in 38 days, although this meant mobilization of some 10,000 laborers. Unlike the old construction methods of T'aejo times, blocks of granite 50 cm by 20 cm were used in neat rows, piled to a height of 6 m with one-third of the lower portion consisting of larger granite blocks cemented with lime for strength and durability. The walls still receded at about 15 degrees, and their thickness tapered from 4 m at the foundation to 1.2 m at the top. The interior of the wall consisted of uncemented blocks. This Korean technology reached its zenith by the time of King Sukchong. Examples of this mature phase abound throughout Seoul. Principal characteristics documented in connection with repair projects that began in 1704 are

1. Stones of square shape, with a side of 65 cm, used in neat rows
2. The surface of the wall solid and vertical (a real innovation)
3. Clefts between big blocks filled with pebbles

[201] *T'aejo sillok*, ch. 9, pp. 1a–3a.
[202] *Sejong sillok*, ch. 15, p. 10b.

4. Gun positions atop the walls for better defense.

This pattern of battlemented wall construction had already begun by 1595, at the time of the construction of Namhansan-sŏng; but its present foundation was laid through repair projects begun in 1626, extending to some eight kilometers of the wall's length. Novelties were also introduced in repairing the gun positions, *tondae* 墩臺 (higher level), *ong-sŏng* 甕城 (fortress wall), and *ammun* 暗門 (emergency gate).[203]

Suwŏn castle wall, built in 1794, may represent the most developed example of this kind of architecture. About five kilometers long, it was fully equipped with watchtowers, signal-fire towers, and a crossbow battery. Plans, with an account of the construction process, a discussion of technical problems, and illustrations, were edited and published in 1800 under the title of *Hwasŏng sŏngyŏk ŭigwe* 華城城役儀軌 (Records and machines of the emergency capital construction service).[204]

BRIDGE-BUILDING TECHNIQUES

Progress in civil engineering was also reflected in the Korean *hyŏnggyo* 桁橋 (girder bridge), *chugyo* 舟橋 (pontoon bridge) and *honggyo* 虹橋 (arched bridge).

Representative examples of the *hyŏnggyo* type include the Sup'yo-gyo (bridge marked with water levels) and Chŏnch'ŏn-gyo (literally, "arrow-pierced bridge"), both still extant in Seoul. Because of its simple structure, the *hyŏnggyo* has been favored from ancient times. The oldest recorded bridge in Japan, built by Paekche engineers at the beginning of the fifth century A.D., was a *hyŏnggyo*.[205] This type of bridge was first made of wood, but gradually stone came into favor. Construction of this kind of bridge included erecting piers and placing girders over them. Sup'yo bridge 水標橋, for example, was constructed of hexagonal stone piers and girders of square section covered with stone. Early bridges of this type were flat;

[203]*Sŏul t'ŭkpyŏlsi sa*, pp. 619–656.
[204]Hong, *Chosŏn kwahaksa*, p. 229.
[205]*Nihon kagaku gijutsushi*, p. 650.

later the center of some such bridges rose a little to form an arch. Chŏnch'ŏn bridge 箭串橋, built during the later part of the fifteenth century, and located in southeast Seoul, is flat. It is famed for having been the longest bridge of the Yi Dynasty.[206] Sup'yo bridge is elegantly arched,[207] a beautiful and advanced form built mainlo within the precincts of royal palaces or temples. Often it incorporated a water outlet of the castle wall. Silla arched bridges, famed for their beauty and construction, include Ch'ŏngun (Dark Cloud) bridge 靑雲橋, Paekun (White Cloud) bridge 白雲橋, Yŏnhwa (Lotus Blossom) bridge 蓮花橋 and Ch'ilbo (Seven Treasures) bridge 七寶橋, all located inside the precincts of Pulguk Temple.[208] These

Fig. 3.36. A typical Korean arch bridge. Yi Dynasty, ca. eighteenth century. Pŏlgyo, south Chŏlla Province.

[206] *Sŏul t'ŭkpyŏlsi sa,* pp. 891–897.

[207] See, for instance, Wada Yūji, "Keijō no suihyōkyō" (Sup'yo bridge in Seoul), *Rekishi chiri,* 1912, *19.* 2: 63–66.

[208] Sekino Tadashi, *Chōsen no kenchiku to geijutsu* (Studies in Korean architecture and art; Tokyo, 1941), pp. 673, 675.

examples interest us because they represent a transitional period in the development from girder bridges to arched bridges, particularly in the steps that help pedestrians across each bridge. Together with the Okch'ŏn bridge 玉川橋 of Ch'anggyŏng palace, the Kŭmch'ŏn bridge 禁川橋 inside Ch'angdŏk palace is typical of stone bridges built inside a royal palace. These bridges are distinguished by their arched piers, resembling the five- or seven-arched bridges that existed near the Tongdaemun 東大門 (East Gate), Seoul, and one of the which, the Hwahongmun 華虹門 (Suwŏn castle arched gate), still remains at Suwŏn, near Seoul. These arched bridges were built by planting two arched abutments in the river, above which were laid a stone surface with stone railings. Although it is not known exactly when arched bridges were introduced to Korea, it is thought that the construction method is at least as old as Samnang-sŏng 三郞城 castle, which reputedly was built before the first century A.D. on Kanghwa Island. The gate of this castle is a brick arch. Another arched bridge still extant is the famous one at Pŏlgyo-ŭp, Posŏng-gun, south Chŏlla Province (Fig. 3.36). It is notable for having been built privately. The three-arched bridge, with its simple beauty, seems to epitomize the aesthetic preferences of the Yi Dynasty.

There is no clear or firm evidence of how the earliest pontoon bridges looked. A reference appears in the *Chŏngjo sillok* to the effect that in 1789 one of them was thrown over Noryangjin ford, Seoul (Figs. 3.37 and 3.38).[209] According to the *Chugyo chŏlmok* 舟橋節目 (Construction manual for the pontoon bridge) of the time, a bridge was made by commandeering 38 official and private boats over which 1,000 planks were laid; and a dozen vessels were connected right and left of the pontoon bridge.

[209] *Chŏngjo sillok*, ch. 28, p. 61a. Chŏng Yag-yong made the plan for this pontoon bridge, according to the records.

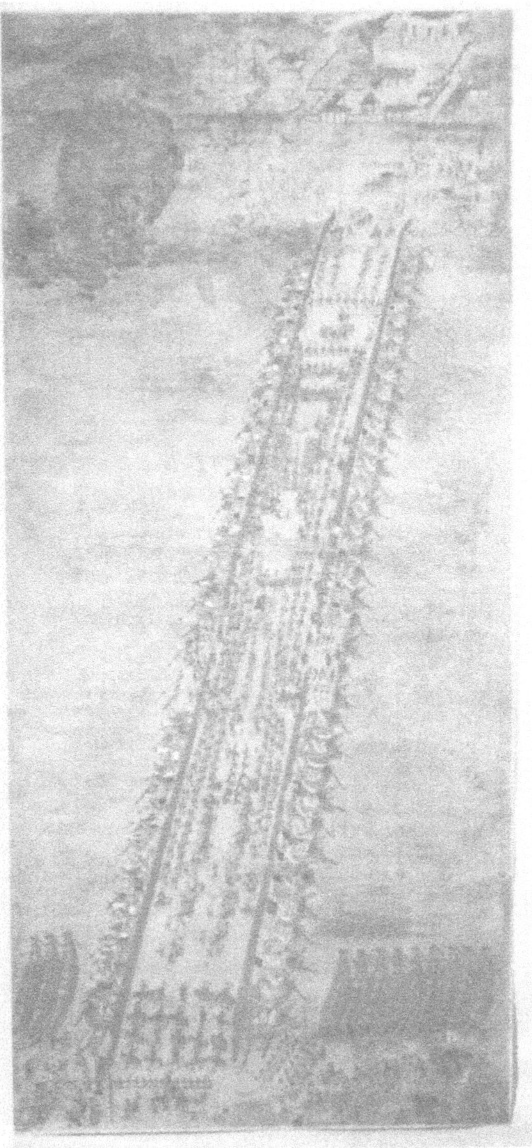

Fig. 3.37. Painting of the pontoon bridge built over Noryangjin ford (Han River), Seoul, in 1789.

Fig. 3.38. Painting of the pontoon bridge built over Noryangjin ferry in 1789.

4

Chemistry, Chemical Technology, and Pharmaceutics

Metal Handicrafts and Mining

BRONZE AGE AND IRON AGE

The Korean bronze age started around the tenth century B.C. with the emergence of the plains pottery people. The first bronze culture, which was introduced into the northeastern and northwestern parts of Korea, produced pottery and bronze similar to those found in Minusinsk, Siberia. It is therefore believed that the north Korean bronze culture was an offshoot of one farther north.[1]

Iron metallurgy, which made its debut during the age of civil wars in China, and the Scythian bronze culture mingled in the Liaotung area before entering Korea in the third or fourth century B.C., from where they played a role in forming the iron culture of Japan.

In 108 B.C., the Han Dynasty set up Lo-lang, a colony in the northwestern part of Korea, and transplanted Chinese metallurgy there. It was around this time that the Koreans were developing their mold-casting techniques (Figs. 4.1 and 4.2). Until the third century B.C., they cast bronze daggers and mirrors in sandstone molds,[2] but they had worked out a wax-mold casting technique by the first century B.C.[3]

On the lower Naktong River, at the time when the Han culture found its way into north Korea through Lo-lang, the

[1]In the past, some Japanese scholars believed that Korea had passed directly from the Neolithic Age to the Iron Age around the fourth to third centuries B.C. But Bronze Age remains were found in north Korea in the 1950s. See Kim Wŏn-yong, *Han'guk kogohak kaeron* (An outline of Korean archeology; Seoul, 1967), pp. 27 ff.

[2]Most of the sandstone molds found in Korea are from sites dated around the third century B.C.

[3]The delicate linear beauty of several bronze implements in the form of dagger handles excavated in the vicinity of Taejŏn in July 1967 is a good example.

Fig. 4.1. Bronze implement in the form of a dagger handle, third to second century B.C., 22.7 cm long, 9 cm and 12.3 cm wide. Excavated at Taejŏn, south Ch'ungch'ŏng Province. National Museum of Korea, Seoul. Photo courtesy National Museum of Korea.

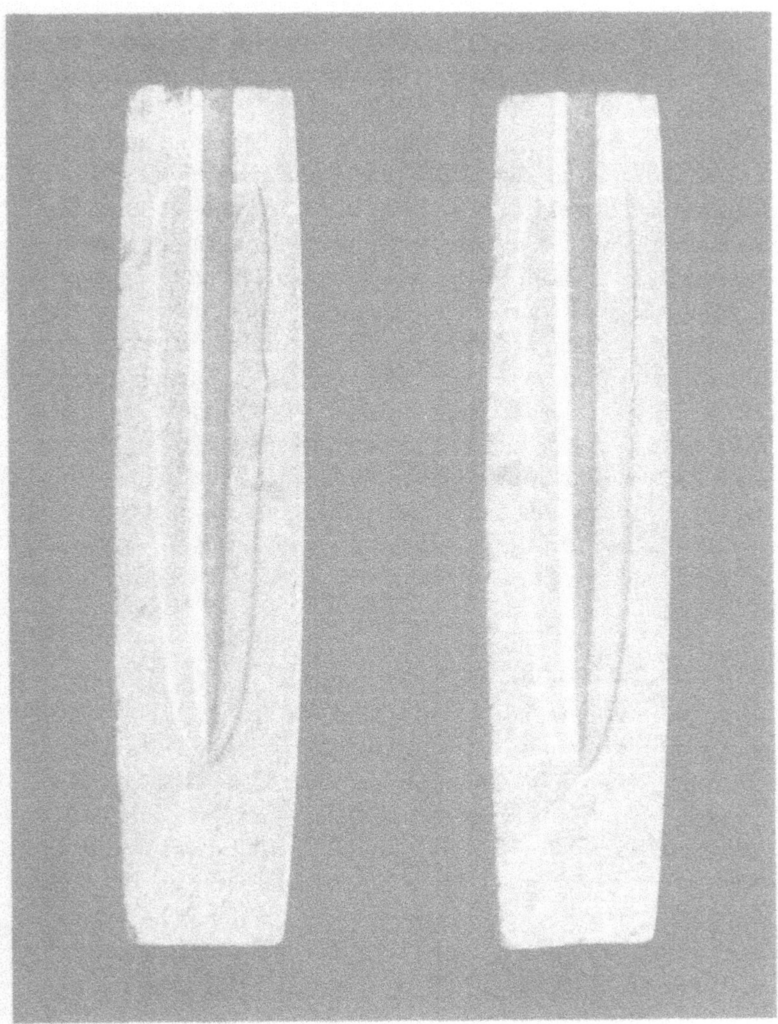

Fig. 4.2. Stone mold for bronze dagger. Third to second century B.C. Two daggers could be molded by one set of molds. The dimensions are 22.5 cm × 5.2 cm (4.2 cm both point). Found at Yongin, Kyŏnggi Province. National Museum of Korea, Seoul. Photo courtesy National Museum of Korea.

new dolmen culture of south Korea rapidly developed new techniques under the influence of the northern iron culture. This new Kimhae culture, as it is usually called, was based on the production of iron in the area. The smelting technique it developed was so successful that even Lo-lang and Japan purchased iron from this region, even though at this time (100 B.C.–300 A.D.) the smelting and production of iron was widespread throughout the Korean peninsula. The raw material used here was iron sand, as was the case in China.

The oldest written records concerning the Iron Age of Korea are seen in the Chinese *San kuo chih* 三國志 (History of the Three Kingdoms period) and the *Wei chih* 魏志 (History of the Wei Dynasty). According to the *Tung i chuan* 東夷傳 (Memoir on the eastern barbarians) of the *Wei chih* (end of the third century), Koreans regarded gold and silver as priceless and knew how to use iron. It noted that in Chinhan 辰韓, iron was produced and supplied to neighboring countries, including Japan.[4]

METALWORKING IN ANCIENT KOREA

Beginning in the fourth century, we can trace the development of mining, smelting, and working gold, silver, bronze, and iron. According to the *Nihon shoki* 日本書紀 (Chronicle of Japan), King Kŭnch'ogo 近肖古王 of Paekche presented to a Japanese envoy forty iron plates. The king also told the envoy, presenting him with a sword and a mirror: "The Kongna 谷那 iron mine is in the west of Paekche. Distant though it is, we will dig iron from the mine and continually send it to you." The sword, 75 cm long and of special shape, is said to be the one preserved at the Isonokami 石上 Shrine in Tenri, Japan. Inscribed on both sides of the blade are sixty-one characters, which say that the sword was made of steel on the sixteenth day of the fifth lunar month, A.D. 369, that it was the best available in Paekche, and that the sword was presented

[4]See, for instance, *Han'guksa* (A history of Korea; Seoul, 1959), vol. 1, p. 314, and Takashi Hatada, *A History of Korea* (trans. and ed. W.W. Smith and B.H. Hazard; Santa Barbara, Calif., 1969), p. 11.

as a gift to the Japanese king so that he might hand it on to his posterity.[5] Japan dispatched troops to Korea in the latter half of the fourth century to secure its iron resources.[6]

From the fifth to the sixth century, Korean metalworking made remarkable progress; ornaments and decorations excavated from Silla tombs, especially golden crowns and earrings, command our admiration. Metallurgy, given patronage by the royal family and the aristocracy, gained momentum in the sixth and seventh centuries with the flourishing Buddhist culture, which encouraged the casting of gilt bronze Buddhist images. Cobalt-60 radiography conducted on bronze artifacts of the period in the Museum at Kyŏngju has led to inferences, based on inner structure, about casting methods employed at that time. A gilt bronze image of Buddha from Koguryŏ (539?) has no air cells. Most small Buddhist statues cast during Unified Silla show signs that several holes left by spacers which supported the interior molds of heads and bodies were filled in. A middle-sized gilt bronze statue of Maitreya in the National Museum, 80 cm in height, was made in Silla ca. 600 by welding together the head and body, which had been cast separately. Its interior mold was supported by thin iron nails. In many cases, craftsmen used wax molds to turn out delicate and slender lines. The lost wax process was inherited from foundrymen of Paekche, who taught the techniques of casting Buddha images in Japan in the sixth century. Such molds were made by carving a mixture of rosin and beeswax, with inlet and outlet channels of wax attached. The mold was coated with refractory clay mixed with clay slip. The mold was then dried in the sun and baked at red heat so that the wax melted away.[7]

Usually, bronze Buddhist images were plated with gold by use of a mercury amalgam made with gold leaf. The amalgam

[5]Kobayashi Yukio, *Kodai no gijutsu* (Ancient technology in Japan; Tokyo, 1962), p.186.

[6]Many Japanese historians assert this as a fact, but most Korean historians believe that the information is not reliable. See, for instance, Yi Pyŏng-do, *Kuksa taegwan* (An outline of Korean history; Seoul, 1958), pp. 66–70.

[7]Kobayashi, *Kodai no gijutsu*, pp. 207–208.

was applied to the surface of the bronze object, which had been washed with plum vinegar or some similar acid. The object was then heated to 350°C in order to evaporate the mercury, and polished. The process had to be repeated several times.[8]

Thanks to positive government support, several masterpieces of bronze temple bells were cast in the eighth century.[9] The Great Bell of Pongdŏk Temple (now in Kyŏngju Museum), 333 cm in height and 227 cm in diameter, was cast with 120,000 *kŭn* (ca. 72 tn) of brass in 770. The bronze bell of Sangwŏn Temple at Mt. Odae, 167 cm in height and 90 cm in diameter, was cast in 725 with 3,300 *yang* (ca. 124 kg) of brass (Chinese *t'ou* 鍮). Those two big bells are the oldest and biggest surviving examples of Silla work. The images of flying devas on the left and right sides of the bodies of the bells are beautiful examples of craftsmanship, and the bells have a fine rolling tone.[10]

METAL HANDICRAFTS OF KORYŎ

Inheriting the metal craftsmanship of Silla and influenced by the techniques of Sung and Yuan, Koryŏ achieved further development in the crafting of large bronze bells, bronze mirrors, and incense burners.

Typical of Koryŏ bronze bells is that of Ch'ŏnhŭng Temple, cast in 1010, which imitated the decorative patterns of Silla. According to the *Koryŏsa*, the golden tower of the Hŭngwang Temple was made with 427 *kŭn* (ca. 256.2 kg) of silver and 144 *kŭn* (ca. 86.4 kg) of gold in 1067. A thirteen-story tower

[8]*Oju sŏjong pangmul kobyŏn* (Oju's book on the investigation of phenomena), p. 1080; Kobayashi, *Kodai no gijutsu*, p. 196. The great gilt bronze statue of Buddha at Nara, Japan, was cast by a descendant of Paekche craftsmen. See, for instance, *Meiji zen Nihon kōgyō hattatsu shi* (A history of the development of Japanese mining before the Meiji era; Tokyo, 1958), p. 31.

[9]In 754, a craftsman cast a huge bell for the Hwangyong Temple, which is said to have been 497,581 *kŭn* (ca. 29.7 tn) in weight, 1 *chang*, 3 *ch'on* (ca. 2.2 m–3 m) in height, and 9 *ch'on* in thickness. Records have it that King Kyŏngdŏk had a craftsman by the name of Kanggo-naemal, who made the "Medicine Buddha" image at the Punhwang Temple, 306,700 *kŭn* in weight, in 755. See, for instance, *Han'guksa*, vol. 1, pp. 696–697.

[10]See, for instance, *Han'guksa*, p. 697.

and several vases were made in 1223 with 144 *kŭn* of gold.

In the early part of the twelfth century, Koryŏ produced a number of excellent incense burners, among them the bronze incense burners at P'yoch'ung (cast 1177) and Kŭmsan temples (cast 1178, now of the Hōryūji Temple, Nara, Japan). Both of them are famous for their simple and balanced shapes and the graceful patterns of the silver inlays. The bronze mirrors of Koryŏ were mere imitations of Chinese implements. Most were based on Han and T'ang mirrors. Later there were imitations of Sung and Yuan mirrors, but the quality of bronze used and the skills employed were inferior. The metalcrafting techniques of Koryŏ advanced only in those areas based on the skills of Silla.

In addition, tributes to China increased year by year. Officials of Yuan, in particular, came to Koryŏ to exploit its supply of gold and silver ores. As a result, the Koreans greatly limited the mining of gold and silver ores, gradually even spreading the impression that Korea produced little gold and silver.

As already discussed, Korea was famous in Japan and China for its iron and copper even in ancient times. Yuan had Koryŏ produce arms, swords, and even raw iron and submit them as tribute. As is well known, iron coins were cast and used in 996 and a sort of silver coin came into use in 1098.[11] Toward the end of Koryŏ, when bronze movable type and firearms were cast, there was no need to import the raw materials. Because of periodic wars and poor administration of mining and metallurgy beginning halfway through the Koryŏ period, however, Korean metal technology began to retrogress.

MINING IN THE YI DYNASTY

From the beginning, the Yi Dynasty promoted the exploitation of valuable metals. Eighty soldiers were mobilized to dig gold ores at Tanch'ŏn, Hamgyŏng Province, in May 1398. The effort was intensified under the third king, T'aejong, with

[11]See, for instance, Kim Sang-gi, *Koryŏ sidae sa* (A history of the Koryŏ period; Seoul, 1961), pp. 62, 207–208.

300 miners digging and mining silver as well. Soldiers sought silver in 1401 and in the winter of 1406. But they could get only 8 *ton* 돈 (匁) (ca. 30 g) of silver.[12] King T'aejong was so disappointed in the projects that he suspended the enterprise in 1407. The *T'aejong sillok* 太宗實錄 (Veritable records of the King T'aejong era) gave as the cause of the failure "ignorance of skills in mining."[13] Dismayed by this lack of success, King T'aejong continually appealed to Ming China to remit Korea's "tributary" contributions of gold and silver. At the same time, he urged his officials to see to the production of mineral ores.

A second series of projects was launched in 1411. In December, the king dispatched the Minister of Works, Pak Chach'ŏng 朴子靑, to a mine near Seoul to encourage and supervise silver mining. But only 1 *yang* (ca. 38 g) of silver could be obtained. Undaunted, King T'aejong kept the laborers digging for mineral ores. In March 1412, he ordered them to dig for silver, lead, and gold. In December 1417, the king gave a disquisition on the methods of exploiting supplies of gold and silver in the State Council. In May of the same year, he ordered provincial governors to visit all silver and iron mines in areas under their jurisdiction and encourage the digging and refining of gold and silver ores. King Sejong also encouraged the digging of gold and silver ores in 1421, by setting up a system of rewards.[14]

The demand for bronze sharply increased after the Office of Typefounding was set up in February 1403, and, a little later, firearms began to be cast. The king also had technicians learn skills of refining copper from Japanese metal craftsmen in August 1418—a change in the customary direction of transmission.[15] The demand for copper increased further during the reign of King Sejong because of the recasting of bronze printing types in October 1422 and the casting in quantity of firearms and coins. In September 1424, King Sejong sent

[12]Ko Sŭng-je, *Kŭnse Han'guk sanŏpsa yŏn'gu* (A study of the industrial history of the Hermit Kingdom of Korea; Seoul, 1959), pp. 217–219.
[13]*T'aejong sillok,* ch. 13, p. 11b.
[14]*Sejong sillok,* ch. 11, p. 14b.
[15]Ibid., ch. 1, p. 7a.

officials to Chŏlla, Kyŏngsang, Hwanghae, and P'yŏngan provinces to try to produce copper ores.[16] In March 1426, officials proposed that a quota of copper be submitted to the court every year from Ch'angwŏn 昌原 in Kyŏngsang Province, and Suan 遂安 and Changyŏn 長淵 in Hwanghae Province, on the grounds that prospects of long-term importation of copper from Japan would be unreliable. This proposal must have been based on the conclusion that copper ores produced at the three places the previous year in experimental refining were of the best quality available.

The Treatise on Geography of the *Sejong sillok* records the existence of 54 iron, copper, and lead mines. There were 17 refineries which refined mineral ores during the agricultural off-season and submitted their products to the court. The one at Ulsan supplied 12,500 *kŭn* (ca. 7.5 tn) of minerals to the court. Year-round miners did not emerge even during the early Yi Dynasty, when the demand for minerals was great: farmers were still mobilized. Four hundred ninety-three blacksmiths (Fig. 4.3) made up 32 percent of all technicians working at refineries in provincial areas, but they mainly fabricated a variety of arms and farming implements. There were only three each of specialist iron and lead craftsmen.

Korea was finally exempted from submitting tribute gold and silver to China in 1430, but this merely led to neglect of gold and silver mining. The Yi Dynasty even lost its self-sufficiency in copper. For example, whereas the monthly demand for copper was 4,050 *kŭn* (ca. 2.4 tn) and annual demand was 48,600 *kŭn* in 1424, only 4,011 *kŭn* of copper was in the hands of the government at that time.[17] The *Ch'ao-hsien fu* 朝鮮賦, written by the Chinese Tung Yueh 董越, who came to Korea in the late fifteenth century, says that copper was produced in the Yi Dynasty in greater quantity than any other mineral. But despite its efforts the government could not meet the demand.

Under the rule of King Sejong, Korea hired Japanese iron

[16]Ibid., ch. 25, p. 23a.
[17]Ibid., ch. 23, p. 6a.

Fig. 4.3. Painting of a traditional Korean blacksmith's shop by Kim Hong-do (1760–?). National Museum of Korea, Seoul.

technicians in 1438.[18] In 1443, Yi Ch'ŏn had Jürchen technicians, who could process wrought iron from cast iron, teach their metallurgical skills to Korean technicians.[19] In July 1444, King Sejong put up rewards for those who could teach mining and refining skills.[20] But the results were negligible. For about half a century from the beginning of the Yi Dynasty to the era of King Sŏngjong (1470–1494), the mining of gold and silver ores was almost suspended, and the Yi Dynasty's mining was all but confined to exploitation of copper ores. Records have

[18]Ibid., ch. 84, p. 19b.
[19]Ibid., ch. 106, p. 18b.
[20]Ibid., ch. 109, p. 1a.

it that 584 *kŭn* of iron were dug and refined in 1424, at Sŏhŭng 瑞興, Hwanghae Province,[21] a mine with most abundant ores. Although it is not clear whether the figure meant output per month or year, the total was not very great. The production of iron declined too in later years; there were few soldiers who wore armor in P'yŏngan and Hamgyŏng provinces in 1502.[22] The shortage of iron was not relieved until the era of King Sŏnjo (1568–1608). As the situation was aggravated following the Hideyoshi invasion, 5,000-odd *kŭn* (ca. 3 tn) of iron was collected from all over the kingdom in February 1605. The Koreans could not produce enough arms to meet their needs until 1607.[23]

From this account it is evident that the conservative policy adopted for the purpose of avoiding excessive tributes of gold and silver to China was not the whole cause of the desuetude of mining in the Yi Dynasty. Although it remained a government monopoly, there was no long-term policy, and no attempt to free mining from the limitation of seasonal labor.

Mining of precious metals, suspended during the reign of King Sejong, was resumed by King Sŏngjong. Because of poor techniques, however, the new project was also given up before long. A breakthrough came when Kim Kam-bul 金甘佛, together with Kim Kŏm-dong 金儉同, an official slave, discovered an epoch-making new method of refining silver from lead ore in 1503,[24] to be explained in the following section. Thanks to this discovery, refining was started at the Tanch'ŏn 端川 lead mine, where a trial run produced 4 *ton* (ca. 15 g) of silver from 2 *kŭn* (ca. 1.2 kg) of lead, and at Yŏnghŭng 永興 lead mine, from which they produced 2 *ton* of silver. For the first time, the government permitted private management of the Tanch'ŏn silver mine and had the manager submit part of his product to the government as tax. The mining tax rate was 2 *yang* (ca. 75 g) of silver per miner for two days, which indicates that

[21]Whether it was a monthly or annual amount is not known.
[22]Ko, *Kŭnse Han'guk sanŏpsa yŏn'gu,* pp. 236–237.
[23]Ibid., pp. 237–241.
[24]See, for instance, ibid., p. 258.

output under the new method was several hundred times improved. The momentum, however, was lost under the reign of King Chungjong because of new Chinese tribute demands. In 1542, consequently, the Yi Dynasty had to import 15,000 *yang* of silver from Japan,[25] to which it had formerly exported the metal.

The exploitation of silver and copper was revived under King Sŏnjo as a means of replenishing the treasury, exhausted after the Hideyoshi invasion. Chinese refining methods, learned by a craftsman on the Manchurian border, were employed this time.[26] Private silver mining was again permitted following the Hideyoshi invasion. In 1651, several private silver mines were licensed and taxed directly for the first time.[27] But there was no consistent policy; management jumped from private to public hands repeatedly. The silver tax rate of 1,000 *yang* per year was halved in 1702. The number of silver mines at one time reached 68, but they were gradually deserted. The management of the 23 remaining silver mines was delegated from the Board of Taxation to provincial officials in 1775, but their abandonment could not be stopped. Exploitation of new silver mines was banned in 1798. In 1807, there remained only three silver mines. Among them, Unp'a silver mine, which paid 1,500 *yang* of silver in tax in 1764, paid only 140 *yang* in 1807.[28]

The mining and refining of copper were started in 1751 but were later suspended because of poor techniques. Except for this brief experiment, copper was entirely imported from Japan.[29]

Thus the metal handicrafts of the Yi Dynasty fell short of Koryŏ standards of quantity and quality. An examination of gilt bronze images of Bodhisattvas from the middle of Koryŏ

[25] Ibid., p. 267.
[26] Ibid.
[27] Kawasaki Shigetarō, *Kobunken ni arawaretaru Chōsen kōsanbutsu* (Mineral resources of Korea which appear in ancient documents; Seoul, 1935), pp. 21–22.
[28] *Man'gi yoram* (Handbook of economics and military affairs), Economics, part 4.
[29] See, for instance, Kawasaki, *Chōsen kōsanbutsu*, p. 21.

until the Yi Dynasty by radiography has given this result: "Changes in quality of materials used as well as in manufacturing methods are conspicuous. The alloy is not consistent, nor can a consistent composition be found in molds. Each was made by a different method."[30]

Methods of Refining Metals in the Yi Dynasty

REFINING OF GOLD

The refining methods for gold and some other precious metals employed during the Yi Dynasty will be explained, mainly based on the *Oju sŏjong pangmul kobyŏn* 五洲書種博物考辨 (Oju's book on the investigation of phenomena).[31]

Gold dust was purified by fusion, and further purified by cupellation with borax, with the separated silver reclaimed by fusing lead in the extraction crucible. The pieces of gold were then wrapped in clay mixed with salt, and baked over fire to produce excellent gold platelets (*yŏpchagŭm* 葉子金). Several layers were baked at a time, separated by sand and tightly bound by wire, whereupon the gold became thinner and turned reddish.

This method of refining gold, as explained by Yi Kyu-kyŏng, does not involve much difficulty. It is thought to date from the era of the Three Kingdoms. In Korea there were three classes of genuine gold: pure gold (*sipp'um-gŭm* 十品金), of the highest quality; platelet gold (*yŏpchagŭm*), of medium quality; and ingot gold, of lower quality. It is construed that *yŏpchagŭm* meant gold in the form of flat pieces like leaves (much thicker than what is now called gold leaf), and that *sipp'um-gŭm* was pure gold of reddish color.

This classification of gold is almost identical with that provided in the great Chinese technical encyclopedia of 1637,

[30]Ko Chong-kŏn and Ham In-yŏng, "Pangsasŏn t'ugwapŏpe ŭihan komisulp'umŭi chosa" (Examination of ancient art objects by gamma radiography), pt. 2, *Misul charyo,* 1964, *9*: 17.

[31]*Oju sŏjong pangmul kobyŏn,* pp. 1074–1075.

T'ien kung k'ai wu 天工開物 (The exploitation of the works of nature), from which Yi Kyu-kyŏng apparently was quoting.[32]

TANCH'ŎN SILVER REFINING METHOD

As noted in the preceding section, Kim Kam-bul and Kim Kŏm-dong discovered a method of refining and separating silver out of lead ores at Tanch'ŏn, northeastern Korea, late in the fifteenth century.[33] The method was actively employed by the Yi Dynasty government from the early sixteenth century. The *Yŏnsan'gun ilgi* 燕山君日記 (Veritable records of the King Yŏnsan era, literally, "Diary of Prince Yŏnsan") has a brief account of it,[34] and Yi Kyu-kyŏng gave a relatively detailed description in his book.[35]

A furnace was constructed with a forced-air blast and a small ashpit underneath. Lead pieces were first put into the furnace, and a layer of crude silver was placed on top. The lead melted first and sank down, while the crude silver bubbled and fused. The lead separated and sank into the ash. Water was sprayed onto the lead to solidify it, and it was then melted in the furnace to recover the pure lead. Silver produced at Tanch'ŏn, it is said, was of much higher purity than its counterpart in China or Japan.[36]

[32]See, for instance, E-tu Zen Sun and Shiou-chuan Sun (tr.), *T'ien-kung k'ai-wu: Chinese Technology in the Seventeenth Century* (by Sung Ying-hsing; University Park and London, 1966), pp. 236–237.

[33]Their discovery can be evaluated when it is recalled that a similar silver-refining method began to be used in Europe in the first half of the sixteenth century.

[34]*Yŏnsan'gun ilgi* (Diary of Prince Yŏnsan), ch. 49, p. 28b. They refined 2 *ton* (ca. 3.75 g) of silver from 1 *kŭn* (ca. 600 g) of lead ore.

[35]*Oju sŏjong pangmul kobyŏn*, p. 1089.

[36]Yi Kyu-kyŏng called the method the "Tanch'ŏn silver-refining method," undoubtedly because Kim Kam-bul and Kim Kŏm-dong tried the method first at Tanch'ŏn. The refining method was different from its counterparts in China and Japan in several respects, but all are ash-blowing methods. In his discussion of silver refining, Yi first took up the silver-refining method given in the *T'ien kung k'ai wu* and then explained the Tanch'ŏn silver-refining method. See also *Imwŏn simyukchi* (Sixteen treatises on science and technology), ch. 51, pp. 51a–51b (photo reprinted., vol. 2, p. 583).

REFINING OF COPPER AND LEAD

The method of refining copper during the Yi Dynasty was as follows:[37]

A furnace was made, connected through a horizontal opening in its left side with the metallurgical bellows. The four walls were built on a small ridge, and flat stones were laid inside like the *ondol* floor of a Korean house. After ash was spread over the stones, the surface was pasted with thin mud. The flat stones were laid to make the middle concave and the edges high, and a hole was made in one of the upper flat stones. Hard charcoal was spread on the stone floor and burned in the air blast for five to six hours. The copper ore was than charged and fired. The melted ores flowed down to the stone floor, and the impurities bubbling on the surface flowed beneath the floor through the hole in the stone. Then the copper was taken out after cooling by spraying it with water. The refining method for lead was the same.

This is a dry method of producing copper and belongs to the so-called floor furnace type. According to the *Yijo sillok*, Mun Kae 文益 and other craftsmen learned the method of refining copper from a Japanese technician in 1566, and the Japanese technician was rewarded with a government post.[38] The Japanese method of refining copper is believed to have been introduced into Korea at that time.

Copper Alloys and Other Metal Compounds

BRONZE AND BRASS

It is well known among historians of the fine arts that Korea's metalworking techniques reached a high level.[39] From ancient

[37] *Oju sŏjong pangmul kobyŏn*, p. 1098.
[38] *Myŏngjong sillok*, ch. 32, p. 35a.
[39] Two books concerning Korean arts (including metalworking) have recently been published in English; Chewon Kim and Won-yong Kim, *The Arts of Korea: Ceramics, Sculpture, Gold, Bronze and Lacquer* (London, 1966), also published as *Treasures of Korean Art: 2000 Years of Ceramics, Sculpture and Jeweled Arts* (New York, 1966); and Evelyn McCune, *The Arts of Korea: An Illustrated History* (Rutland, Vermont, and Tokyo, 1962).

times, the quality of Korean copper alloys was recognized in China. Their reputation is verified by the records in old books. In the *Pen-ts'ao kang mu* 本草綱目 (The great pharmacopoeia, 1596), Li Shih-chen 李時珍 maintains that Persian bronze (brass) is suitable for mirrors but that Silla copper (bronze) is superior for casting bells.[40] *Tongguk yŏji sŭngnam* 東國輿地勝覽 (Geographical conspectus of the Eastern Kingdom, Korea, completed 1482) records that of the five metals (gold, silver, copper, iron, and tin), copper was produced in the largest quantity in Korea. Korean copper was known for its hardness and somewhat reddish color. Spoons, chopsticks, and tableware were made of Korean bronze, which is called *kao-li t'ung* 高麗銅 in China.[41] Tung Yueh 董越 in the *Ch'ao-hsien fu* 朝鮮賦 also praised the high quality of Korean copper.[42] T'ang and Sung China purchased Korean brass and bronze for use in casting coins and manufacturing tableware.[43] According to the *Koryŏsa,* Koryŏ copper (probably bronze or brass) was bought by the Later Chou Chinese in 958–959. In 1262, Yuan sent special envoys to Koryŏ to obtain brass.[44]

Chemical and physical research on Korean bronze is just beginning.[45] So far it has given rise to a new classification of Korean bronze into three major kinds: bronze, which includes

[40]Kawasaki, *Chōsen kōsanbutsu,* p. 17.

[41]*Tongguk yŏji sŭngnam* (Geographical conspectus of the Eastern Kingdom), ch. 1, pp. 13b–14a.

[42]Son Po-gi, "Han'guk insoe kisulsa" (A history of Korean printing technology), *Han'guk munhwasa taege* (Seoul, 1968), p. 1047.

[43]*Haedong yŏksa* (Historical treatises on Korea; modern reprint edition), ch. 26, p. 3; Ch'oe Nam-sŏn, *Urinara yŏksa* (A Korean history; Seoul, 1952), p. 59; Kim, *Koryŏ sidae sa,* pp. 925–926.

[44]*Koryŏsa* (History of the Koryŏ Dynasty), ch. 2, pp. 27b–28a.

[45]It is recorded in the *Samguk sagi* (History of the Three Kingdoms) and in *Samguk yusa* that these bells were made of brass, but no chemical analysis has ever been made to determine their ingredients. Beginning in 1920, Japanese scholars launched full-scale archeological, chemical and physical research on ancient Chinese bronze which reached a climax in the 1930s. In recent years, Chinese scholars have also conducted such research. Professor Martin Levey and others have analyzed some Korean bronzes and found them to be over 24 percent tin. This is approximately the amount found in modern cymbals, giving a rare martensitic structure.

copper, tin, and lead as its principal ingredients; a bronze resembling brass, which includes only copper and tin; and Korean bronze of copper, zinc, tin, and lead.

It is presumed that Koreans began casting copper alloys as early as the tenth to seventh century B.C. Perhaps Koreans learned casting techniques from the Chinese, but it should be noted that the bronze that Korean metalworkers produced was in most cases different in composition from that cast by the Chinese. According to the recent result of chemical analysis,[46] bronze daggers made in Korea around the third to second century B.C. are composed of copper, tin, and lead in an average ratio of 75 : 16 : 9. In the case of Chinese bronze mirrors, Chikashige's analysis shows a ratio of 70 : 20 : 10, while Komatsu and Yamanouchi list a ratio of 75 : 15 : 10 for Korean bronze mirrors. The latter is almost identical with the composition of the Korean daggers. In other words, Korean bronze, an alloy of copper, tin, and lead, contains less tin but much more lead than the Chinese product.[47]

It is especially significant, in view of the specialization of

[46]Ch'oe Sang-jun, "Urinara wŏnsi sidae mit kodaeŭi soebuch'i yumul punsŏk" (Report on the chemical analysis of ancient Korean bronze and iron wares), *Kogo minsok*, 1966, *3*: 44. This source gives the following percentages of copper, tin, and lead in five bronze daggers:

Cu	Sn	Pb
78.20%	17.12%	4.32%
73.13	19.17	6.39
70.30	14.84	14.22
78.09	14.39	8.39
75.94	15.08	9.45

[47]See Komatsu Shigeru and Yamanouchi Yosito., "Kokyō no kagakuteki kenkyū" (A chemical and metallographic study of ancient Far Eastern mirrors), *Tōhō gakuhō* (Kyoto), 1937, *8*: 19–22; and Chikashige Masumi, "Tōyō kodōki no kagakuteki kenkyū" (Chemical analysis of ancient Oriental bronze wares), *Shirin*, 1919, *3*. 2: 74–76.

Cu	Sn	Pb	Sb	As
65.37	21.90	9.68	2.60	0.44
72.29	11.76	15.14		—
74.63	14.18	9.59	0.33	0.32

alloys in China, that there is no remarkable difference in ratio between Korean daggers and mirrors. This implies that it might have been technically difficult to regulate that ratio of principal ingredients without running into problems when casting.

Korean metalworkers were more enterprising in the use of new ingredients. An analysis of bronze wares of around the tenth century B.C., found in North Korea,[48] suggests that they were pioneering efforts in the addition of zinc to bronze. This process is so difficult that even in China it was not used until after the Sung Dynasty. According to the excavation report, these bronzes were used for ornamental and ceremonial purposes. Furthermore, the report points out that they must have been made where they were found.[49] The copper alloy

However, spectrochemical analysis of an eleventh-century Koryŏ bronze mirror by J.W. Mellichamp, Institute for Exploratory Research, U. S. Army Electronics Command, Fort Monmouth, N. J., shows a different composition: Cu major, Pb 5%, Sn 4 %, As 0.1 %, Sb 0.1 %, Bi 0.1 %, Ni 0.1 %, Zn 0.05 %, Ag 0.02 %, Fe 0.05 %, Si 0.01 %, Mg and Al traces. See also Noel Barnard, *Bronze Casting and Bronze Alloys in Ancient China* (Canberra, 1961), p. 192, which gives a Cu: Sn: Pb ratio of 68: 21: 4 for mirrors established as being from the same period. According to Chikashige's analysis, in the case of Chinese bronze mirrors, the mean ratio of Cu, Sn and Pb is 68: 23.6: 5.6. The maximum percentage of Pb is 9.0, minimum, 4.2. In the case of Korean bronze mirrors, according to the analysis of Komatsu and Yamanouchi, the maximum percentage of Pb is 9.7, minimum, 7.8.

[48]Ch'oe, "Urinara wŏnsi sidae mit kodaeŭi soebuch'i yumul punsŏk," pp. 43–44.

	Composition (%)					
Item	Cu	Sn	Zn	Pb	Fe	Others
Ornament	59.93	22.30	13.70	5.11	1.29	3.67
Mirror with fineline pattern	42.19	26.70	7.36	5.56	1.05	—
Ceremonial hatchet	40.55	18.30	24.50	7.50	1.05	—

[49]The remains are believed to be from about the tenth century B.C. An excavation report points out that it is not certain whether the bronze wares are from the same period (Chŏng Paek-un and To Yu-ho, *Najin Ch'odo wŏnsi yujŏk palgul pogo* [Report on the excavation of primitive ruins in Ch'odo Island, Najin, North Korea; P'yŏngyang, 1956]). On this basis, the bronze is estimated to have been from the tenth century B.C. (pp. 16, 44–46).

includes a maximum 24.50 percent of zinc.[50] Pongsan-gun, Hwanghae Province, where the remains were unearthed, is listed as a production site of smithsonite in a Korean geographical treatise of the fifteenth century.[51] Therefore, it is thought that primitive men living in the region happened to find out that the quality of the copper alloy changes when smithsonite is added. It was far from easy to find a method of adding zinc to bronze, since the boiling point of zinc is 900°C.

Korean documents record brass, known as *yusŏk* 鍮石 (*t'ou shih,* a word distantly descended from Persian *tutiya*) or *hwang-dong* 黃銅 (*huang t'ung,* yellow copper) from the early eighth century. The use of a copper-tin-zinc alloy in casting great bronze bells reflected both the facts that Korea produced more zinc than tin, and that when tin was used skillfully and in moderation it provided better liquidity of solution, thus facilitating casting.

But such a copper alloy was not suitable for making tablewares because of the poisonous elements of zinc and lead. Therefore, Koreans used a conventional copper-tin bronze (75 : 25 or 80 : 20) from the Koryŏ period.[52] In the 1940s, bronze tableware was still as widely used in Korea as the white porcelain variety. The traditional name for the copper alloy was *not* 놋, which was written in Chinese characters as *yu* 鍮 (*t'ou*) *yudong* 鍮銅 (*t'ou t'ung*) or *yuch'ŏl* 鍮鐵 (*t'ou t'ieh*), all of which originally referred to brass. On this account, this common bronze was taken for brass until recently in Korea. In the *Oju sŏjong pangmul kobyŏn,* it is explained that *yu* is a resonant bronze made by mixing copper with tin: to produce 1 *kŭn* of *not soe* 놋쇠 (*t'ou t'ieh,* literally, "brass-iron"), 1 *kŭn* (ca.

[50]In Persia brass cups were used in the fifth century B.C. See Charles Singer (ed.), *A History of Technology,* vol. 2 (Oxford, 1945), p. 54. According to *Sejong sillok,* the Yi government had 5,083 *kŭn* (ca. 3 tn) of zinc in 1424 (ch. 23, p. 34a).

[51]*Sejong sillok,* ch. 152, "Chiriji," p. 4a.

[52]According to the *Koryŏsa* (ch. 39), in September 1357 the king approved a recommendation to encourage his subjects to use bronze (or brass) tableware freely. Masumi Chikashige, *Oriental Alchemy—Alchemy and other Chemical Achievements of the Ancient Orient* (Tokyo, 1936), pp. 71, 211.

600 g) of copper and 4 *yang* of tin were alloyed. Korean metalworkers employed this 4 : 1 ratio until recently.

When bronze type was first made in 1403, it was cast from various kinds of bronze wares which had been collected and melted together. It was not an alloy of controlled composition; chemical analysis puts it in the category of Korean bronze. Photographs of a piece of bronze type of the eighteenth century, enlarged 100 times,[53] prove that it was an alloy of higher quality, but it has not been possible to reconstruct the process by which it was made. There is, however, information on the making of the best copper alloys in the *Oju sŏjong pangmul kobyŏn*.

Yi Kyu-kyŏng says that the best-quality brass can be obtained when 6 *kŭn* (ca. 3.6 kg) of red copper and 4 *kŭn* (ca. 2.4 kg) of zinc are melted in a stoneware jar and cooled. He goes on to say that 1.5 *kŭn* (ca. 900 g) of brass can be obtained when 1 *kŭn* of copper is mixed with 1 *kŭn* of smithsonite in the same way. He stresses that brass made from smithsonite should be hammered while red-hot, but when zinc is used, it should be hammered cold. Brass becomes dull white in color and unresonant after hammering. It becomes yellow and its metallic sound becomes clear after it is filed and polished. Yi further says that bronze becomes very hard when the melted mixture of copper and tin is cooled with water.[54] Although such records are fragmentary, it is thought that the secret of producing quality copper alloy lies in hammering, reheating and cooling. Starting in the Koryŏ period, government artisans responsible for the production of copper alloys were classified as experts on brass ware, red copper, bronze ware, metalworking, mirrors, and paktong[55] (to be discussed in the following section).

PAKTONG

Paktong (Chinese *pai-t'ung* 白銅, literally, white copper) is an alloy of copper and nickel, but often also contains zinc.

[53]Son, "Han'guk insoe kisulsa," p. 1051.
[54]*Oju sŏjong pangmul kobyŏn*, p. 1115.
[55]Kim, *Koryŏ sidae sa*, pp. 925–926.

Originally, paktong incorporated arsenic for the purpose of giving copper a silver color.[56] There is a passage in the *Samguk sagi* on paktong bells, evidence that paktong was produced before the seventh century.[57] In addition, the *Koryŏsa* states that there were metalworking specialists called paktong artisans; and that a special envoy of Koryŏ presented paktong wares to the Sung emperor in 1124.

The *Sejong sillok* writes that Koreans used copper and nickel in making paktong. In 1438, King Sejong ordered the governor of Kyŏngsang province to produce it for royal use. The governor had his technicians use copper from Ch'angwŏn and "white iron" (nickel) of Ulsan in its manufacture.[58]

Meanwhile, Yi Kyu-kyŏng asserted that copper containing pure arsenic was so hard that it could not be hammered. In order to make the metal soft, "white tin" or nickel was used. This leads us to interpret that nickel was combined with copper-arsenic alloy for this purpose alone. He said that five *p'un* to one *chŏn* (*ton*) of nickel was combined with one *yang* of paktong to soften the latter.

Paktong was produced by the following method: A furnace was built and a round wall constructed around the smelt. Arsenic ore containing nickel was placed in the furnace and copper was laid on it. Then the furnace was heated and a blast of air introduced from a double-acting bellows installed outside the wall. The metals combined into an alloy and began to flow into the pot outside the wall. When the pot cooled off, paktong was removed. The round wall was designed to protect metalworkers from the poisonous vapor of copper. Hazards involved in the production of paktong were far greater than those in producing brass.

[56]According to the *T'ien kung k'ai wu* (The exploitation of the works of nature), the Chinese also refined paktong with copper and arsenic, while the Japanese used copper and nickel. What Koreans called *wae-paektong* (Japanese paktong) was an alloy of copper, zinc, and tin. See *Oju sŏjong pangmul kobyŏn*, p. 1108.

[57]According to ch. 40 of the *Samguk sagi* concerning government positions, junior officials held paktong bells, while senior officials carried golden and silver bells and subordinates iron bells.

[58]*Sejong sillok*, ch. 81, p. 11a.

Paktong made this way was further refined by heating until it was almost melted, and polished with great care. During refining, arsenic powder was put on the surface with a view to making the paktong whiter. It is said that Korean metalworkers often added brass or zinc to make the work easier, and the Korean product used to be colored with a yellow tint. Yi Kyu-kyŏng maintained that Korea's paktong was softer and less white than that of China, because arsenic stone was rare in Korea and substitutes were used.[59]

The popularity of paktong wares (especially spoons) was second only to that of silver utensils in Korean households from the Yi Dynasty period to 1945, when stainless steel became popular for household utensils. The word paktong itself may have been derived from the Korean pronunciation "*paektong*" rather than the Chinese "*pai-t'ung.*"

MERCURY AND ITS COMPOUNDS

It is believed that Koreans had already acquired knowledge of mercury and its compounds from the Chinese around the fourth century, in light of the fact that cinnabar and vermilion are contained in the red pigment of mural paintings in the old tombs of Koguryŏ. Their knowledge must have expanded with the introduction of books by the alchemists Ko Hung 葛洪 and T'ao Hung-ching 陶弘景 in the Three Kingdoms period (ca. 300–650).[60]

In the early sixth century, Koguryŏ artisans developed a plating technique using gold amalgam and manufactured beautiful gilt bronze figures of Buddha. At that time the Koreans began to plate their bronze figures of Buddhas by this process. But it appears that gilding did not make rapid progress until after the fifteenth century, because Koreans until then

[59]The mineral ores, Yi Kyu-kyŏng says, were brownish yellow and looked somewhat like lead or iron. See *Oju sŏjong pangmul kobyŏn*, p. 1106. Even today, arsenic is considered a rare mineral ore in Korea.

[60]Kim Tu-jong, *Han'guk ŭihaksa* (A history of Korean medicine; Seoul, 1966), p. 34.

used imported mercury, mainly from China and other countries.[61] According to the *Koryŏsa*, merchants from Arabia, China, and Japan brought mercury to Koryŏ.[62] The *Yijo sillok* had it that Kim Chung-bo 金仲寶 discovered a method of making mercury from cinnabar in 1492. In the *Tongguk yŏji sŭngnam*, there are records concerning deposits of cinnabar which were used in the production of mercury. It seems that a large quantity of mercury was produced in Korea around this time (early sixteenth century). A total of nine methods of preparing mercury are explained in the *Oju sŏjong pangmul kobyŏn*. They include common methods also known from such Chinese sources as *T'ien kung k'ai wu*, *Tan yao pi chueh* 丹藥秘訣 (Secret formulas for the elixir), and *Tan fang ching yuan* 丹方鏡源 (Source-mirror of elixir formulas).[63]

We will take an example, typical in the simplicity of its apparatus: First, cinnabar was ground into powder, wrapped in blue-colored hemp cloth, and put into a steamer (*siru* 시루). The bottom of the steamer was then covered with a white porcelain bowl and luted with salt-seasoned clay so that no steam could leak out. The implement was heated from above on a charcoal fire. The mercury vapor moved downward through the steamer and mercury was condensed as it cooled at the bottom. This method is very close to one known to have been used in Ise, Japan; both depend on the downward movement of mercury vapor.[64]

Mercury compounds that were used as medicines or paints included calomel, corrosive sublimate, cinnabar, and vermilion.

LEAD COMPOUNDS, ARSENIC, ANTIMONY, AND OTHERS

Lead compounds produced at the time included white lead,

[61]In Korea, mercury is scarce even these days. The shortage of ores must have limited production of mercury in ancient Korea.
[62]Kim, *Han'guk ŭihaksa*, pp. 130–131.
[63]*Oju sŏjong pangmul kobyŏn*, pp. 1177–1179, and, for instance, Nathan Sivin, *Chinese Alchemy: Preliminary Studies* (Cambridge, Mass., 1968), p. 69.
[64]Yoshida Mitsukuni, "Chūsei no kagaku (rentan-jutsu) to senjutsu" (Medieval chemistry [alchemy] and the arts of immortality), *Chūgoku chūsei kagaku gijutsushi no kenkyū* (Tokyo, 1963), p. 236.

minium, and lead acetate, or litharge, used in the prescriptions of Korean folk medicine in the fourteenth century. Preparation methods are the same as those in the *T'ien kung k'ai wu*.[65] White lead was made before the seventh century. In 692, a Korean monk in Japan was rewarded for preparing white lead.[66]

Koreans did not distinguish arsenic and antimony minerals. Yi Kyu-kyŏng writes in his book on arsenic stone that there are two kinds of arsenic stone, "Chinese" and "local." He further calls the products of sublimation (As_2O_3) "Chinese arsenic frost 唐砒霜" and "local arsenic frost 鄕砒霜," respectively. He gave a detailed preparation for both, quoting the *T'ien kung k'ai wu* on the method for Chinese arsenic oxide. The "local" arsenic oxide was prepared in two different ways. One of them is as follows:[67]

Korean arsenic stone was put into an urn and covered with pine needles. It was them inverted over another stoneware urn and the joint tightly luted. The lower half of the combination was then buried in the ground. The upper jar was heated from above with faggots and then cooled. The melted fluid flowed to the jar below. Remnants in the upper urn were mixed with the next charge and re-treated.

This method, which unlike the Chinese process does not involve sublimation, would be of very little value in purifying arsenic oxide. The secret of its "local" success is that the Korean "arsenic stone 砒石" actually contained considerable antimony oxide, which would indeed be purified by gravity upon fusion. One can see traces of the artisans' original confusion (if not the compiler's incomprehension) in a second "local" method, in which a kiln and a pot for condensing sublimate are set up, but the two-part reaction vessel is again tightly sealed so that no vapors can escape! It is not surprising that Korean-made paktong has contained considerable anti-

[65]Sun and Sun (tr.), *T'ien-kung k'ai-wu*, p. 256.

[66]Kazuo Yamasaki, "Pigments Employed in Old Paintings of Japan," in Martin Levey (ed.), *Archeological Chemistry* (Philadelphia, 1967), pp. 350–351.

[67]*Oju sŏjong pangmul kobyŏn*, pp. 1207–1208.

mony since the eleventh century. Arsenic and its compounds were detected by silver in medical jurisprudence from the Koryŏ period (tenth to fourteenth century).

Korean Alchemy and Medicine

ORIGIN AND HISTORY

Korean alchemy dates back to the fifth century, when, under the influence of Taoism, efforts were made to manufacture an elixir of immortality.[68] But Koreans had a good deal of chemical and pharmacological knowledge well before the fifth century. According to ancient Chinese documents Koreans produced medicine and poisons before the birth of Christ.[69]

Beginning in the sixth century, the influence of Taoist ideas gradually became widespread. King Chinhung 眞興王 (540–575) of Silla had a high regard for the idea of immortality. During the period from 624 to 643, Taoists came to Koguryŏ from T'ang and propagated the *Tao te ching* 道德經. It was around this time that the *Chou hou pei chi fang* 肘後備急方 (Easy emergency prescriptions) of Ko Hung 葛洪 (283–343) and the *Pen-ts'ao ching chi chu* 本草經集註 (The Shen-nung pharmacopeia with collected annotations) by T'ao Hung-ching 陶弘景 (451–536) were introduced into Korea. The *Pen-ts'ao ching chi chu* recorded that Koguryŏ's gold was well refined and could be taken medicinally.[70] Judging from the fact that the teachings of the Yŏlban 涅槃 Buddhist sect in-

[68]According to Korean belief, the Realm of the Immortals, where those who achieved perfection spent their endless lives, was a part of Koguryŏ territory. For the Chinese complex of immortality conceptions, see J. R. Ware, *Alchemy, Medicine, Religion in the China of A.D. 320, the Nei P'ien of Ko Hung* (Cambridge, Mass., 1966). For further references on this and other aspects of Taoism, see M. Soymié and F. Litsch, "Bibliographie du Taoisme. Études dans les langues occidentales," *Dokkyo kenkyū*, 1968, *3*: 249–318; 1971, *4*: 225–290.

[69]Kim, *Han'guk ŭihaksa*, pp. 22–24.

[70]The book introduced eleven kinds of Korean-made medicines, which included gold and silver, both of which were powdered for internal use.

corporated the Taoist idea of longevity, we can guess the extent of the influence of Taoism.[71]

The *Pao p'u tzu nei p'ien* 抱朴子 (Esoteric writings, ca. 320) by Ko Hung must have exerted exceptional influence. The famous painting Three Mountains, pictures of Taoist immortals, which are presumed to have been drawn in the late sixth or early seventh century in Koguryŏ tombs, and drawings of female immortals plucking magic mushrooms with their right hands while holding medicine bowls in their left hands—all these represent Taoist ideas. The toad, symbol of the moon, was also used as an emblem of the idea of immortality. The Korean acceptance of the idea of immortality can probably be traced back to the early currency of prayers wishing eternal life for the dead.[72]

Unfortunately, we lack materials on the methods and equipment Korean Taoists, influenced by Chinese alchemy, used in making elixir. We have no choice but to believe that Chinese books which contain elements of Taoist practice—especially medical books—largely determined the Korean alchemical style. One of these, widely known in Korea, was the *Ch'ien chin fang* 千金方 (Prescriptions worth a thousand, between 650 and 659) by Sun Ssu-mo 孫思邈. We also know that Kim Ka-gi 金可紀, a man of letters of the mid-ninth century, went to T'ang and lived the rest of his life as a Taoist, and that Ch'oe Ch'i-wŏn 崔致遠, a distinguished scholar and once a civil official of T'ang China, relinquished his public post, wandered around the country, and finally retired to his hermitage in Mt. Kaya. Their lives attest to the fact that some people of Silla were deeply submerged in the Taoist milieu and preferred to live in the midst of nature, aloof from profane life. But Taoist thought was also destined to influence the medical ideas which bore on daily life. According to the *Ishinhō* 醫心方 (Tamba no Yasuyori's collected prescriptions, 982) of Japan, one of the great Chinese-style medical com-

[71]Yi Ki-baek, *Han'guksa sillon* (New interpretation of Korean history; Seoul, 1967), p. 80.

[72]See, for instance, Kim, *Han'guk ŭihaksa*, p. 34.

pendia of its time, the Silla masters' two secret methods on the drugs for sexual intercourse came from the Taoist belief that regulating sexual behavior leads to longevity.[73]

MEDICINE IN SILLA AND KORYŎ

By about the eighth century, the folk pharmacological knowledge of Silla had been systematized into an academic discipline under Chinese influence. Eleven medicines of Korean origin were included in T'ao Hung-ching's pharmacopoeia. By the ninth century another 22 kinds were known to Japan and China.[74] These medicines were mostly herbs; powdered gold and powdered silver, which were known during the sixth century, were the only mineral drugs. This suggests that Koreans were satisfying a Chinese demand rather than being greatly enamored of mineral remedies themselves. Around this time, Silla's ginseng and "cow bezoar" (bovine bilestones) were valued highly for their qualities in China and Japan.

Medical knowledge among the people of Silla must have been expanded with the influx, by way of T'ang China, of medicines made in Southeast Asian countries, India, Persia, Arabia, and Rome.[75]

In the tenth century, Koryŏ's medicine was further subjected to the influence of Indian medicine with the rise of Buddhism. Koryŏ was able to build a foundation for the development of its own pharmacology by combining the local variant of T'ang medicine which it inherited from Silla and the pharmacological information it got directly from Sung China. Koryŏ established two state-run medical schools and added a specialization in medicine to the state civil service examinations which were inaugurated in 958. In this the government was following the lead of T'ang China.

Beginning in the early eleventh century, Koryŏ began a positive policy of introducing Sung medicine. Five or six

[73] *Ishinhō*, ch. 28, Drugs 26.

[74] Kim, *Han'guk ŭihaksa*, pp. 79–82; No Chŏng-u, "Han'guk ŭihaksa" (A history of Korean medicine), in *Kwahak kisulsa, Han'guk munhwasa taege*, pp. 770–771.

[75] Kim, *Han'guk ŭihaksa*, p. 83.

medical books, including the *T'ai-p'ing sheng hui fang* 太平聖惠方 (Great compendium of therapeutics, 992), *T'u ching pen-ts'ao* 圖經本草 (Illustrated pharmacopoeia) and *Huang-ti chen ching* 黄帝鍼經 (The acupuncture classic of the Yellow Emperor) were officially imported.[76] Many Sung medical doctors were invited to Koryŏ. In 1058, eight Chinese medical books, including such basic classics as the *Huang-ti nei ching* 黄帝內經 (The inner classic of the Yellow Emperor) and *Shang han lun* 傷寒論 (On febrile diseases) by Chang Chung-ching 張仲景 were reprinted in Koryŏ, accelerating the study of Chinese medicine.[77]

In 1079, Sung sent Koryŏ some 100 medicinal substances for use in curing the illness of King Munjong 文宗. The medicines, sent at the request of Koryŏ, included mineral, animal, and vegetable medicines: cinnabar, refined Borneo camphor, lead acetate, realgar, iron powder, copper crystal, benzoin, gypsum, glue, elecampane, and others.[78] These gave new impetus to Koryŏ medicine.

In the twelfth and thirteenth centuries, several medical books were written by Koreans. Among others, the *Hyangyak kugŭppang* 鄉藥救急方 (First-aid measures with local medicines) is outstanding and worthy of notice. The book dealt mainly with prescriptions using native medicines. In a sense, it was an epitome of the traditional medical knowledge of Korea up to that time, and probably reflected much of the Korean alchemical knowledge of the period. The book, which was published around 1236, contains the nomenclature of 180 kinds of Korean medicines, their properties, and the methods of collection. Included among the medicines are stove deposit (a mixture of metallic oxides and silicates), "cow bezoar," deer antler, talc, bear gall, minium, musk, glue, lime, and magnetite. The book provides much precious information on various aspects of Koryŏ's drug manufacturing techniques and alchemical work.

[76]Miki Sakae, *Chōsen igakushi oyobi sitsubyōshi* (A history of Korean medicine and disease; Osaka, 1955), pp. 23–28.
[77]*Koryŏsa*, ch. 8, p. 11b.
[78]Ibid., ch. 9, p. 37a.

It was Koryŏ which broke from the tradition of relying upon Chinese pharmacology and launched an independent study of medicines. Arabic medical knowledge, especially of the distillation of alcohol, also reached Koryŏ during the Yuan period, when a distilled rice wine called *soju* 燒酒 (literally, burned liquor) began to be made. The vapor distillation technique was later extended to a variety of medical applications. Modern Koreans use a domestic wine brewed from grains in worshiping their ancestors, an application for which *soju*, because of its foreign origin, is still considered unsuitable.

MEDICINE IN THE YI DYNASTY

The Yi Dynasty developed its medicine on the strength of the medical heritage of Koryŏ. In 1398, the *Hyangyak chesaeng chipsŏngbang* 鄕藥濟生集成方 (Collected lifesaving prescriptions of native Korean medicines) was compiled out of the Korean medical books popular in the late Koryŏ period.[79]

In the fifteenth century, the place of medicine as an academic discipline under government control was made firmer. King Sejong urged on a comparative study of Korean and Chinese medicines and had scholars compile the *Hyangyak ponch'o* 鄕藥本草 (Native Korean pharmacopoeia). The king dispatched No Chung-ye 盧重禮 and others to China for further research, which produced in 1433 the *Hyangyak chipsŏngbang* 鄕藥集成方 (Great collection of native Korean prescriptions), in which a total of 703 Korean native medicines—109 mineral, 220 animal, and 374 herbal medicines—were included.[80] The book, is classifying and describing the medicines, followed the pattern of the Chinese *Cheng ho ching shih cheng lei pei yung pen-ts'ao* 政和經史證類備用本草 (The pharmacopoeia of 1249).[81] As in Chinese medical books, medicine and alchemy are inter-

[79] *Yangch'onjip* (Collected works of Yangch'on = Kwŏn Kŭn), ch. 17, preface to the *Hyangyak chesaeng chipsŏngbang*.
[80] Yi Tŏk-pong, "Han'guk saegmulhaksa" (A history of Korean biology), in *Kwahak kisulsa, Han'guk munhwasa taege*, pp. 383–384.
[81] Kim, *Han'guk ŭihaksa*, p. 219.

mingled.[82] The supplement to chapter 75 of the book contains about 70 prescriptions for immortality drugs. Many of these were taken from the *Ch'ien chin fang* 千金方 by Sun Ssu-mo and *T'ai-ping sheng hui fang* 太平聖惠方.

A similar relationship is found in *Ŭibang yuch'ui* 醫方類聚 (Classified collection of medical prescriptions), a large encyclopedia completed in 1445. Its prescriptions were taken from 153 Korean and Chinese medical treatises, including all the major medical treatises of renowned Chinese alchemists.

The influence of Taoism, especially alchemy, on Yi Dynasty medicine is also seen in the *Tongŭi pogam* 東醫寶鑑 (Precious mirror of Eastern medicine), which ranks with the *Ŭibang yuch'ui* as one of the greatest medical encyclopedias of traditional East Asia. In this vast work, begun in 1596 and completed in 1610, Hŏ Chun 許浚, the most eminent physician of sixteenth-century Korea, writes:

> Taoists uphold purity and serenity as their principles, while physicians practice medicine, acupuncture, and moxibustion. Thus the Taoist makes use of the essence 精, and the physician makes use of the crude residue 粗.
>
> According to the books about cultivating one's nature 養性書, there are established rules and principles for mental training and hygiene. In general, one must not deplete his seminal essence, must not waste his spirit and must cultivate his mind. These three elements are the very norms of the Taoists.[83]

The *Tongŭi pogam* includes some 1,420 classified medicinal substances, including 143 minerals, 451 animal products, and 746 herbal medicines, compared with 1,882 substances in the contemporary Chinese *Pen-ts'ao kang mu*. Among them only 102 kinds were imported from China. In other words, about 1,300 kinds of Korean drugs were used in prescriptions in the sixteenth century.[84] Of the 143 mineral medicines, 27

[82]Regarding the relation between alchemy and medicine in China, see, for instance, Sivin, *Chinese Alchemy*.

[83]*Tongŭi pogam* (Precious mirror of Eastern medicine), vol. 1, ch. 1; Yi, "Han'guk saengmulhaksa," p. 388.

[84]Yi, "Han'guk saengmulhaksa," pp. 390–394.

were of Chinese origin. Mercury compounds made up the largest portion of these, which suggests that in the sixteenth century Koreans produced few mercury compounds for use. The *Tongŭi pogam* includes some 20 prescriptions for epilepsy, then regarded as the most serious disease of children. The main ingredients were cinnabar and realgar. One of the prescriptions recommended that cinnabar be administered in its pure state. This, I believe, shows the influence of alchemical prescriptions on sixteenth-century Korean medicine, partly because of the Taoist tendencies of Hŏ Chun, the compiler of this immensely influential collection. But although the goal of immortality took firm root in Korea, alchemical means were not considered indispensable by most people. Like some Chinese Taoists who preferred to use certain rare natural substances, Koreans put great faith in the efficacy of ginseng, a Korean product, as an elixir of life. Many Koreans believe even today that ginseng and deer horn are superior tonics whose optimum use results in eternal youth.

There are many folk tales in Korea about men who spent their lives wandering deep in the mountains and valleys in search of several-hundred-year-old ginseng. There is no doubt that these stories have a philosophical background in Chinese alchemy.

ALCHEMY AND KOREAN TAOISTS

Taoists in Korea as well as China called the search for eternal life *sŏndo* 仙道 (*hsien tao*, literally, way of the immortals). Koreans believed that one could become an immortal by mental training, physical discipline, and by taking artifical elixirs or certain rare natural substances.

Yi Su-gwang 李睟光, one of the initiators of the Yi Dynasty practical learning (*sirhak* 實學), presents his view on the Taoist idea of the immortal in the *Chibong yusŏl* 芝峰類說 (Classified writings of Chibong). He quotes the *Su wen* 素問 section of the Inner classic of the Yellow Emperor (*Huang-ti nei ching*), the fountainhead of Chinese medical theory, and the writings of the lengendary Taoist Keng-sang Ch'u 庚桑楚, to affirm that the immortal can travel freely between heaven and earth, see and hear things far away, and know everything without cogitation. The Taoist elixir can be obtained when one eschews exaggeration, self-torment, and rage, and stays aloof from other people's business. If one cleaves to this in his daily life, then a divine spirit will guide him and serve as his elixir.[85]

The concept of longevity among the people of the Yi Dynasty thus leaned toward spiritual and physical self-restraint rather than toward the use of alchemical preparations. King Sŏnjo 宣祖 (1568–1608)—to whom Hŏ Chun, the author of the *Tongŭi pogam*, was court physician—gave a famous royal lecture against the theory of immortality. He said it had been foolish for kings before him to seek it. Yi Chun-min 李俊民, one of his subjects, responded, to the satisfaction of King Sŏnjo, that immortals do exist, taking Wŏn Hon 元混 as an example. Wŏn Hon, he said, could be called an immortal, for he maintained his youthfulness and wisdom until the age of ninety through an ascetic and meditative life.[86]

Among those who were known for taking elixirs in the sixteenth and seventeenth centuries were 20 persons mentioned in the *Haedong ijŏk* 海東異蹟 (Miracles in the Eastern country).[87] The *Chibong yusŏl* also refers to some of them.[88]

The sixteenth century marked a peak in the development of Taoist alchemy in Korea.[89] Yun Kun-p'yŏng 尹君平, an army officer, learned the *Huang t'ing ching* 黃庭經 (which in China was the basis of physiological, not alchemical, disciplines) from a Chinese Taoist when he visited in China, and became

[85]*Chibong yusŏl,* (Classified writings of Chibong = Yi Su-gwang), vol. 2.

[86]Ibid.

[87]Yi Nŭng-hwa, *Han'guk togyosa* (A history of Taoism in Korea; Seoul, 1956), pp. 299–304.

[88]*Chibong yusŏl,* vol. 2.

[89]The Yi Dynasty in the period of the sixteenth and seventeenth centuries witnessed the Hideyoshi invasion and the Chinese invasion. The land was reduced to smoldering ruins, the nation's strength was at its lowest point, and the people endured great hardship. What is more, the struggle for power among the *yangban* class was such that bloody purges occurred one after another. Consequently, many sought exile in remote mountainous areas.

an expert in alchemical preparations. According to his son, Yun died at the age of 90. The son said his father ate heated pieces of "cold iron," which is presumed to be a kind of elixir. Fisherman who lived in the southern coastal region had a legend that there were immortals on several islands. It would not be difficult to imagine that if they existed they were Taoists. The Taoists to whom Koreans referred all followed special dietary regimens, and their physical conditions were always said to be excellent.

Alchemy in Korea, as with the case of China, was associated chiefly with the "way of the immortals" and was a branch of Taoist learning. For Koreans, as for Chinese alchemists, on the whole the transmutation of base to precious metals was not important. The aim of the alchemical Taoists was to become immortal. But the comparative lack of popularity of alchemy in Korea can be explained only partly by its foreign origin. Perhaps Koreans were in general more satisfied with mundane life, and less inclined to escape into immortality.

Papermaking Technique

KOREAN PAPER

The craft of papermaking is believed to have been introduced into Korea from China in the fourth and seventh centuries,[90] and must have been accelerated greatly by the development and growth of culture in the Three Kingdoms era. The paper which was produced at the time is recognized as being of exceptional quality by today's standards for handmade paper. Silla paper was called "white hammered paper" and earned the high reputation in China of being the best paper in the world. The "white hammered paper" was made of pith from the paper

[90]There is no established theory as to when and how papermaking was transmitted from China to Korea. But it is well known that Tamjing (579–631), a renowned Buddhist monk of Koguryŏ, disseminated it in Japan in 610. This enables us to presume that paper was produced in Koguryŏ by about 600 at the latest. On the other hand, Paekche had already compiled the Historical Records (*Sasŏ*) in the second half of the fourth century.

mulberry shrub (*Broussentia papyrifera*).[91] Other paper, excavated recently in Koguryŏ's historical remains in North Korea, was proved to have been made of hemp fiber.[92] Indications are that Korea introduced and adapted from China at roughly the same time methods of producing mulberry paper and hemp paper.

Beginning in the Koryŏ Dynasty, Korean paper was made of paper mulberry. Yi Dynasty craftsmen made more paper of paper mulberry than of hemp because it was easier to raise the shrub and to process pulp from it.[93]

Starting in the twelfth century, demand for paper increased drastically as more Buddhist scriptures and books on history were printed. The Koryŏ Dynasty sponsored an incentive drive among farmers to grow mulberry shrubs from 1145 to 1188. The dynasty also encouraged private papermaking. The government set up an Office of Papermaking (Chiso 紙所) and helped methods of manufacture to develop. Thus, the Koryŏ Dynasty was able to produce quality paper, hard, thick, and smooth on both sides, suitable for printing and calligraphic work. The so-called Chosŏn (Korean) paper, in use from the Yi Dynasty on, thus was given its unique characteristics in Koryŏ days. The papermaking craft, which reached maturity in the Koryŏ period, evolved large-scale production during the Yi Dynasty when the demand for paper increased with the revival of movable metal type. In 1415, the Paper Manufactory (Chojiso 造紙所) was established as a government-run organization responsible for papermaking technology and rational

[91]In the *T'ien kung k'ai wu* there is a passage which reads: "It is not known what the white hammered paper of Korea is made of."

[92]The Koguryŏ method of bleaching and of homogenizing the quality by beating the fiber made hemp paper very white, and its fiber uniform and tight. See *Chōsen bunkashi* (Cultural history of Korea; Tokyo, 1966), vol. 1, p. 52.

[93]According to *Yijo sillok* (Veritable records of the Yi Dynasty), Korean paper craftsmen sent to China in the fifteenth century learned how to make hemp paper and propagated the technique on their return. Yi Dynasty craftsmen, upon hearing that the Chinese used bamboo and paper mulberry fiber, experimented with it. See Yi Kwang-nin, "Yijo ch'ogiŭi chejiŏp" (Paper manufacturing in the early Yi Dynasty), *Yŏksa hakpo*, 1958, *10*: 9 ff.

production management. Considerable efforts were made to improve the quality of paper and cut down on production costs.[94]

In Seoul, a central paper mill was set up and run by 2 directors in charge of technical affairs, 1 overseer (*saji* 司紙), 4 assistant overseers, 85 papermaking craftsmen, and 95 laborers. A total of 698 papermaking craftsmen were assigned to paper factories in the provinces.[95] All of them were given privileged treatment. The Yi Dynasty stepped up its efforts to introduce advanced papermaking methods from foreign countries. In 1414, Korean craftsmen were advised to learn the characteristics of Chinese paper. In 1428, they learned of the Japanese technique of papermaking from a report submitted by Korean envoys to Japan.[96] In 1475, the dynasty dispatched papermaking experts to China and had them study techniques for producing hemp paper. But the main raw material used in Korea remained the mulberry shrub. In 1439 (twenty-first year of King Sejong), the dynasty imported Japanese mulberry seeds as a part of its plan to improve the Korean mulberry shrub.[97] In addition to those made of mulberry and hemp, rice-straw papers were produced, as well as a paper in which fine shreds of sea-laver were intermixed, in order to conserve mulberry resources and produce a greater variety of papers for diverse uses. Oyster lime, available at a lower cost, was used as a substitute for wood ash in dissolving the pith and bleaching the product beginning in 1457 (third year of King Sejo).[98]

METHOD OF MANUFACTURING KOREAN PAPER

Basically, the method of manufacturing Korean paper is quite similar to that used for Chinese or Japanese paper for one thousand years. The traditional method of producing mulberry paper, a typical Korean paper, is as follows:

[94]Yi, "Yijo ch'ogiŭi chejiŏp," pp. 20–23.
[95]*Kyŏngguk taejŏn* (National code), ch. 1, Taxation Code; ch. 6, Works Code.
[96]Yi, "Yijo ch'ogiŭi chejiŏp," p. 9.
[97]*Sejong sillok*, ch. 84, p. 4b; Marugame, "Chōsen no katsuji juzōsho ni tsuite" (Type foundries in Korea), *Chōsen gakuhō*, 1953, *4*: pp. 114–115.
[98]Yi, "Yijo ch'ogiŭi chejiŏp," p. 8.

In Korea, the mulberry shrub is harvested in the late autumn. If the tree is cut at a spot slightly above the root, a new shoot grows in the spring. The mulberry is put into large kettles and steamed until the bark can be removed. It is then dried. When this black bark, as it is called, is ready to be bleached, it is soaked in water in a tub or stream for 24 hours until it is softened. Then the pith is removed with the feet or hands or by using a knife and bleached in the sun for several days. It is then called white bark. The white bark is again submerged in water until it is completely swollen, mixed with lime at the ratio of 10 *kwan* (ca. 83 lb) to 1 *mal* or 2 *kŭn* (ca. 2.6 lb), and heated to boiling for three to four hours. It is then put into a bag and placed in a stream for about a week until excess lime and impurities are eliminated. The pulp is then sun-bleached, a process that usually takes two days in the winter and one in the summer. Care is needed at this point to keep the pulp free from dust and nodes of cellulose. Then the bleached pulp is pounded to very fine fibers on a flat stone or wooden board. This pulverizing is more extreme than the maceration employed in other Far Eastern countries. Unlike the Chinese method of milling the pulp, the Korean and Japanese pounding leaves long fibers and tiny pieces of stalk, which are often visible in the paper. But this remnant is useful in ensuring the durability of Korean paper.

The hammered pulp is poured into a wooden pulp tank and water is added to suspend the pulp, along with a mucilaginous liquid, usually taken from the root of a kind of yam (Japanese *tororo-aoi*) or pith of makino, which assures a uniform suspension. When the paper-making screen is lowered into the pulp tank and agitated to lay down a thin, even layer of pulp, the product is a sheet of paper. This process of laying pulp on the screen is one of the most important phases in making Korean paper. On the uniqueness of the Korean screen, Hunter says:[99]

Korean papers have always had their own special charac-

[99]Dard Hunter, *Papermaking, The History and Technique of an Ancient Craft*, 2d ed. (New York, 1947), p. 94.

teristics, largely due to the moulds on which they were formed. Like the common "laid" mould of China, the Korean mould consists of four separate parts: the frame, the laid-cover, and the two deckle sticks. While bamboo is the most common material used in making Korean "laid" moulds, there have been instances where a tall Korean grass (*miscanthus sp.*) has been found suitable for the purpose. In India grass has long formed a useful material for the making of "laid" mould-covers.

All Korean moulds as well as the paper made on them may be distinguished by several marked characteristics: the "laid-lines" run the narrow way and the "chain-lines," often narrowly spaced and irregular, run the length of the mould.

The thin layers of pulp are placed one by one on a hot rack, then stacked and pressed dry. Then each damp sheet is placed on a drying board and dried in the sun.[100]

Papermaking methods using other materials are the same as those described in the Chinese *T'ien kung k'ai wu* (1637).[101]

Gunpowder

GUNPOWDER MAKING IN ITS EARLY STAGES

As was mentioned in the previous chapter, gunpowder was introduced into Korea from China in the early fourteenth century.[102] In Korea, Ch'oe Mu-sŏn succeeded in manufacturing gunpowder, after long years of painstaking effort, with the help of a Chinese. Ch'oe wrote a treatise on gunpowder making, but the book has been lost. Ch'oe Mu-sŏn's method was inherited by his son, Ch'oe Hae-san.[103]

Between the fifteenth and sixteenth centuries, gunpowder-making methods, under government sponsorship, grew in scale. But it appears that few changes were made in the chemical process itself. *Tongŭi pogam* quoted a passage from *I-hsueh*

[100]Kim Hwa-ja, "Sŏsa chaeryorosŏŭi Han'gukchi palchŏne kwanhan yŏn'gu" (A bibliographical study on the development of Korean paper; unpublished master's thesis, Ehwa Women's University, 1968), pp. 118–125.
[101]Sun and Sun tr., *T'ien-kung k'ai-wu*, pp. 223–231.
[102]See Chapter 3, p.186.
[103]See Chapter 3, p.187.

ju-men 醫學入門 (Introduction to medicine, 1575) concerning the preparation of saltpeter. This method is similar to that of the *T'ien kung k'ai wu.*[104] Gunpowder-making methods were also included in *Ch'ongt'ong tŭngnok* 銃筒謄錄 (Complete records of firearms, first half of the fifteenth century), which has not survived.

Between the the sixteenth and early seventeenth centuries, important changes began to take place in the Korean gunpowder manufacturing technique. Yi Sŏ's *Sinjŏn chach'ui yŏmch'obang* 新傳煮取焰硝方 (New preparation of saltpeter), written in 1635, is an important source of detailed information on how Koreans made gunpowder at the time. Yi describes fifteen new processes discovered through experiments by Sŏng Kŭn 成根, an army officer.

The main raw material which Sŏng Kŭn used in preparing saltpeter was soil collected from the ground under the kitchens and heated floors of houses. According to Yi Sŏ 李曙, the soil should taste salty, sour, sweet, or bitter. The soil was mixed with urine (actually the major source of the nitrate) and ash and covered with ashed horse-dung (another rich source). The ingredients were thoroughly mixed, put into a wooden bucket, and filled with water. The mixture was heated to a boil three times. When the solution cooled, saltpeter was crystalized out. The remaining solution and soil were saved for reuse.[105]

Yi maintained that three technicians and seven laborers were able to produce 1,000 *kŭn* (ca. 1,320 lb) of saltpeter per month by his method. His treatise does not mention the way black gunpowder was made from the saltpeter. Therefore it is likely that Yi followed the conventional method.

NEW METHOD IN THE SEVENTEENTH CENTURY

The year 1698, about half a century after Sŏng Kŭn, saw great strides in the manufacturing of saltpeter and gunpowder. Such progress was made possible when Kim Chi-nam 金指南

[104] *Tongŭi pogam*, "Medicine," ch. 3; Sun and Sun, tr., *T'ien-kung k'ai-wu*, pp. 268–271.

[105] *Sinjŏn chach'ui yŏmch'obang* (New preparation of saltpeter), pp. 1a ff.

(1654–?), an interpreter, engaged in extensive research on the basis of confidential Chinese information on gunpowder making, which he obtained during a trip to China as a member of an official mission. His *Sinjŏn chach'obang* 新傳煮硝方 (New preparation of saltpeter) presents a detailed process for manufacturing saltpeter and then gunpowder in ten steps.[106]

Kim did not confine soil collection to specific places. Any kind of soil was acceptable, if acrid, sweet, or bitter; salty soil easily absorbed moisture. Another ingredient was ash. According to him straw and mugwort ash were ideal. Ashes of weeds or bushes could be used, but pine-tree ash was not suitable. Soil and ash were mixed in equal amounts (in the text, the ratio was prescribed with 10 *mal*, ca. 100 l, of each). When the soil was highly viscous, 1 additional *mal* of ash was added to the mixture.[107] When sandy soil was used, 1 *mal* of ash was deducted from the 10-*mal* unit. In a jar with several small holes in its base, tree branches were stacked in a grid and covered with a grass screen. The mixed materials were spread over the screen; water was poured on it, and the solution obtained from this filtration process was boiled in a kettle. When the solution began to boil, glue was added. This manufacturing process is not chemically an innovation; its significance lies in the simplification of earlier methods.

The book also explains the manufacturing method for black gunpowder. According to the treatise, gunpowder was made by mixing 1 *kŭn* (ca. 600 g) of saltpeter with 3 *yang* (ca. 110 g) of willow ash, and 1 *yang* 4 *ton* (ca. 50 g) of fine powdered sulfur. The ratio of saltpeter ash and sulfur was 78 : 15 : 7. This ratio may be compared to that given in *T'ien kung k'ai wu,* 70 parts saltpeter, an unspecified ratio of ash, and 30 parts sulfur for an explosive mixture, and to the ratio used for gunpowder today, roughly 6 : 1 : 1. The ingredients were made into a dough by using rice water, and milled to form a thick paste. The process

[106] *Sinjŏn chach'obang* (New preparation of saltpeter), pp. 1a ff.

[107] *Mal* is a unit of dry measure for grain, similar to the Anglo-Saxon bushel. The present-day *mal* in Korea is nearly equivalent to a half-bushel.

is quite similar to the modern method for manufacturing black gunpowder.

The author of the *Sinjŏn chach'obang* insists that gunpowder made by this process was free from moisture, and that materials such as soil and ash were reduced by one-third when compared with the earlier method. It is certain that the quality of gunpowder was excellent.[108] The saltpeter manufacturing process was much closer to medieval European practice than to the contemporary Chinese method, which involved merely purifying natural surface deposits.

[108]See Yoshida Mitsukuni, "Mindai no heiki" (Weapons in the Ming Dynasty), *Tenkō kaibutsu no kenkyū* (Tokyo, 1953), pp, 181–182.

5

Geography and Cartography

Geography in Ancient and Medieval Times

GEOGRAPHICAL KNOWLEDGE IN ANCIENT KOREA

Very few materials are available with which to judge whether "scientific" geography existed from the Three Kingdoms period to the time of Unified Silla. The fact that the development of geography in Korea in these times cannot be clearly separated from the development of astronomy and divination would in any case make the present day discipline a poor yardstick for the evaluation of ancient Korean geography.

The *Chiu t'ang shu* 舊唐書 mentions a map of Koguryŏ, the *Pong'yŏkto* 封域圖, presented to the T'ang court by a Koguryŏ envoy in 628.[1] That maps existed before this date may also be seen in a detailed map of Liaotung found on the wall of a Koguryŏ tomb.[2] The map details the structure and facilities of the city and the surrounding terrain. Walls and major buildings are included, even to specification of type of private dwellings.[3] Rivers, streams, mountains, and roads are keyed in red, blue, purple, and white shades. The style of this fourth-century map is quite similar to that of the watercolor maps of the Yi Dynasty.

Paekche had easy sea access to China and Japan, and may have drafted primitive charts for navigational purposes.[4] Of the Three Kingdoms, Paekche maintained the closest relations

[1] *Samguk sagi* (History of the Three Kingdoms), ch. 20, chronicle of Koguryŏ, 8; *Haedong yŏksa* (Historical treatise on Korea), ch. 42; Hong I-sŏp, *Chosŏn kwahaksa* (A history of Korean science; Seoul, 1946), p. 45.

[2] The tomb was found and unearthed in Sunch'ŏn-gun, north P'yŏngan Province, north Korea, in 1953.

[3] Yi Chin-hi, "Kaihōgo Chōsen kōkogaku no hatten—Kokuri hekiga kofun no kenkyū" (The postwar development of Korean archeology—Tombs with wall-paintings of the Koguryŏ Dynasty), *Kōkogaku zasshi* 45.3, 1959: 52–53.

[4] *Samguk yusa* (Memorabilia of the Three Kingdoms), ch. 2. The record also suggests that there was an original map of Paekche and a geographical treatise on the kingdom.

with Japan and on occasion sent scholars and books on astronomy to that country.

In A.D. 487 Silla established postal stations in all parts of the country and repaired roads for official use.[5] The renowned Silla monk, Hech'o 慧超, authored a record of his extensive travels in India, the *Wang o-Ch'ŏnch'ukkuk chŏn* 往五天竺國傳, which included discussions of the political situation, social conditions, and other matters such as food, clothing, local products, and climate. Regarded today as one of the best available records of India and Central Asia of the eighth century,[6] this work seems not to have had influence on his homeland.

It is possible that the *K'uo ti chih* 括地志 (Comprehensive geography), authored by Li T'ai 李泰 of the T'ang in 638 and quoted in many Korean geographies, may have been introduced in Korea toward the end of the Unified Silla period.[7]

KORYŎ MAPS

It was not until the early Koryŏ period, in the eleventh century, that a generally accurate outline of Korea and its natural features became known. The two earliest full peninsular maps extant are the *Chao-hsien t'u* 朝鮮圖 (Map of Korea) in the *Kuang yü t'u* 廣輿圖 of the Ming cartographer Lo Hung-hsien 羅洪先,[8] and the *P'altodo* 八道圖 (Map of the eight provinces) prepared by Yi Hoe 李薈 in the early Yi period.[9] The *Kuang yü t'u* copied Chu Ssu-pen's 朱思本 fourteenth century *Yü ti t'u* 輿地圖; the Korean map in the *Kuang yü t'u*

[5]Hong I-sŏp, *Chosŏn kwahaksa*, p. 86.

[6]Ko Pyŏng-ik, "Hech'o *Wang o-ch'ŏnch'ukkukchŏn* yŏn'gu sosa" (Brief history of studies on Hui-ch'ao's *Wang Wu-t'ien-chu-kuo chuan*), in *Paek Sŏng-uk paksa hoeryŏk kinyŏm pulgyo nonch'ong* (Commemorative papers on Buddhism for the sixtieth birthday of Dr. Paek Sŏng-uk; Seoul, 1959), pp. 299–316.

[7]In 1091, Sung requested Koryŏ to obtain the copies of the *K'uo ti chih* and *Yü ti chih* which were preserved in the Koryŏ libraries.

[8]*Kuang yü t'u*, ch. 2, pp. 81b, 82a.

[9]*Yangch'onjip* (Collected works of Yangch'on = Kwŏn Kŭn), ch. 2; Aoyama Sadao, "Richō ni okeru nisan no Chōsen zenzu ni tsuite" (About some general maps of Korea made during the Yi Dynasty), *Tōhō gakuhō* (Tokyo), *9*, 1939: 143–172.

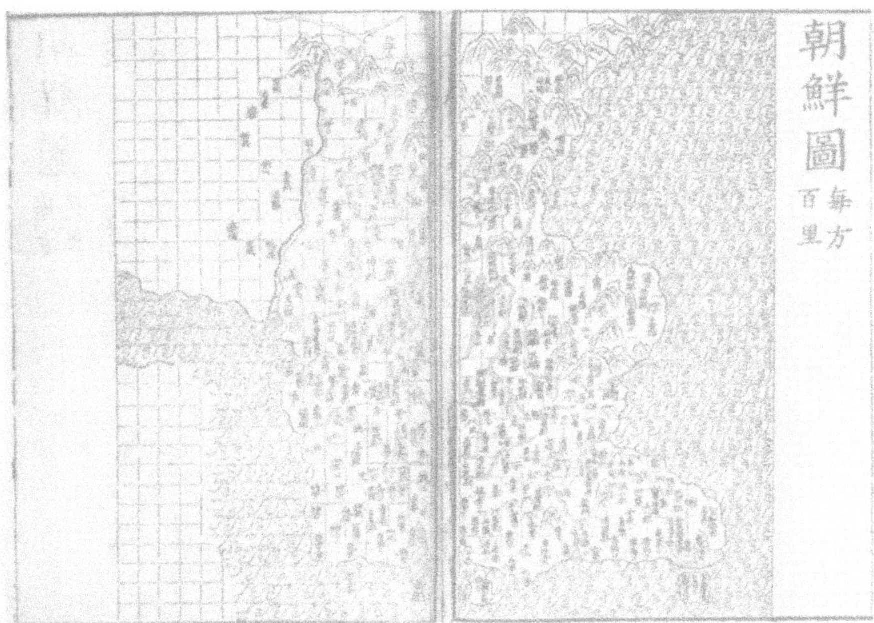

Fig. 5.1. The map of Korea in the *Kuang yü t'u* of the Ming cartographer Lo Hung-hsien. The *Kuang yü t'u* was copied from Chu Ssu-pen's fourteenth-century *Yü ti t'u*. The Korean map in the *Kuang yü t'u* is believed to be a copy of a Koryŏ map prepared before the thirteenth century.

is believed to be a copy of a Koryŏ map prepared before the thirteenth century (Fig. 5.1).[10] A detailed examination of the *Ch'ao-hsien t'u* shows it to be identical with the Koryŏ map described in the preface to the Map of the Three Kingdoms in the *Tongmunsŏn* 東文選 (Selections from Korean literature).[11] The map not only presents a complete outline of Koryŏ but includes mountains, rivers, towns, and cities as well. Yi Hoe's map of 1402 was particularly accurate on these points,[12] and one can speculate that he may have used late Koryŏ maps (Fig. 5.2). Na Hŭng-yu 羅興儒, a noted geographer of the mid-

[10]Aoyama Sadao, "Gendai no chizu ni tsuite" (On maps of the Yuan Dynasty), *Tōhō gakuhō* (Tokyo), *8*, 1938: 105–109.
[11]*Tongmunsŏn* (Selections from Korean literature), ch. 92.
[12]*T'aejong sillok*, ch. 3, p. 27a.

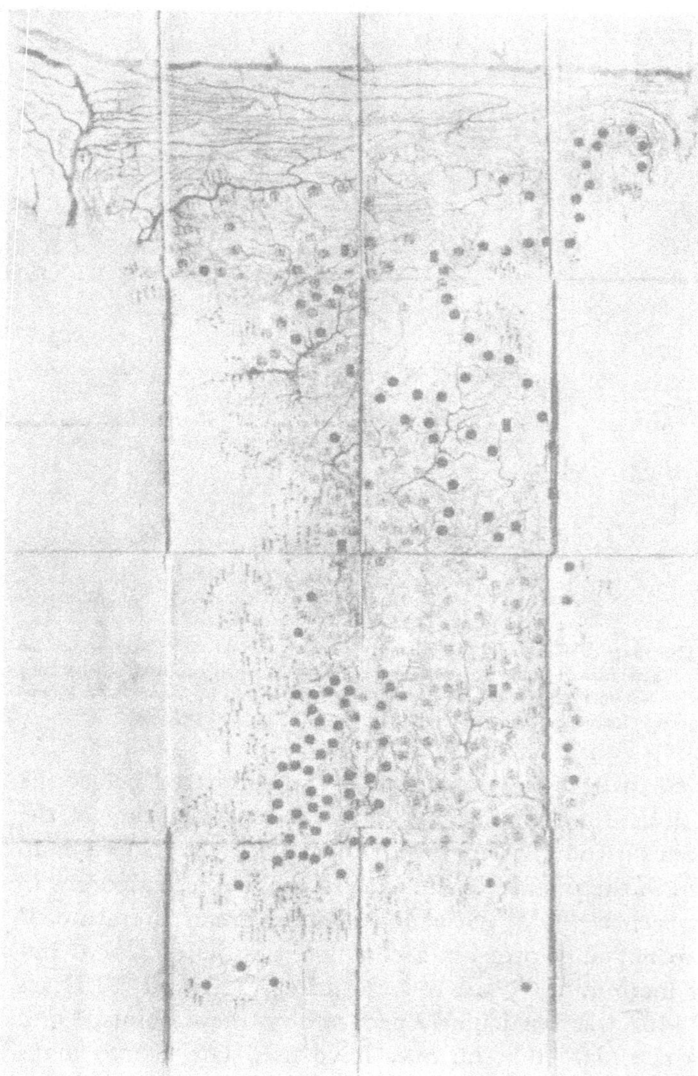

Fig. 5.2 A map of Korea, MS. The map is believed to be a copy of Chŏng Chŏk's map, which was prepared in the early Yi period. It must also be derived from late Koryŏ maps. It is preserved at the Naikaku Bunkō in Tokyo. Its size is about 5 ft × 3 ft. Photo from Aoyama, "Richō ni okeru nisan no Chōsen zenzu ni tsuite."

fourteenth century, is reported in several sources to have prepared maps of Koryŏ and China.[13] Yun Po 尹誧 (d. 1329) wrote an account of India, accompanied by maps and entitled *O-Ch'ŏnch'ukkukto* 五天竺國圖, based on the Chinese *Ta T'ang hsi yü chi* 大唐西域記 of Hsuan-chuang 玄奘 (602–664). Yun's work helped Koryŏ officials broaden their perspectives to include India and Central Asia.[14]

Kim Pu-sik 金富軾 was the first to write a treatise on geography, in 1145. The treatise, contained in the *Samguk sagi* 三國史記 (History of the Three Kingdoms), is a historical geography placing stress on the conditions, historical and geographical, that accompanied the separation or integration of local administrative regions in the Three Kingdoms. The treatise contains historical accounts of the capitals of the Three Kingdoms, as well as of local cities, and good descriptions of their locations, sizes and administrative districts, but it omits treatment of natural geographical features.[15] The importance of Kim Pu-sik's work is found in its contributions to the study of ancient topographical nomenclature.

There are no other Koryŏ geographical materials extant, but it is evident that the treatise on geography of the *Koryŏsa,* compiled in the early Yi Dynasty, must have used reference materials from the Koryŏ period. In view of a Sung request in 1091 for copies of the *K'uo ti chih* and *Yü ti chih* 輿地志,[16] we may suspect that many Chinese geographical works were preserved and available in Koryŏ. Geographical works obtained from China may have included the Sung *Yü ti chi sheng* 輿地紀勝 of Wang Hsiang-chih 王象之; the *Fang yü sheng lan* 方輿勝覽 of Chu Mu 祝穆; the *Ta Yuan i t'ung chih* 大元一統志, a comprehensive geography of the Yuan Empire, and the *Ti li t'u* 墜理圖 of the Southern Sung, among others. Some scholars think that Islamic cartographic technique, introduced through

[13]*Chōsen bunkashi* (Cultural history of Korea; Tokyo, 1966), vol. 1, p. 188.
[14]Hong I-sŏp, *Chosŏn kwahaksa*, p. 113.
[15]On geographical names, see Sin T'ae-hyŏn, *Samguk sagi chirijiŭi yŏn'gu* (A study of the treatise on geography of the *Samguk sagi*; Seoul, 1958).
[16]*Koryŏsa* (History of the Koryŏ Dynasty), ch. 10, p. 23b.

the Yuan, exerted a great influence on Koryŏ; but materials remain only from the early Yi period, and further research on this subject is necessary.

It is known that Koryŏ had contacts with Islamic merchants and naturalized Mohammedans serving as officials with the Yuan. In Songdo (the capital of Koryŏ, now Kaesŏng) Moslem merchants were known as *hoe hoe abi* 回回아비 and were widely popular. In the *Koryŏsa* it is recorded that a group of about one hundred Moslems came twice, in 1025 and 1026, bearing gifts for the Koryŏ king.[17] A group which visited in 1016 brought with it many kinds of medicine, including mercury. Such exchanges provided opportunities for Koryŏ to make contacts with neighboring countries and obtain knowledge about foreign medical science, culture, and customs. The *Koryŏsa* contains a record of a visit of Thai and Cambodian diplomats (traders?) who brought gifts to the king.[18] Emissaries were at times received from Okinawa. Exchanges with Southeast Asians provided Koryŏ with samples of spirits, mercury, sugar, sulfur, black pepper, alcohol, and incense.[19]

P'UNGSU (*FENG-SHUI*) GEOGRAPHY IN SILLA AND KORYŎ

Toward the end of the Silla period, Tosŏn 道詵 (A.D. 826–898) formulated a theory of geomancy ("*p'ungsu*" 風水, literally wind and water) based on the principles of yin and yang and similar to those current in China.[20] His theory was based on the analogy of natural and geographical conditions with the human body and its functions. Geography based on the *p'ungsu* theory was originally very similiar to that dealing with natural topography. During the Koryŏ and Yi Dynasty periods, however, *p'ungsu* geography became connected with shamanism and thus lost much of its theoretic content. During the Koryŏ period in particular, Tosŏn's ideas mingled with

[17]Ibid., ch. 5, p. 4a.

[18]Ibid., ch. 6, p. 6a. See also notes 16 and 17.

[19]Kim Tu-jong, *Han'guk ŭihaksa* (A history of Korean medicine), pp. 146–148.

[20]See, for instance, Yi Ki-baek, *Han'guksa sillon* (New interpretations of Korean history; Seoul, 1967), pp. 128–129.

those of Buddhism, becoming a basis for the political ideals of King T'aejo.[21]

According to the *yin-yang* theory on which geomancy is based, changes in geographical and natural phenomena are correlated. Those who adhered to this theory believed that the destiny of a nation or an individual could be foretold by observing and interpreting peculiar natural or geographical phenomena. Thus what was known as *ŭmyang sunyŏk* 陰陽順逆 geomancy (the geomancy of yin and yang alternation) maintained that geographical and natural phenomena included both favorable and unfavorable aspects, which when understood properly could be harnessed and applied in practical matters.

Tosŏn's *p'ungsu* theory was composed of two theories. The theory of *chiri soewang* 地理衰旺 ("decline and flourishing of the land") held that the strength and capacity of the land, thought of as its specific energy (*chigi* 地氣) rises at some times and declines at others. Dynasties or individuals who are settled on land where this energy is flourishing will prosper, but when the energy of the land weakens and declines they will be ruined. The theory of *ŭmyang pobi* 陰陽補裨 held that a deficiency in geographical conditions could be supplemented and improved through human efforts. For example, terrain in which strata are weak or ill-formed can be converted, by means of cutting rocks or heaping up soil, to an area in which advantageous elements can thrive.

P'ungsu theory wielded a great influence over Korean thought and society until late in the Yi Dynasty. It exerted a powerful influence throughout the country on the construction of palaces, temples, and private dwellings and on the location and layout of cities and graveyards.[22] In awakening interest in the survey and observation of geographical features, *p'ungsu* thought contributed greatly to the development of a uniquely

[21]One of the best studies on *p'ungsu* and related thought in Korea is Yi Pyŏng-do's *Koryŏ sidaeŭi yŏn'gu* (Studies in the Koryŏ period; Seoul, 1948, 1957). See also Yi Ki-baek, *Han'guksa sillon,* p. 129.
[22]Yi, *Koryŏ sidaeŭi yŏn'gu,* pp. 3–85, 371.

Korean map style. In Korean regional maps, for example, one notes many instances in which mountains are represented not individually but as part of a system or ridgeline; *p'ungsu* theory emphasizes long-distance flow of terrestrial energy. This feature is absent from Chinese maps, except for those of graveyards, and reveals the influence of *p'ungsu* grave mapping techniques.

Geography in the Early Yi Dynasty

P'UNGSU (FENG-SHUI) GEOGRAPHY

The study of geography had its closest connection with political events at the start of the Yi Dynasty, when it began to develop under government sponsorship and service. From the establishment of Tosŏn's geomantic concepts in Koryŏ times under King T'aejo, *p'ungsu* geography underwent a broadening transformation into a widely followed system of geomancy. Many high-ranking ministers and scholars were convinced that Koryŏ had fallen because of the neglect of geomantic factors, and it was clear that geomancy would play an important role in locating the new Yi capital.[23] In the deliberations surrounding the selection of a propitious site for the new capital, a number of Koryŏ geomantic works were consulted. Scholars completed a compendium of geomantic theories, the *Tongguk yŏkdae chehyŏn pirok* 東國歷代諸賢秘錄 (Secret records of the worthies of successive eras in the Eastern Kingdom), in the third year of King T'aejo (1394).[24]

The book played a great role in the selection of the new site for the Yi capital, setting out fully the various geomantic concepts held during the Koryŏ period. With a view to consolidating these theories, T'aejo established in 1394 a Supervisorate of Geomancy (Ŭmyang Sanjŏng Togam 陰陽刪定都監), recruiting for service all of the scholars who had par-

[23]The *Haedong pirok* (Secret record of Korea), the *Haedong kohyŏn ch'amgi* (Prophecies of great Korean sages), and the *Tosŏn myŏngdanggi* (Tosŏn's geomantic theories) were among those used.

[24]*T'aejo sillok*, ch. 5, p. 5b. *Tongguk yŏkdae chehyŏn pirok* is also known as the *Chiri pirok ch'walyo* (Elements of the secret record of geography).

ticipated in the compilation of the *Tongguk yŏkdae chehyŏn pirok*. This office was to further theoretical research.[25] The development of *p'ungsu* as an established doctrine was halted by the Confucianist and rationalist policies of T'aejong, who instituted strict controls over Buddhism and proscribed geomantic writings.[26]

AN EARLY YI DYNASTY MAP OF THE WORLD

The second chapter of the *Yangch'onjip* 陽村集, the collected

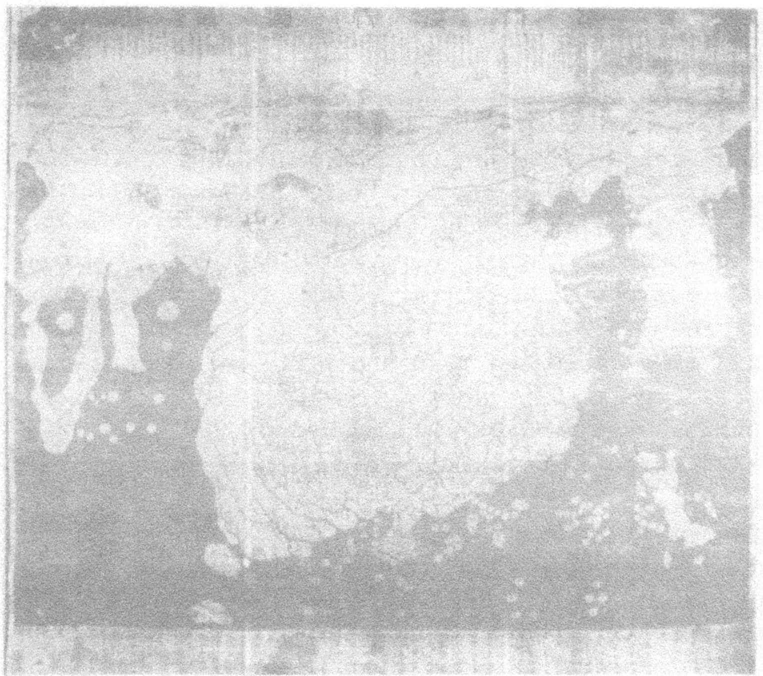

Fig. 5.3. The Korean world map of 1402, *Honil kangni yŏktae kuktojido,* by Kim Sa-hyŏng, Yi Mu, and Yi Hoe. It was the first highly accurate world map drawn in East Asia until the appearance of Matteo Ricci's world map of 1602. Its size is 5 ft × 4 ft. A copy is now preserved in the Library of Ryūkoku University in Kyoto, Japan.

[25] *T'aejo sillok,* ch. 6, p. 6a. The office compared in function with the Sanch'ŏn Sinbo Togam 山川裨補都監 of Koryŏ.
[26] Yi, *Koryŏ sidaeŭi yŏn'gu,* p. 437.

writings of the scholar Kwŏn Kŭn 權近, is devoted to information about the drafting of a world map, the *Honil kangni yŏktae kuktojido* 混一疆理歷代國都之圖, by Kim Sa-hyŏng 金士衡, Yi Mu 李茂, and Yi Hoe 李薈 (Fig. 5.3). It was revised by Kim Sa-hyŏng in 1399 under the influence of two Chinese maps which he had acquired, the *Sheng-chiao kuang-pei t'u* 聲教廣被圖 of Li Tse-min 李澤民 and the *Li-tai ti-wang hun-i chiang-li t'u* 歷代帝王混一疆理圖 of the monk, Ch'ing Chün 清濬 (1329–1392). (These two maps omitted a great part of the area east of the Liaotung Peninsula.) When a map was brought from Japan in 1401 by Pak Ton-ji 朴敦之, Tsushima and Iki Islands were also included.

A copy of the *Honil kangni yŏktae kuktojido* has been preserved in the library of Ryūkoku University in Kyoto, Japan. Ogawa Takuji and Aoyama Sadao note that the copy, which measures five feet by four feet, bears the same name as the *Yü t'u* 輿圖 prepared by Chu Ssu-pen 朱思本 in 1320, leading to the conclusion that it was prepared with the aid of contemporary maps.[27] The section of the map dealing with the Western world draws our particular attention. It contains some one hundred European and thirty-five African areas. The Sahara and Gobi deserts are colored black. Alexandria is marked as a port and distinguished by the likeness of a Pharoah. One of the conspicuous features of the map is the influence of Islamic cartographic techniques. The Nile River is marked in the same manner as on the terrestrial globe of Jamal al-Din brought to Peking in 1276. Seas are colored green with waves in black. Although the Mediterranean is given a clear-cut coastline, it shows no waves, probably because the Korean cartographers did not yet realize that it is a sea. Indochina is missing.[28] The map is similar in many respects to

[27]Ogawa Takuji, "Shina chizugaku no hattatsu" (Development of Chinese cartography), in *Shina rekishi chiri kenkyū* (Studies in Chinese historical geography; Kyoto, 1928); Aoyama Sadao, "Gendai no chizu ni tsuite," pp. 103–152.

[28]Needham, *SCC*, vol. 3, pp. 555–556. But it is certain that the map is based on the world atlas prepared under the influence of Ptolemy. See Takahashi Tadashi, "Tōzen seru chūsei Isuramu sekaizu" (The introduction of

the *Kuang yü t'u* 廣輿圖 of Lo Hung-hsien 羅洪先 (1504–1565), which is based in turn on Chu Ssu-pen's *Yü t'u*.

The Korean map conveys more precise information than contemporary Chinese maps, especially concerning the outlines of Western Europe, Africa and Southwestern Asia.[29] The portion dealing with Korea is far more accurate than the *P'alto ch'ongdo* 八道總圖 (Complete map of the eight provinces) of the *Tongguk yŏji sŭngnam* 東國輿地勝覽, and is presumed to be based on the *P'altodo* 八道圖 (Map of the eight provinces) prepared by Yi Hoe. The Chinese *Kuang yü t'u* of Lo Hung-hsien contains a rough map of Japan by Li Tse-min and a flat Korean map which was probably drawn from the map of Korea in the *Tongguk yŏji sŭngnam* (Fig. 5.4). The *Honil kangni*

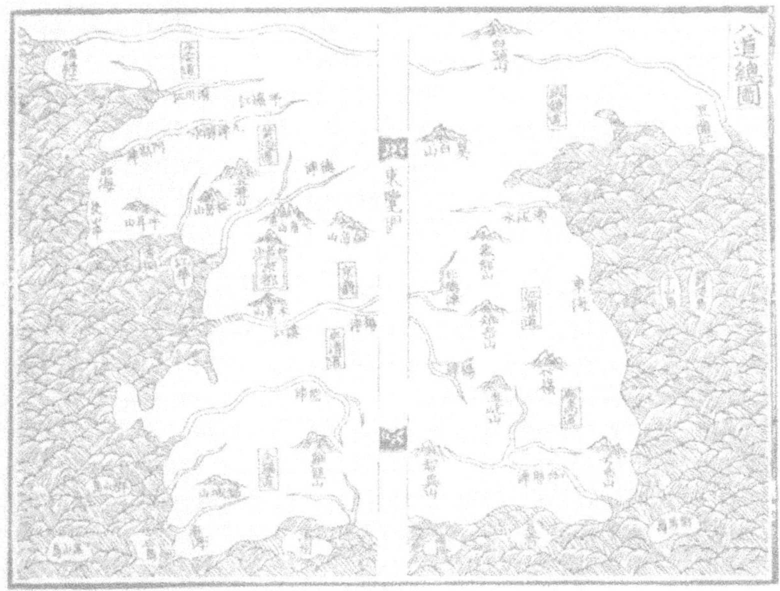

Fig. 5.4. Map of the eight provinces, *P'alto ch'ongdo*, in the *Tongguk yŏji sŭngnam*. It was first published in the fifteenth century.

Islamic world maps to China and Korea; on the Korean world map of 1402), *Ryūkoku University Theses*, no. 374, p. 86.

[29]Takahashi Tadashi, "Tōzen seru chūsei Isuramu sekaizu," p. 86.

yŏktae kuktojido of 1402 is apparently based on the most up-to-date and detailed Korean and Japanese maps available.

It can be seen that scholars of the early Yi Dynasty had a cartographic knowledge of Western Europe and Africa perhaps exceeding that of contemporary Western Europeans concerning the Far East. The *Honil kangni yŏktae kuktojido* was the only world map prepared by a Yi Dynasty scholar, and it remained unsurpassed until the arrival of Matteo Ricci's second world map of 1602. Aoyama has called the Yi Dynasty map the first highly accurate world map drawn in East Asia.[30] Its format was Sinocentric and Korea was disproportionately emphasized; an overexpanded China and Korea dwarf the remainder of the Asian continent, Europe, and Africa. Nevertheless, as Needham and other scholars have pointed out, this Korean effort far excels the Catalan map of 1375 prepared with roughly contemporary materials.[31]

Another map of the world was apparently completed in 1469, but fragmentary records in the *Yejong sillok* 睿宗實錄[32] are insufficient to tell us whether it was a traditional Sinocentric map, or part of a wood-block engraved atlas such as those which are extant today.

OTHER WORLD MAPS IN THE YI PERIOD

During the Yi period, there were two types of world map; one was similar to the *Honil kangni yŏktae kuktojido* of 1402 described in the preceeding section and the other was a wheel-type map known as a *ch'ŏnha ch'ongdo* 天下總圖, or "complete world map." Copies of this latter type, variously known as the *Ch'ŏnha yŏjido* 天下輿地圖 (World map) or *Taemyŏng ch'ŏnhado* 大明天下圖 (Great Ming world map), were mentioned in a memorial of Yang Sŏng-ji 梁誠之 of 1482 requesting the preservation of maps[33] and in the *Ch'ŏnha yŏjido,* prepared by the Hongmungwan 弘文館 (Office of Special Counselors) in 1511.[34] These

[30]Aoyama, "Gendai no chizu ni tsuite," p. 556.
[31]Needham, *SCC,* vol. 3, p. 556.
[32]*Yejong sillok,* ch. 6, pp. 15ab.
[33]*Sŏngjong sillok,* ch. 138, p. 10b.
[34]*Chungjong sillok,* ch. 14, p. 21a. According to the same record, ch. 36, p.

may have been similar to the map of the Ming donated to the Honmyōji Temple by Katō Kiyomasa after the Hideyoshi invasion. This map was prepared in Korea, according to one scholar, and included China, Japan, and Korea. The axis of Japan did not incline to the right but was comparatively vertical. Honshū and Shikoku were treated as islands.[35] This treatment corresponds to that of the map of Japan contained in the *Haedong chegukki* 海東諸國紀 (Geographical treatise on the countries in the East [Japan and Okinawa]), and could have been copied from the Hongmun'gwan map of 1511 mentioned above.

An atlas based on Buddhist views of the world and known as the *Ch'ŏnhado* or the *Ch'ŏnha ch'ongdo* was compiled in either the early Yi period or late Koryŏ times. Copies which remain today are wood-block engravings presumed to date from the period 1717–1719.[36] In the unique format of this map, mainland China with Mt. K'un-lun in its center is located in the midst of a great sea and surrounded by the other continents. To both east and west are two islands. The map is round and covers some four hundred million *ri* in circumference (a *ri* is about a third of a mile). The distance represented between the eastern and western extremes is more than one hundred million *ri*. Eighty-four thousand countries are indicated schematically in all directions on the map.[37] It is identical to

---

75b, the *Taemyŏng yŏjido* (Map of the great Ming) was preserved in the palace by the order of King Chungjong in 1519.

[35]Akioka Takejirō, *Nihon chizushi* (A history of Japanese cartography; Tokyo, 1955), pp. 80–81.

[36]The atlas contains the *Ch'ŏnha ch'ongdo* (World map), eight Korean provinces in its 11 pages. The provincial maps are the same as those in the *Tongguk yŏji sŭngnam*. The maps of each province include brief notes on the military bases, castles, the number of households, population, and ports. The author's conclusion on the atlas's publication date (1717–1719) was made on the basis of these notes. There are many maps similar to that in the atlas, including the copies made during the reign of King Yŏngjo.

[37]The atlas consists of a world map on a reduced scale of 1 : 60,000,000, as well as a map of China (1 : 25,000,000), a map of Korea (1: 2,500,000), maps of each Korean province (1 : 800,000), map of Japan (1 : 2,500,000), and a map of the Ryūkyū Islands (1 : 500,000). Differences in scale were believed to have been due to adjustments of the map sizes for accomodation

the world map prepared by Jen Ch'ao 仁潮 in 1607 which is contained in the *Fa chieh an li t'u* 法界安立圖, and is obviously influenced by the "wheel maps" (so-called T-O maps) of the European middle ages and by contemporary Islamic maps.[38]

THE *HAEDONG CHEGUKKI*

The world map of 1402 gave Koreans cartographic knowledge of the Middle East, Africa, and Western Europe, but concrete and detailed knowledge of the natural and social conditions of these and many other areas was lacking. The lack of knowledge was the more acute with respect to Japan, considering the long history of contact between the two countries. Following the cessation of coastal raids by the Japanese pirates in the early years of the dynasty, the two countries made efforts to restore friendly relations, and it became important for Korea to possess more accurate information about Japan. The *Haedong chegukki* 海東諸國紀 (Geographical treatise on the countries in the East [Japan and Okinawa]), produced by Sin Suk-chu 申叔舟 in 1471, was to meet this need in part. In the preface, Sin said, "In approaching a neighboring country and establishing friendly relations through exchange of envoys, we should be [kept] informed of that country's customs and traditions. By so doing, we demonstrate our tact, and this would further display our sincerity. . . . The author has compiled this book for presentation to His Majesty after studying various reference materials on Japan. The book includes a map of Japan, with information about her climate and geographical conditions and details of protocol that our country has to observe. . . ." The work included a history of exchanges

in the atlas. See Mok Yŏng-man, *Chido iyagi* (The story of maps; P'yŏngyang, 1965), pp. 126–127. In the map of Japan, a certain compass bearing is 180 degrees off, as on the map of Japan in the world map of 1402. See Akioka, *Nihon chizushi*, pp. 34–37. Judging from this, one may assume that the original text of the atlas could have been drawn up during the early Yi Dynasty or before. This possibility also serves as a guide to a series of Korean efforts to probe into the question of when the wheel map made its debut in Korea.

[38]But the T-O map was oriented to place East at its upper corner, while the wheel map in the *Ch'ŏnhado* places East on the right side.

Fig. 5.5. A map of Japan in the *Haedong chegukki* (Geographical treatise on the countries in the East), by Sin Suk-chu, published in 1471. It is the oldest printed map of Japan in the world.

between the two countries as well as a history of the Ryukyus. Six maps were appended, the *Haedong cheguk ch'ongdo* 海東諸國總圖 (General map of the countries in the East), the *Ilbon pon'gukto* 日本本國圖 (Map of Japan, Fig. 5.5), and maps of Western Kyushu, Iki, Tsushima and the Ryukyus.

In the preface, Sin discussed the general geographic situation of Japan: "When one looks at the Eastern Sea areas [one finds that] there are several countries in the Eastern Sea. Japan appears to be the oldest country. She is also the largest country. Her territory stretches from the north of the Amur to the southern tip of Cheju Island and borders on the Ryukyus." Sin's work is not, strictly speaking, a geographical journal, but may be regarded as a fairly successful attempt to consolidate contemporary Korean knowledge of Japan. It remained unsurpassed until the later part of the Yi period, and several editions were published. The *Haedong chegukki,* containing as it does what is perhaps the oldest printed single map of Japan in the world, is a measure of how deeply Koreans were interested in Japan and may provide valuable assistance to those interested in researching Korean-Japanese relations and Japanese history and geography.

The Yi court had access to several other maps of Japan in the fifteenth century. One of these was a replica of the map brought from Japan by Pak Ton-ji during the reign of King T'aejong; others were the maps of Japan and the Ryūkyūs drawn up by the Japanese monk, Dōan, and brought to Korea in July of 1453. It is presumed that the fifteenth-century Japanese maps brought to Korea were based on such maps as the *Gyōki zu* 行基圖, owned by the Niwa temple, and the map of Japan in the Chōdoku temple. Both the *Haedong cheguk ch'ongdo,* contained in the *Haedong chegukki,* and the map of Japan in the 1402 world map of Kim Sa-hyŏng, Yi Mu, and Yi Hoe are similar to the one in the Niwa temple, and it is believed that the maps prepared by Sin Suk-chu 申叔舟 were based on these Japanese maps. Both Sin's maps and the *Samp'o chido* 三浦地圖 of Nam Che 南悌 employed methods similar to those seen in the *Tongguk yŏji sŭngnam* and are important materials for the study of Yi Dynasty cartography.

Geographical Compendia of the Early Yi Dynasty

THE *SINCH'AN P'ALTO CHIRIJI*

As the Yi court consolidated its political position and pushed forward with the work of compiling records, geography quickly became an important area of documentation. It was apparent to many that a geographical survey of the country, dependent as it was on local elders and records, should be undertaken as soon as possible, and in 1414 the court appointed Pyŏn Kye-ryang 卞季良 chief editor of the project.[39] This was the first major state venture into geography in the Yi period and laid the groundwork for further study. The work was

[39]*Sejong sillok,* ch. 26, p. 25a.

completed by Maeng Sa-sŏng 孟思誠, Kwŏn Chin 權軫, Yun Hoe 尹淮, and Sin Chang 申檣 in 1422, seven years after it was begun, under the title *Sinch'an p'alto chiriji* 新撰八道地理志 (Newly compiled geography of the eight provinces).[40]

The method of compilation is well illustrated in the preparation of the *Kyŏngsang-do chiriji* 慶尙道地理志 (Geography of Kyŏngsang Province).[41] Pyŏn Kye-ryang conducted an extensive study of state and privately-owned materials, while the Board of Rites sent questionnaires to each province and had the provincial authorities in each case send their documents to the Bureau of State Records (Ch'unch'ugwan). The twelve items listed in the preface of the Kyŏngsang Province geography are typical. They were numbered as follows: (1) the history of place names in each province; (2) the history of prefectures, cities, and counties; (3) the current number of prefectures, cities, and counties; (4) mountains in the province, mileage of rivers, borders, important topographical and other features; (5) provincial products sent to the capital for court use; (6) hot springs, caves, cattle stations, horse stations, and mines; (7) locations of military camps and naval ports and the number of officers, troops, and ships at each location; (8) royal tombs, shrines, and the tombs of renowned local scholars; (9) land fertility, depth of wells, temperature, climate, and customs; (10) distances between islands and the mainland; (11) amounts of taxes, ports for shipping, canals, and land transportation routes; (12) population and the number of households.

The survey on these items for Kyŏngsang Province was completed by the governor, Ha Yŏn 河演, in 1425 after a year of work and submitted with prefatory remarks to the Bureau of State Records. Other provinces had finished the survey by 1426. The provincial survey reports and materials prepared previously by Pyŏn Kye-ryang were edited by Maeng Sa-sŏng, Kwŏn Chin, Yun Hoe, and Sin Chang at the Bureau of State Records and became the basis for the *Sinch'an p'alto*

[40]Ibid., ch. 55, p. 7b.
[41]*Kyŏngsang-do chiriji* (Geographic treatise on Kyŏngsang Province), pp. 1a ff.

chiriji. Since no complete copies of this compendium are extant, it is difficult to tell precisely what it was like, but some idea of the contents and approach may be obtained from the geography of Kyŏngsang Province and from the Treatise on Geography of the *Sejong sillok*. This treatise contributed immensely to the academic development of geography in the Yi period and became the basis for all later geographies. The compilation of the *Sinch'an p'alto chiriji* marked the first step in academic geography and was an emancipation from the conceptual categories of geomantic and cosmological theories of geography.

THE *SEJONG SILLOK CHIRIJI*

Compilation of the Veritable Records, or *Sillok*, for the reign of Sejong was begun in the first year of Munjong's reign (1451) and completed three years later, in the second year of Tanjong's rule, more than twenty years after the completion of the *Sinch'an p'alto chiriji*. Under the influence of Chinese practice in compiling standard histories, the Veritable Records for the reign of Sejong included a number of treatises, among them one on geography. Because of the lack of geographical writings in the interim, the Geographical Treatise of the *Sejong sillok* was based closely on the information contained in the *Sinch'an p'alto chiriji*, although some information was added for newly established border towns. The treatise was an encyclopedic compendium of regional information, covering local history and geographical situations as well as political, social, financial, economic, industrial, military, and transportation organizations. Each of eight chapters dealt with one of the eight provinces: Kyŏnggi, Ch'ungch'ŏng, Kyŏngsang, Chŏlla, Hwanghae, Kangwŏn, P'yŏngan and Hamgil. Specifically, each chapter discussed the geography and history of the province and included the names of the governor and his subordinates, a table of provincial organizations, provincial boundaries, famous mountains and rivers, ferry points, river origins, and tributaries. There were also details concerning the number of households, military camps, naval forces, acreage of government-owned paddies and fields and their production, points of scenic interest, and stations on the main roads. Similar information was given under headings for cities, countries, and villages in each province. In these sections was given information concerning local customs and products, fish storage points, salt warehouses, ceramics factories, town fortifications, historical remains, pavilions, post stations, and beacon stations used for long-distance alarm signals.[42]

The *Chiriji* of the *Sejong sillok* was an important landmark in the development of Yi Dynasty geographical studies, providing a foundation in its empirical approach for the development of geographical science, despite a relative weakness in the area of theory. As a topographical work, it demonstrated independence from Chinese models, which, since the *Fang yü sheng lan* 方輿勝覽 (Conspectus of geography) of the Sung period with its inclusion of poetry, had undergone a decline in substance. Its matter-of-fact approach undoubtedly stemmed from its official character. All in all, one is inclined to concur in the judgment of Katsushiro Sueji: "The book should not be regarded as a mere geography book but as a pioneering [effort] in modern geography."[43]

THE *P'ALTO CHIRIJI* OF 1478

The *P'alto chiriji* 八道地理志 (Treatise on the geography of the eight provinces) of Yang Sŏng-ji 梁誠之 was commissioned in 1455, shortly after the enthronement of Sejo.[44] The purpose of the new geography was to correct various defects of the *Sejong sillok chiriji* and provide supplementary information. Like the *Kyŏngsang-do sokch'an chiriji* 慶尙道續撰地理志 (Supplementary geographical treatise on Kyŏngsang Province), which was a regional effort to document changes in administrative districts

[42]See, for instance, *Sejong sillok chiriji* (Treatise on geography of the *Sejong sillok*; modern reprint, Seoul, 1936), pp. 1–14, Katsushiro Sueji's commentary.

[43]Ibid.

[44]*Sejo sillok,* ch. 2, p. 7a. However, according to the *Nulchejip* (Collected writings of Nulche = Yang Sŏng-ji), it began by the order of King Tanjong in 1453. *Nulchejip,* ch. 3, p. 75b.

and government organizations since the 1430 compilation of the *Sinch'an p'alto chiriji,* Yang's work was methodologically similar to its predecessor. Under Yejong 睿宗, Yang was able to have orders issued that each province compile its own topography as means of expediting the project as a whole. These were commonly researched in much greater detail then the previous topographies.[45] Yang combined the provincial studies with the results of his own extensive research. The *P'alto chiriji* is unfortunately lost, but an idea of its quality may be gotten from the text of the supplementary treatise on Kyŏngsang Province mentioned above.[46] Inasmuch as it took into account administrative and other changes which had taken place since the compilation of the *Sokch'an chiriji* 續撰地理志, and incorporated additional studies, the *P'alto chiriji* was undoubtedly superior in many respects to the *Sokch'an chiriji* and may be viewed as a consolidation of Yi Dynasty achievements in the study of geography to its time.[47]

THE *TONGGUK YŎJI SŬNGNAM*

The compilation of the *Tongguk yŏji sŭngnam* 東國輿地勝覽 (Geographical conspectus of the Eastern Kingdom) was apparently stimulated by the introduction from China of the *Ta Ming i t'ung chih* 大明一統志. Its style and title were influenced by the Sung work *Fang yü sheng lan* 方輿勝覽. The original draft, a synthesis of Sŏ Kŏ-jŏng's 徐居正 *Tongmunsŏn* 東文選 (Selections from Korean literature) and the *P'alto chiriji* 八道地理志, was prepared by No Sa-sin 盧思慎, Kang Hi-maeng 姜希孟, Sŏng Im 成任 and Sŏ Kŏ-jŏng under the orders of Sŏngjong 成宗 in 1481. Revision for publication began the following year and the book went through several versions before completion in 1487. Revision was the responsibility of a committee established in the Office of Special Counselors. The final version, titled *Sinch'an tongguk yŏji sŭngnam* 新撰東國輿地

[45] *Kyŏngsang-do sokch'an chiriji,* pp. 1a ff.

[46] Ibid.

[47] Yang's treatise appears in the list of the geographical books in the *Taedong unbu kunok* (A dictionary of arts and sciences of the Yi Dynasty).

勝覽, contains the fifty chapters edited by No Sa-sin and the others and five chapters added by Kim Chong-jik 金宗直 of the Office of Royal Decrees.[48]

The *Tongguk yŏji sŭngnam* made use of all available materials, including Yang Sŏng-ji's *P'alto chiriji*.[49] It may be regarded as an improvement over existing compilations. Its academic value is perhaps diminished by the inclusion of only purely literary works and its faithful adherence to the values of the traditional political context. In its inclusion of information of historical, administrative, and religious importance it reflects the influence of Chinese encyclopedic geographies and hence represents a decline in a trend toward independence seen from the time of T'aejong through the reign of Sejong.

The Use of Surveys

THE *P'ALTODO* (MAP OF THE EIGHT PROVINCES) OF YI HOE

One of the earliest Yi Dynasty maps of Korea was Yi Hoe's 李薈 *P'altodo* 八道圖,[50] which was copied in the 1402 world map, the *Honil kangni yŏktae kuktojido* 混一疆理歷代國都之圖.[51] The accuracy of this map leads us to conclude that it was compiled on the basis of actual observation. It is valuable for the information which it thus provides on early Yi Dynasty mapmaking.

TONGGUK CHIDO 東國地圖 OF CHŎNG CH'ŎK 鄭陟 AND YANG SŎNG-JI 梁誠之

The possibility of improving the accuracy of maps grew as the result of astronomical observations became available in the time of Sejong. Through the computation of latitudes for Mt. Paektu, Mt. Mani on Kanghwa Island, and Mt. Halla on Cheju Island, calculations could be made for the length of the peninsula, the distance between the east and west coasts,

[48] *Tongguk yŏji sŭngnam*, postscript; *Chōsenshi* (Chronological history of Korea; Seoul, 1932–1938), pt. 4, vol. 5, p. 570; *Sŏngjong sillok*, ch. 200, p. 6a.
[49] The format employed is the same as the one Yang used, with a separate page and a different color given to each province.
[50] Completed in May 1402. See *T'aejong sillok*, ch. 3, p. 27a.
[51] *Yangch'onjip*, ch. 2, description of the world map of 1402.

the polar altitude of Seoul, the distance between Seoul and Mt. Paektu, and the distance between Seoul and Cheju Island. This was essential groundwork for the drawing of complete and accurate maps of Korea.

Preparations were begun in 1424 (sixteenth year of King Sejong) for the drafting of new maps.[52] As the first step, Chŏng Ch'ŏk was sent to the northern provinces of Hamgil, Pyŏngan, and Hwanghae to draw up maps on the basis of observation to replace existing maps of these areas, which were incomplete and contained errors.[53] Chŏng later completed a map of the eight provinces, the first to be completely prepared on the basis of actual survey.[54]

When the Yi boundaries were expanded to the east of the Yalu River, reaching to just south of the Tuman River during the period from T'aejo to Sejong, Chŏng Ch'ŏk's map had to be revised to include the new territories. This revision, based upon direct observations and surveys conducted by local administrators on the order of the court,[55] published in 1451, was titled the *Yangge chido* 兩界地圖.

It was during the reign of Sejong that a third-century Chinese hodometer carriage known as the "mile-counting drum carriage" was put into wide use.[56] In 1441, a survey of major roads was conducted and milestones were laid or trees planted every thirty *ri*.[57] In 1450 the court ordered each province to conduct surveys on the distance between major cities and counties and to use the materials in drawing up new maps.[58] The preparation of maps was regarded as an important aspect of administration at this time, and efforts were made to select experienced and expert men to assist. For example, when the court wanted maps of the capital city and each province drawn up in 1453, Yang Sŏng-ji, who had

[52]*Sejong sillok*, ch. 64, p. 4b.
[53]Ibid., ch. 71, p. 9a.
[54]Aoyama Sadao, "Richō ni okeru nisan no Chōsen zenzu ni tsuite."
[55]*Munjong sillok*, ch. 7, p. 47a.
[56]*Chōsenshi*, pt. 4, vol. 3, p. 638.
[57]*Sejong sillok*, ch. 93, pp. 26ab.
[58]*Munjong sillok*, ch. 5, p. 22b.

worked on the treatise on geography of the *Koryŏsa,* was appointed.[59] The field methods used are described in the Veritable Records of Tanjong: "Prince Suyang 首陽大君 climbed the Pohyŏn hill of Mt. Samgak together with Chŏng Ch'ŏk, Kang Hi-an 姜希顏, and Yang Sŏng-ji and drafted a map of Seoul. Chŏng Ch'ŏk was well informed on the terrain, Kang Hi-an was versed in drawing, and Yang Sŏng-ji was an expert on geography. . . ."[60] Yang's administrative map, titled *Hwangguk ch'ip'yŏngdo* 皇國治平圖, was completed in January of 1454. Further work under Tanjong 端宗 began in Kyŏnggi Province in the spring of 1455. Findings underwent extensive analysis, and further work was started in August of the same year in Kyŏngsang, Chŏlla, and Ch'ungch'ŏng Provinces. These may have relied as well on maps of Kyŏngsang Province and Nŭngch'ŏn 熊川 presented to the court by Hwang Su-sin 黃守身.[61] The results of these efforts were edited and compiled as the *Tongguk chido* 東國地圖 in 1463. This map is no longer extant, but from what we have seen of the compilation process we can guess that it must have been one of the most precise of its time.[62] It is believed that the *Tongguk chido* of Chŏng Sang-gi 鄭尚驥 was largely based upon this map.

THE USE OF TRIANGULATION

Even before his accession to the throne, Sejo 世祖 had a deep interest in the accuracy of maps. His interest eventually led to his invention in 1467 of a triangular surveying instrument which he called the *kyuhyŏng injiŭi* 窺衡印地儀. Among the many scholars to whom he explained his method, only Yu Hi-ik 兪希益 and Kim Yu 金紐 were reported to have understood it.[63] In the *Munhŏn pigo,* the instrument is explained simply as a device for measuring distances. Yi Yuk 李陸 said of it: "The *kyuhyŏng* 窺衡 is made of bronze with twenty-four

[59]*Tanjong sillok,* ch. 8, p. 21b. ch. 10, pp. 24b–25a.
[60]Ibid., ch. 11, p. 3a.
[61]*Chōsenshi,* pt. 4, vol. 4, p. 403.
[62]*Sejo sillok,* ch. 31, p. 25b.
[63]Ibid., ch. 41, pp. 20ab.

cardinal points. Its center is sustained by a bronze pillar; by turning the round instrument, one can determine the direction."[64] We know from this that the instrument was similar to what we use today, and that it employed the principle of equivalent triangular proportions. Sejo had scholars prepare a map of the capital city using triangulation.[65]

SURVEYING FOR THE MAP OF KOREA IN THE *TONGGUK YŎJI SŬNGNAM*

The maps contained in the *Tongguk yŏji sŭngnam*, published in 1487,[66] are in many cases inaccurate and incomplete. The formulae of the Korea map for P'yŏngan and Hamgyŏng are particularly poor, and since it is evident that they were prepared from Yang Sŏng-ji's maps in his *P'alto chiriji* (1478), the question arises whether the maps drawn up on the basis of on-the-spot observations such as Yang conducted were really reliable. When one compares the overall map for Korea (the *P'alto ch'ongdo* 八道總圖) with the various individual provincial maps, one finds great differences in terrain, these being especially acute in the cases of Hamgyŏng and P'yŏngan provinces. The individual provincial maps contained in each of the various chapters are on the whole far more accurate than the *Ch'ongdo* 總圖, and more accurate as well than Yi Hoe's *P'altodo.* Some distortions in the *Ch'ongdo* are the result of the size of paper used for the map; proportions were apparently sacrificed to enable the map to be attached to the *Tongguk yŏji sŭngnam.*

The separate provincial maps were naturally more detailed than the *Ch'ongdo.* The *Ch'ongdo* included names of provinces and major mountains, rivers, and islands. The provincial maps included in addition the names of cities, towns, mountains, rivers, and most of the islands in the provinces. Mountains were marked topographically and the seas were drawn with black waves as in the world map of 1402. The method of representing each mountain separately compares closely with

[64] *Munhŏn pigo,* ch. 2, pp. 32ab.
[65] *Sejo sillok,* ch. 44, p. 9b.
[66] *Tongguk yŏji sŭngnam,* postscript.

that in the *Ta Ming i t'ung chih* 大明一統志 (Comprehensive geography of the Ming Empire).[67]

Korean cartography placed heavier emphasis on towns and terrain to the time of Sejo. During the reign of Sŏngjong, special emphasis was given to coastal lines and the surrounding seas, and it was this period that saw the development of charts for use in maritime transport.[68]

Treatises on Geography in the Sixteenth Century

Two months after the publication of the newly compiled *Tongguk yŏji sŭngnam*, in early 1487, Kim Chong-jik 金宗直 presented to the king a list of deficiencies in the work and requested correction. Kim maintained that details taken from the testimony of local residents were often in error and that distances were incorrect because no survey had ever been made. In particular, local products had been poorly reported out of fear of taxation. Accordingly, work was begun on a revision early in the reign of Yŏnsan 燕山 but stopped in 1499. A major role in this revision was played by Sŏng Hyŏn 成俔, Im Sa-hong 任士洪 and Yi Tŏk-sung 李德崇, but their works were not published. The final revision was completed only in 1530 through the efforts of Yi Haeng 李荇 and others. It was published in November of that year under the title *Sinjŭng tongguk yŏji sŭngnam* 新增東國輿地勝覽 (Newly augmented *Tongguk yŏji sŭngnam*).[69]

At the end of the sixteenth century, Han Paek-kyŏm 韓百謙 published his *Tongguk chiriji* 東國地理誌 (Geography of the Eastern Kingdom), which was comparable in scope to the government-edited geography.[70] Han's treatise contained excerpts from the *Ch'ien Han shu* 前漢書 and the *Hou Han shu* 後漢書 as well as other Chinese works, adding his own comments. His effort was noteworthy in that it did not simply follow Yang Sŏng-ji's *P'alto chiriji* or the Chinese official geography. As the

[67]Aoyama, "Richō ni okeru nisan no Chōsen zenzu ni tsuite," p. 149.
[68]*Chōsenshi,* pt. 4, vol. 5, p. 405.
[69]*Tongguk yŏji sŭngnam,* postscript.
[70]It was first published in 1640.

first "positivist" approach to the study of Korean historical geography, his work was a pioneering one in the development of *sirhak* (practical learning) in Korea.[71] His method of study greatly influenced later works on Korean geography, such as Sin Kyŏng-jun's 申景濬 *Kanggego* 疆界考 (A study on the frontier, 1756), *Torogo* 道路考 (A study on roads, 1770), and *Sansugo* 山水考 (A study on mountains and rivers). Other evidences of his influence may be found in Chŏng Yag-yong's 丁若鏞 *Kangyŏkko* 疆域考 (Studies in Korean geography, 1800–1834) and *Taedong sugyŏng* 大東水經 (Korean water classic, 1800–1834), in Han Ch'i-yun's 韓致奫 *Sok-chirigo* 續地理考 (Further study on Korean geography), in the *Haedong yŏksa* 海東繹史 (Historical treatises on Korea, continued, 1823) and in Yi Ik's theory of natural geography as developed in the *Sŏngho sasŏl* 星湖僿說 (Sŏngho's detailed discourses).[72]

The sixteenth century also saw the publication of Yun Kyŏl's 尹潔 *Yugu p'ungsokki* 琉球風俗記 (Folk customs of the Ryūkyūs) and Yu Tae-yong's 柳大容 *Yugu p'ungt'ogi* 琉球風土記 (Natural history of the Ryūkyūs). Both were based on accounts from Pak Son 朴孫 and his colleagues who returned to Korea in February 1546 after living for four years in the Ryūkyūs.[73] The books related the peculiar travel experiences of the ship's crew.

The T'aengniji 擇里誌 of Yi Chung-hwan 李重煥

The *Sejong sillok chiriji* and Yang Sŏng-ji's *P'alto chiriji* were the most magnificent achievements in Korean scientific geography of the fifteenth century. Inasmuch as they were government-sponsored publications, the scholars working on them, uninterested in the inadequate geomantic theories of the time, devoted little attention to the interrelation between man and his surroundings, but focussed instead on comprehensive topographic geographical description. The astronomers assigned to the Sŏun'gwan were well versed in *p'ungsu* 風水 theory, but more conventional scholars in general disregarded them.

[71]There are many theses on *sirhak* available but Yi Ki-baek's *Han'guksa sillon* contains an excellent summary (pp. 268–274).

[72]Hong I-sŏp, *Chosŏn kwahaksa*, pp. 250–252.

[73]*Myŏngjong sillok*, ch. 3, pp. 19ab. *Chōsenshi*, pt. 4, vol 8, p. 168.

In the early eighteenth century, geography began to take on the character of a discrete academic subject. Developments at this time were an outgrowth of the massive disruptions in every sphere of life brought on by the Hideyoshi and Manchu invasions. The *T'aengniji* (Notes on choosing a domicile) of Yi Chung-hwan (1690–1753) was written at this time. It must be admitted that Yi in his search for a "livable place" (可居地) was not entirely free in his methodology from *p'ungsu* theories. His work is important because he adhered generally to a methodology which logically attempted to analyze the history of mankind with a view to producing more reliable conclusions. The *T'aengniji* was essentially a programmatic work, concerned with the welfare of the people. In both its practical purposes and academic methodology, then, it represents a departure from establishmentarian *p'ungsu* theory studied and developed in the Soun'gwan.

CONTENTS OF THE *T'AENGNIJI*

The preface contains general remarks on the geography of Korea in two parts, the "Samin ch'ongnon 四民總論" (Introduction to the Korean people) and the "P'alto ch'ongnon 八道總論" (General remarks on the eight provinces). In the second of these, Yi described the territory of the country in this way:

A branch of the K'un-lun mountain range stretches eastward along the south side of the Great Desert and forms a basin of the Lo-shan mountains, located west of Liaotung. From here the Liaotung Plain starts. Across the plain is Mt. Paektu, or Pulamsan, referred to in the *Shan hai ching* 山海經. The energy current of the mountain range reaches as far as one thousand *ri* to the north, crossing the two rivers, and creates Ningkut'a (in Ki-lin Province, cradle of the Ch'ing Dynasty) in the south. One of the branches in the rear constitutes the Korean mountains. There are eight provinces in Korea; P'yŏngan Province adjoins Mukden, Hamgyŏng Province neighbors Jürchen, and Kangwŏn Province borders the south of Hamgyŏng Province. Hwanghae Province is adjacent to the south of Hamgyŏng Province, Kyŏnggi Province is to the south of Kangwŏn and Hwanghae, Ch'ungch'ŏng Province and Chŏlla Province are

to the south of Kyŏnggi Province, and Kyŏngsang Province is to the east of Chŏlla Province.[74]

In describing geographical features of the country, Yi wrote:

The peninsula is surrounded by seas. Its main road to the north leads to Mukden through Jürchen. The terrain is mountainous and paddies and fields are scarce. The people are obedient. The peninsula's length is three thousand *ri* and its waist is one thousand *ri*. The land of Chosŏn (Korea) is situated between China and Japan.[75]

In the chapters concerning the eight provinces, Yi dealt with each separately in the order above, which was based on the arrangement of mountain ranges, rather than in the order conventionally used by court officials. He touched on the location of each province, terrain, climate, natural conditions, industry, and towns. He attempted to go beyond the mere listing of facts and correlate natural features with human life. His work is, needless to say, an epoch-making venture in placing Korean geography on a scientific foundation.[76]

Making the best use of his general statements, Yi proceeded to develop concretely his geographical approach to finding a "livable place." In the chapter dealing with residence, Yi said that one must consider, in choosing his residence, geography, practical benefits, and surroundings. In accordance with these prerequisites, he offered the criteria of geomantic adequacy (*yangt'aek* 陽宅), productive geography (*saengni* 生利), rivers (*kangha* 江河), and tenor of the town (*insim* 人心). Yi listed six preconditions for the formation of a community: water resources, fields, mountains, fertile soil, irrigation, and extension of the mountains and rivers. The ideal place would be one in which fields and paddies were fertile and productive of foods and agricultural products with which to make clothing, and where there was easy access to (preferably water) trans-

[74] *T'aengniji*, tr. No To-yang (Seoul, 1968, p. 37). There are other editions available.

[75] Ibid., pp. 37–38.

[76] Yi Ch'an, "Han'guk chirihaksa" (A history of Korean geography; *Kwahak kisulsa*. in *Han'guk munhwasa taege* series, Seoul, 1968), pp. 706–707.

portation. He called a "good neighborhood" one in which every man had a pleasant character.

Yi strongly believed that terrain affected man's spirit. In his chapter on *sansuron* 山水論 (theory of mountains and rivers), he divided his discussion into an introduction and six sections, discussing one by one famous mountains, rivers, islands, and other points of scenic interest. He maintained that it was geomantically preferable to live in a mountain valley than on a riverbank.

Yi Chung-hwan's *T'aengniji*, with its consistency and format, its systematic presentation, and its accurate observation of the relationship between man and natural conditions, contributed greatly to the development of Korean geography.[77]

Introduction of New Maps of the World

WORLD MAPS BASED ON WESTERN MAPS

Western maps introduced into China by Matteo Ricci helped the Chinese modify their Sinocentric geography and exercised an influence in turn on Korean cartography (Figs. 5.6 and 5.7). Ricci's map was initially accepted in China because he combined European data with traditional Chinese geographic notions. His first effort, the *Yü ti shan hai ch'üan t'u* 輿地山海全圖 of 1584, and the considerably revised *K'un yü wan kuo ch'üan t'u* 坤輿萬國全圖 of 1602 helped the Chinese become aware of the five known continents and the western classification of climate into five groups according to latitude.[78]

The Yi court was eager to obtain such knowledge and customarily sent an astronomer and geographer to China each year along with the regular mission. Ricci's map, which had been drafted jointly with Li Chih-tsao 李之藻, was brought

[77]Sŏ Su-in, "T'aengniji yŏn'gu sŏsŏl" (Preliminary study of the *T'aengniji*), *Chirihak, 1*, 1963: 88; Yi, "Han'guk chirihaksa," p. 706. Sŏ Su-in indicated in the article just cited that the *T'aengniji* was published before the pioneering work of Karl Ritter and A. von Humboldt. It would seem to deserve a prominent place in the history of geography.
[78]Needham, *SCC*, vol. 3, pp. 583–586.

Fig. 5.6. Terrestial globe from the armillary sphere of 1664–1669. Diameter, 3.5 in. The picture shows part of Asia and America. The map of South America is quite accurate, but North America has a very distorted outline. Koryŏ University Museum.

into Korea by Yi Kwang-chŏng 李光庭 in 1603, one year after its publication in Peking.[79] The following year a revised edition (the *Liang i hsuan lan t'u* 兩儀玄覽圖) was brought into Korea, followed by several copies of maps drawn by Ricci. Only the *Liang i hsuan lan t'u* is extant in Korea, in the Sungjŏn Univer-

[79]*Chibong yusŏl* (Classified writings of Ch'ibong = Yi Su-gwang), ch. 2, geography.

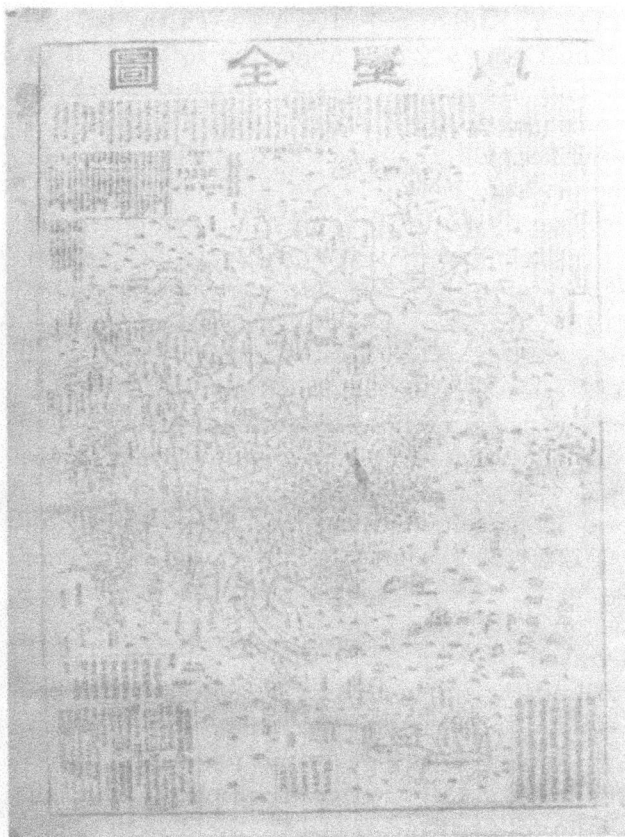

Fig. 5.7. A map of the world, *Yŏji chŏndo*. Late eighteenth century. This is one of the Sinocentric world maps influenced by Chinese tradition and the Korean world map of 1402.

sity Museum, Seoul.[80]

Ricci's map was later copied under government auspices. A typical copy, the *Kŏnsang konyŏdo* 乾象坤輿圖 (Chinese *Ch'ien hsiang k'un yü t'u*), was prepared at the Bureau of Astronomy

[80]Kim Yang-sŏn, "Myŏngmal Ch'ŏngch'o Yasohoe sŏn'gyosatŭri chejakhan sege chidowa kŭ Han'guk munhwasasange mich'in yŏnghyang" (Jesuit world maps published in Ming and Ch'ing, and their influence in Korea), *Sungdae, 6,* 1961: p.35–37.

(Kwansanggam) by Yi Kuk-hwa 李國華, Yu U-ch'ang 柳遇昌, and the artist Kim Chin-yŏ 金振汝 under the instructions of Sukchong 肅宗. This was based on the *K'un yü wan kuo ch'üan t'u,* which was prepared during the period from 1608 to 1610. It was kept in the Pongsŏn Temple in Kyŏnggi Province but destroyed during the Korean War.[81]

When Matteo Ricci died in 1610, Giulio Aleni completed the preparation of the world map *Wan kuo ch'üan t'u* 萬國全圖 and the treatise *Chih fang wai chi* 職方外紀 (On world geography).[82] These were brought to Korea in 1631 by Chŏng Tu-wŏn 鄭斗源 and Yi Yŏng-hu 李榮後. The *K'un yü ch'üan t'u* 坤輿全圖 (World map) of Verbiest was brought to Korea in 1674. This map was an equatorial projection divided into hemispheres. It far excelled that of Matteo Ricci. It was republished in Korea as the *Konyŏ chŏndo* 坤輿全圖 in 1860.[83]

Another world map containing the two hemispheres, the *Chigu chŏnhudo* 地球前後圖 (Diagram of the earth) was made by Kim Chŏng-ho in 1834. This map bears the inscription, "second edition of T'aeyŏnjae 泰然齊." Apparently this was copied and engraved by Kim Chŏng-ho himself, but it has been difficult to find the original. Kim Yang-sŏn 金良善 believes that the original might be a terrestrial globe which was presented to the Chinese court by a British subject in 1793.[84] The *Konyŏ chŏndo* had placed the western hemisphere at the left but Kim Chŏng-ho reversed them, placing the eastern hemi-

[81] *Myŏnggokchip,* ch. 8. The map currently displayed in the Seoul National University Museum, the *Konyŏ man'guk chŏndo* (eight-fold, 170 cm × 533 cm), with an introductory note by Ch'oe Sŏk-chŏng, is a later one. *Chōsenshi,* pt. 5, vol. 6, pp. 711–712.

[82] *Kukcho pogam* (Precious mirror of the Yi Dynasty), ch. 35, ninth year of King Injo.

[83] The Korean version of the world map of Verbiest consists of eight parts. At present, six engraved wooden parts are in the Seoul National University Library, and two descriptive parts are missing. Tabohashi said of this map that, along with the *Taedong yŏjido,* it is one of the best examples of Yi Dynasty woodcut maps. See Tabohashi Kiyoshi, "Chōsen sokuchishi-jō no ichi gyōseki" (A contribution to the history of Korean cartography), *Rekishi chiri, 60.*6, 1932.

[84] Kim Yang-sŏn, "Han'guk kojido yŏn'gu ch'o" (A study of ancient maps in Korea), *Sungdae, 10,* 1965: 80.

sphere in front and depicting the land and sea areas somewhat differently.

Around 1846, Kim prepared a map of the five continents, and in 1848 he drew up a map with Kögler's star map in the center and Asia, Europe, America, and Africa in the four corners. Kögler's planisphere had been drawn up by Ch'oe Han-gi 崔漢綺 and engraved by Kim in 1834. Ch'oe also copied the extremely informative Chinese compendia of European geographic and political information, the *Ying-huan chih lueh* 瀛環志略 (Brief survey of the maritime circuit, 1850) of Hsu Chi-yü 徐繼畬 and Wei Yuan's 魏源 *Hai-kuo t'u chih* 海國圖志 (Illustrated gazetteer of the maritime countries, 1844). In addition to these, there is an eight-fold *Chigu chŏndo* 地球全圖 (Complete map of the earth) in the Sungjŏn University Museum. The 1866 edition of *Yŏjae ch'walyo* 輿載撮要 (Elements of world geography) contains a global map in its preface. These are identical to Japanese maps published at about the same time.

SINOCENTRIC WORLD MAPS

World maps in the Chinese style prepared in Korea all bear titles of a form similar to *Ch'ŏnhado* 天下圖 (World map; Map of all-under-heaven); they are all sinocentric. Some of these have already been discussed under the heading "Geography in the Early Yi Dynasty." The following discussion will deal with those not already mentioned.

At the end of the eighteenth century the Yi government published a sinocentric world map containing many of the European countries. This simple map—the *Yŏji chŏndo* 輿地全圖 lacked the American continent but included the Indian subcontinent and Russia. The portions of the map dealing with Arabia and Africa were similar to the Korean map of 1402, but the portions dealing with Korea and Japan were rather less accurate than earlier maps, leading us to suspect that the map was not based entirely on the 1402 map. The introduction cites Guilio Aleni's *Chih fang wai chi* and *K'un yü t'u shuo* as reference materials.

Maps in the Later Yi Dynasty

CHŎNG SANG-GI'S MAP OF KOREA

The *Tongguk chido* 東國地圖 (Map of Korea) of Yang Sŏng-ji 梁誠之 and Chŏng Ch'ŏk 鄭陟 was the climax of early Yi Dynasty efforts to prepare maps on the basis of actual surveys (Fig. 5.8). There seem to have been few similar projects for some time afterward, excepting the *Sŏbuk chido* 西北地圖 (Map of the northwestern boundaries), which was drawn up by Sŏng Chun 成俊 and Yi Kŭk-kyun 李克均 while they were working on the *Sŏbuk chebŏn'gi* 西北諸蕃記 (Record of the barbarians in the northwestern districts) and presented to the court in

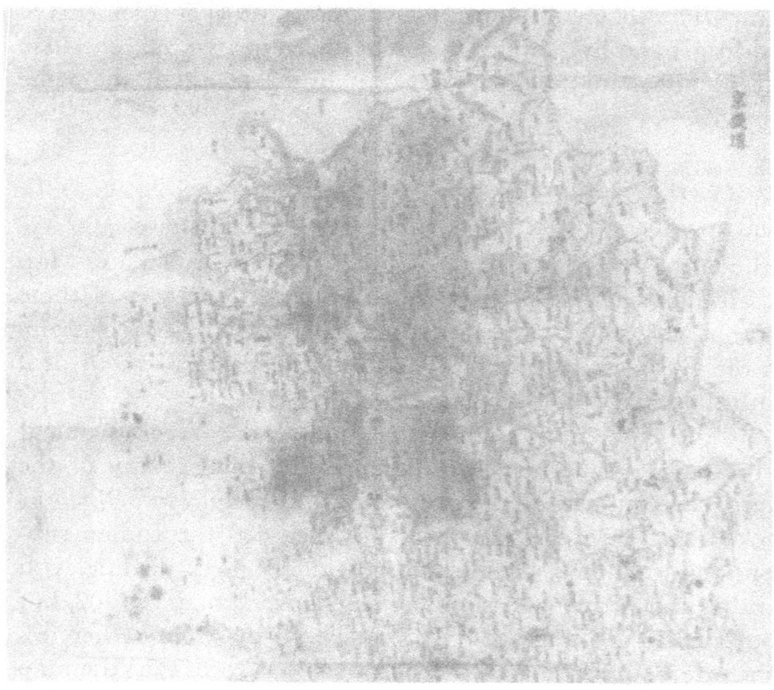

Fig. 5.8. Chŏng Sang-gi's map of Korea, *Tongguk Taejido*, MS. The map shows Kyŏnggi Province. Prepared around 1757. Chŏng's atlas is the best one of its period in Korea, and is well known for its use of a scale of 100 *ri*.

1501.[85] These early attempts were not entirely successful. Particularly conspicuous in early Yi maps is the inaccuracy of border areas adjacent to Hamgyŏng and P'yŏngan Provinces. Several maps available for study show that maps of the early and middle Yi period distorted the border areas. This probably resulted from the use of inaccurate portions of the overall map of Korea in the *Tongguk yŏji sŭngnam* (mentioned above). Another common defect of early maps is the placement of Ullung Island to the east of Usan Island. It is possible that the inaccuracy in depicting terrain in Hamgyŏng and P'yŏngan Provinces in early Yi maps stems from errors in the work of Yang and Chŏng. It is known that new astronomical knowledge available from about the time of Sukchong (1713) would have made it possible for geographers to prepare more accurate maps.

In 1757 Hong Yang-han 洪良漢 submitted to Yŏngjo the *P'alto pundoch'ŏp* 八道分圖帖 (Separate maps of the eight provinces).[86] It is possible that this extremely precise atlas was based on the *Tongguk taejido* 東國大地圖 (Enlarged map of Korea) of Chŏng Sang-gi 鄭尚驥 (1687–1752), a scholar during the reign of Yŏngjo. His poor health forced him to renounce his ambition to become a government officer, and he turned to scholarship. Chŏng Sang-gi was credited by Sin Kyŏng-jun in the *Tongguk munhŏn pigo* 東國文獻備考 with being the first to make a map using scales.[87] Sin also selected Chŏng's map from among all others available in 1770 as the best from which to prepare a "perfect" map for the *Tongguk munhŏn pigo*. Yi Ik 李瀷 highly praised Chŏng's map in the first chapter of the

[85] *Yŏnsan'gun ilgi* (Veritable records of the King Yŏnsan era), ch. 40, p. 20b.
[86] *Yŏngjo sillok*, ch. 90, pp. 8b–9a.
[87] Chŏng used a one-hundred *ri* scale, called the *paengni-ch'ŏk*. Maps reducing 100 *ri* to one *ch'on* had been in use in China, and one of these, Lo Hung-hsien's *Kuang yü t'u*, was well known in Yi Dynasty Korea. Chŏng's scale must have been made under the influence of Chinese map scales. Yi Pyŏng-do, "Chŏng Sang-giwa *Tongguk chido*" (Chŏng Sang-gi and his Map of Korea, *Tongguk chido*), *Sŏji, 1*, 1960: 14–15. As Yi Ch'an pointed out, Chŏng illustrated the concept of reduced scale by the use of a trapezoid. Yi Ch'an, "Han'guk chirihaksa," p. 718.

Sŏngho sasŏl 星湖僿說 (Sŏngho's detailed discourses), and in his inscription for Chŏng's tombstone.

The *P'alto pundo* 八道分圖 (Separate maps of the eight provinces) and the *Ch'ongdo* 總圖 (Complete map) of Chŏng Sanggi currently in the possession of the author are 95 cm × 59 cm in size and in nine folds. These maps appear to have been completed about 1757. The scale on these maps, surprisingly precise, is 3.5 cm to 50 *ri*.

In the eighth part of the *Tongguk taejido*, Chŏng described his work as follows:

There are quite a few maps published in Korea. These maps were proportioned to the size of paper available, resulting in inaccurate distances on the maps. In many cases, short distances were marked as longer than the actual distance, and vice-versa. Incorrect locations were given and terrain was often displaced, leading viewers to great confusion. When one looks at the maps, it seems as if one is left in the darkness. I prepared this map while I was suffering from illness. I made on-the-spot surveys over many areas and employed a method of marking 100 *ri* as one *ch'ŏk* on the map and 10 *ri* as one *ch'on*. Measurement was started from the capital and expanded gradually to all provinces. Each sheet is a map of the province concerned. The eight sections will make a general map of Korea when the provincial maps are combined. Hamgyŏng Province was put on two sheets because of its complex geographical situation and terrain. Kyŏnggi and Ch'ungch'ŏng Provinces, on the other hand, were placed on a single sheet because of their simplicity of terrain and comparatively small size. The northeastern section of the Western boundary region was added to the west of Hamgyŏng Province because those vast areas could not be contained on one single sheet. As for Cheju Island, Ullung, Hŭksan, Hongi, and Kage islands, these were drawn at the end of the sheet and the distances listed. Rivers or mountains situated along the borders of two or more provinces were marked in each sheet. Therefore, if one desires to combine the sheets, he may eliminate overlap from one of the sheets he would like to combine. A reduced scale is used. For example, 100 *ri* was reduced to one *ch'ŏk*. In the places where the terrain is complex and steep the scale of one to 120–130 was used. Kyŏnggi Province was colored yellow; Hosŏ white; Hodong red; Yŏngdong blue and red; Yŏngnam blue; Haesŏ white; Kwansŏ white and black; Kwanbuk black. Mountains were colored green and rivers blue. Red lines were used to mark the waterways; yellow lines denote borders; red dots represent beacon stations. Towns with castles are given a symbol; stations are marked with blue.

It can be seen from Chŏng's description that his *Tongguk taejido* contained many characteristics not found on previous maps, such as the use of a scale of one *ch'ŏk* to one hundred *ri*, the combination of separate provincial maps to form a general map, the clear marking of waterways, the explanation of communications networks, and the distinctive marking of mountains. These creative innovations cause Chŏng's work to be regarded as an important milestone in Korean cartography.

Despite what he said in his discussion, Chŏng's illness makes it unlikely that he was ever able to engage in on-the-spot surveys of key areas. It seems more likely that Chŏng revised older maps, using his own method of reduced scaling. Chŏng himself indicated this when he said that the maps he had used had not shown distances; he did not say that it was impossible to measure the distances. A careful comparison of the distances given in the later (and highly accurate) *Taedong yŏjido* 大東輿地圖 of Kim Chŏng-ho 金正浩 with both the Geographical Treatise of the *Sejong sillok* and with the *Tongguk chido* of Yang Sŏng-ji and Chŏng Chŏk indicates the latter to have been more correct than the former. Taking into consideration that Chŏng Sang-gi was a descendant of Chŏng In-ji, a renowned astronomer during the reigns of Sejong and Sŏngjong, it is possible that he was able to use copies of the *Tongguk chido* possessed bv his ancestor and kept by the family.

THE *TAEDONG YŎJIDO*

The *Tongguk chido* of Yang Sŏng-ji and Chŏng Ch'ŏk had been greatly improved by the work of Chŏng Sang-gi. Further progress was made with the completion in 1834 of Kim Chŏng-ho's *Ch'ŏnggudo* 靑邱圖. The *Ch'ŏnggudo* was a composite map, based on rectangular units, *p'an* 板 (blocks) in *tan* 段 (columns), scaled to represent 70 *ri* × 100 *ri*, respectively. Each unit was drawn up on a sheet 21 cm × 30 cm, and then the sheets were combined. The completed map measured 22 *p'an* by 29 *tan*. This was an improvement on Chŏng Sang-gi's simpler technique. Kim followed Chŏng's method of leaving the upper and lower parts of each map blank, so that there was some duplication of areas when the sheets were combined. The *Ch'ŏnggudo* was drawn up over a ten-year period on the basis of maps prepared by Chŏng Ch'ŏl-cho 鄭喆祚, Hwang Yŏp 黃燁 and Yun Yŏng 尹鍈. Under the influence of contemporary geographical treatises, Kim used blank spaces on the completed map to record information pertaining to the history of each town, and information about old villages, rivers, mountains, castles, walls, warehouses, stations, beacon stations, ships, ferry points, bridges, hills and passes, islands, pasturage, households, troop stations, paddies and fields, crops, embankments, borders, markets, palaces, pavilions, personalities, folk customs, products, shrines, royal tombs, temples, and historical remains.[88]

Aware of the inconveniences inherent in such a sectional atlas, Kim began to prepare a more precise map, the *Taedong yŏjido*, which was completed thirty years after the publication of the *Ch'ŏnggudo*, in 1861. This was both the largest and best map prepared during the Yi period (Figs. 5.9 and 5.10). The 1861 edition was made up of rectangles 20 cm × 30 cm, representing 80 *ri* × 120 *ri* respectively.[89] Another innovation was Kim's use of a map key, which had not been in wide use up to that time. Other improvements were the marking of each 10 *ri* on roads and the marking of county borders with dotted lines.

The map included a concise treatise giving the objectives of the author and, under the title "Chido yusŏl 地圖類說"

[88]The *Ch'ŏnggudo* adopted a method in which a hypothetical (i.e., blank) map was drawn and introductory remarks attached to it. This is the method used in modern atlases. Distances are measured in the *Ch'ŏnggudo* by the use of equidistant, i.e., concentric circles. Enlargement and reduction of maps by means of rectangular projection is explained as in the *Chi-ho yuan-pen* 幾何原本, the early Chinese translation of the works of Euclid. Such a method is still in use, although today often superseded by photography. The *Ch'ŏnggudo* used the Polestar as a guide to latitude, perhaps relying for measured values upon materials available at the Bureau of Astronomy.

[89]Yi Ch'an, "Han'guk chirihaksa," p. 727.

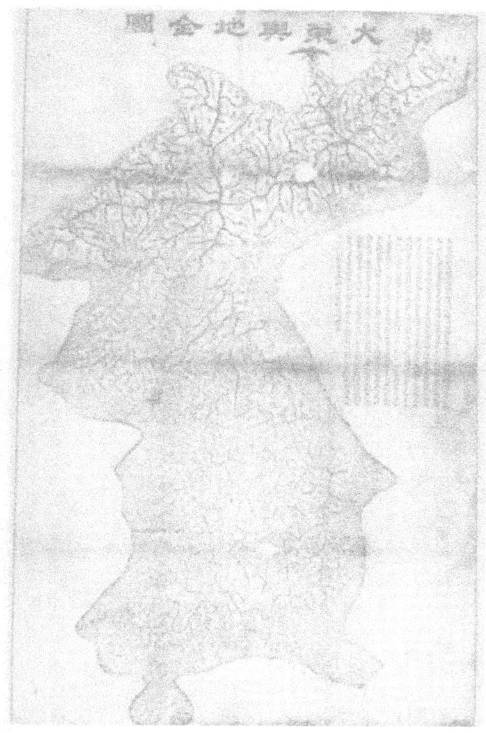

Fig. 5.9. The *Taedong yŏji chŏndo,* Kim Chŏng-ho's general map of Korea. Printed around 1861.

(General description of the map), the principles and history of cartography.[90] Kim's own ideas were developed and illustrated by the use of quotations from older works on geography. He reviewed the origin and historical significance of maps and cartography in China, stressing the indispensability of geography for politics, economics, national defense, and academic

[90]The theory which Kim developed in the general description of the map in connection with the illustration of mountains and rivers was not based on ancient Chinese geographical documents but originated from the Koreans' unique topographical method. The mountains Kim drew are identical to those in illustrations of graves drawn by Korean geomancers.

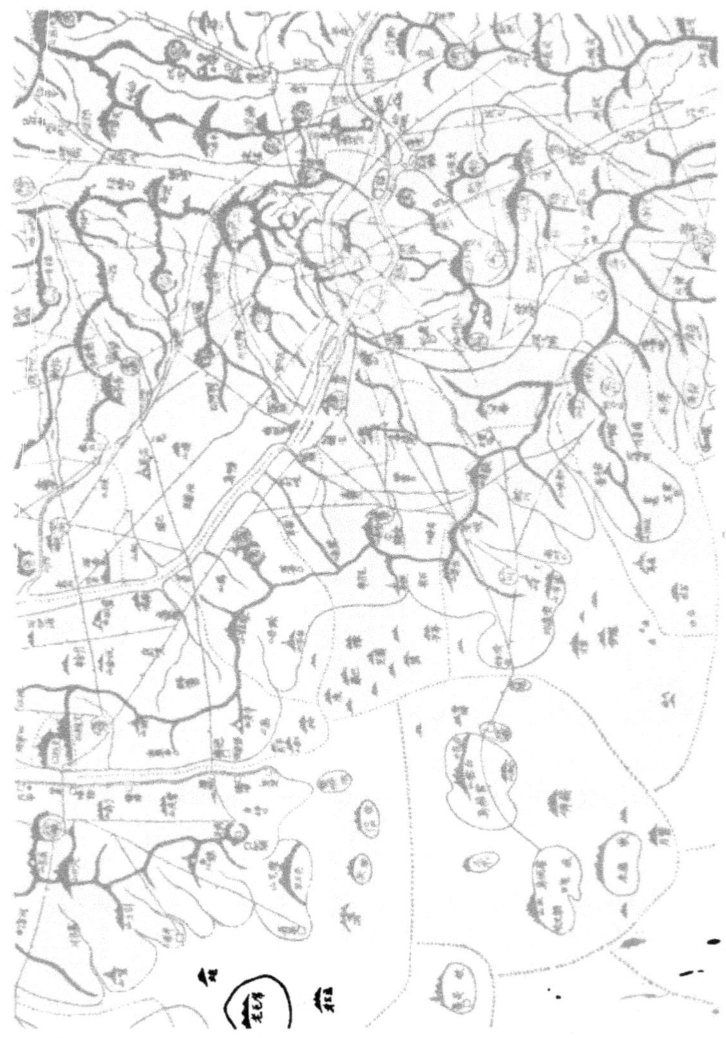

Fig. 5.10. Map of Kyŏnggi Province from the *Taedong yŏjido*, Kim Chŏng-ho's map of Korea, scale 160,000:1. Printed in 1861, in 22 parts of 20 × 30 cm each. Kim's map is the most scientific and accurate made by a Korean before the twentieth century.

research. In particular, he singled out for emphasis the six tenets of the Chinese scholar P'ei Hsiu (224–271):

1. Use of graduated divisions, which are the means of determining the scale to which the map is to be drawn
2. Use of the rectangular grid, which is the way of depicting the correct relations among the various parts of the map
3. Pacing out the sides of right-angled triangles, which is the way of fixing the lengths of derived distances
4. (Measuring) the high and the low
5. (Measuring) right angles and acute angles
6. (Measuring) curves and straight lines.

Kim added the principles of Chinese Han dynasty cartography and of the *Fang yü chi yao* 方輿紀要 of the Sung period, stressing the inclusion of information concerning the size of land areas, their location, the shapes of mountains and rivers, and the number of households. Such information, he felt, was indispensable for the ruler who wished to defend the country and govern it well. Kim further noted that accurate directions and distance are inseparable factors. Kim followed the *Munhŏn pigo* in giving numerical values for land acreage, the lengths of the coastal and border lines, and distances from the northwestern and northeastern corners to the southeastern and southwestern tips. His figures are almost identical to those currently in use.

Kim's ideas are reflected in the organization and content of the *Taedong yŏjido*. Most Yi Dynasty maps before Kim's were designed to meet administrative needs; consequently, 50 to 75 percent of the data pertained to population and administrative centers. The *Ch'ŏnggudo* introduced illustrations and descriptions of natural and topographic features; these were expanded in the *Taedong yŏjido* to about 60 percent of the total notes. About one tenth of the notes in the later map were devoted to rivers and roads. It is the nonurban topographic notes and descriptions in the *Taedong yŏjido* as much as anything else which contribute to the place of the map in the history of Korean cartography.

Perhaps most notable in the *Taedong yŏjido* is the accuracy of coastal lines and the detail in terrain. Mountains are illustrated by ranges, groups, and independent mountains. In putting mountains on his map, Kim used a method of symbolizing shape by profile. More than twenty-eight hundred mountains were entered in the map, including exceptionally high mountains, parallel chains, continuous ranges and overlapping mountains. A variety of symbols was also used for different sizes of cities, villages or administrative centers, and military bases, combined with scales and local features to make them easily recognizable.[91]

In addition to the *Ch'ŏnggudo* and the *Taedong yŏjido*, Kim Chŏng-ho published several general maps of Korea on woodcuts. Among these were the *Taedong yŏji chŏndo* 大東輿地全圖, the *Haejwa chŏndo* 海左全圖 and the *Chosŏn chŏndo* 朝鮮全圖. These were accurate maps on a scale of 1:900,000. The *Taedong yŏji chŏndo* was a reduction of the *Taedong yŏjido*. The two others contained general descriptions of Korean natural geography as well as an outline of regional historical and geographic information. These smaller maps were much closer to modern methods in representing mountains from an overhead "bird's-eye-view" perspective. Despite the various difficulties inherent in the woodblock printing, these maps are among the most accurate Korea ever produced, representing the greatest achievements of Yi Dynasty cartography.[92]

[91]In accordance with this method, valleys and isthmuses were entered on the map. More than 1,160 hills were represented. Various terrains were shown with their names and additional explanations. About 2,400 rivers, coastlines, lakes, ponds, and wells were also entered. The map is especially noted for its excellence in presenting the characteristics of Korean rivers. Mok Yŏng-man, *Chido iyagi*, pp. 21–51.

[92]It should be mentioned that there is much disagreement about the details of Kim's life. It is maintained that he was a commoner and lived in poverty all his life. Some say that he was married to a peddler, others that he was the servant of a military officer. It is questionable, however, given the rigid class system of the time, whether a commoner could have made seven trips to Mt. Paektu and paid his travel expenses. One might also question the likelihood of his access to rare materials on geography and astronomy over a period of thirty years. His map is the result of combined studies—astronomical observations and on-the-spot surveys. Such an achievement would be almost impossible unless he were provided with survey data. However his work was carried out, Kim's contribution to Korean geography was the greatest of the nineteenth century.

GEOGRAPHICAL PUBLICATIONS IN THE LATE YI DYNASTY

From the time the *Tongguk yŏji sŭngnam* was published in the sixteenth century until the publication of the *Tongguk munhŏn pigo* in 1770, a number of geographies were published under government auspices. These are less valuable academically than works published privately in the later part of the dynasty. Private works from this period include Wi Paek-kyu's 魏伯珪 *Hwanyŏngji* 寰瀛誌 (Gazetteer of the maritime circuit), first edited in 1770 and published in 1822; Ch'oe Han-gi's *Chigu chŏnyo* 地球典要 (A handbook of world geography), in six volumes; Kim Chŏng-ho's *Taedong chiji* 大東地誌 (Treatise on Korean geography), published in 1864; and the *Yŏjae ch'walyo* 輿載撮要, published in 1866.

Bibliography

The bibliography that follows is not exhaustive but includes only those works most relevant to the subjects treated in this volume. Primary sources are arranged alphabetically by title; secondary and Western sources are alphabetized in the conventional way, by author.

Primary Sources (Korean, Chinese, and Japanese Sources That Appeared before 1910)

"Chaeigo 災異考"
(Observations of natural disasters and strange events), 1624–1655, MS preserved in Seoul National University Library.

Chamgok p'iltam 潛谷筆談
(Collected jottings of Chamgok = Kim Yuk), 1580–1658. By Kim Yuk 金堉.

Chega yŏksangjip 諸家曆象集
(Collected discourses on astronomy and calendrical science of the Chinese masters), 1445. Edited by Yi Sun-ji 李純之.

Chibong yusŏl 芝峰類說
(Classified writings of Chibong = Yi Su-gwang), preface dated 1614, first published 1633 (modern reprint, Chosŏn kosŏ kanhaenghoe 朝鮮古書刊行會, 1915). By Yi Su-gwang 李睟光.

Ch'i ch'i t'u shuo 奇器圖說
(Illustrated explanations of wonderful machines), 1625. By Johann Schreck and Wang Cheng 王徵.

Chien p'ing i shuo 簡平儀說
(Description of a simple equatorial torquetrum [astronomical instrument]), 1611. By Sabbatino de Ursis (Hsiung San-pa 熊三拔).

Chigu chŏnyo 地球典要
(A handbook of world geography), 1851. By Ch'oe Han-gi 崔漢綺.

Chih fang wai chi 職方外紀
(On world geography), 1623. By Giulio Aleni (Ai Ju-lueh 艾儒略).

Ch'ilchŏngsan naep'yŏn 七政算內篇
(Main part [literally, "inner chapters"; based on the *Shou-shih li* and *Ta-t'ung li*] of the calculation of the motions of the Seven Governors), 1442 (also in the *Sejong sillok*). Officially compiled by Chŏng Hŭm-ji 鄭欽之, Chŏng Ch'o 鄭招, et al.

Ch'ilchŏngsan oep'yŏn 七政算外篇
(Extra part [literally, "outer chapters"; based on the *Hui-hui li* or Islamic tables] of the calculation of the motions of the Seven Governors), 1442 (also in the *Sejong sillok*). Officially compiled by Yi Sun-ji 李純之, Kim Tam 金淡, et al.

Chin shu 晉書
(History of the Chin Dynasty [A.D. 265–419]), 635. Compiled by Fang Hsuan-ling 房玄齡.

Chingbirok 懲毖錄
(Record of difficulties), 1592–1598. By Yu Sŏng-yong 柳成龍.

Chiu T'ang shu 舊唐書
(Old history of the T'ang Dynasty [618–906]), 945. Compiled by Liu Hsu 劉昫 et al.

Ch'ŏnmun yuch'o 天文類抄
(Selected and classified writings on astrology), fifteenth century. Edited by Yi Sun-ji 李純之.

Chŏnsang yŏlch'a punyajido 天象列次分野之圖
(Chart of the constellations and the regions they govern), 1396. Officially compiled by Kwŏn Kŭn 權近 et al.

Chosŏn wangjo sillok 朝鮮王朝實錄
See *Yijo sillok*.

Chou i ts'an t'ung chi 周易參同契
See *Ts'an t'ung chi*.

Ch'ou-jen chuan 疇人傳
(Biographies of Chinese mathematicians and astronomers), 1799. Compiled by Juan Yuan 阮元.

Ch'ubo sokhae 推步續解
(Analysis of celestial motions), supplement 1862. By Nam Pyŏng-ch'ŏl 南秉哲.

Chuhae suyong 籌解需用
(Practical computation), ca. 1780 (modern reprint in *Tamhŏnsŏ*, Seoul, 1939). By Hong Tae-yong 洪大容.

Chŭngbo munhŏn pigo 增補文獻備考
(Comprehensive study of civilization, revised and expanded edition). See also *Tongguk Munhŏn pigo*. 1908. Officially compiled (modern reprint ed., Kosŏ Kanhaenghoe 古書刊行會, 1959)

Chŭngbo sallim kyŏngje 增補山林經濟
(The agricultural handbook, revised and expanded), eighteenth century (modern reprint ed., Chosŏn Kwangmunhoe 朝鮮光文會, 1935). Edited by Hong Man-sŏn 洪萬選.

Chunggyŏngji 中京志
(Gazetteer of the middle capital [Kaesong]), 1824 (modern reprint, Chosŏn Kwangmunhoe, 1911). Compiled by Kim I-chae 金履載.

Ch'ung-chen li shu 崇禎曆書
(Treatises on [astronomy and] calendrical science of the Ch'ung-chen reign period—Jesuit astronomical writings published separately, 1629–1634) Ed. Johann Adam Schall von Bell.

Chungsŏng sinp'yo 中星新表
(New table of meridian transits), 1853. Compiled by Nam Pyŏng-gil 南秉吉, corrected by Yi Chun-yang 李俊養.

Chungsu taemyŏngyŏk 重修大明曆
(Ta Ming calendrical treatise, revised), ca. 1450. Officially compiled by the Sŏun'gwan 書雲觀.

Fang yü sheng lan 方輿勝覽
(Conspectus of geography [Sung geography]), 1240. Compiled by Chu Mu 祝穆.

Haedong chegukki 海東諸國紀
(Geographical treatise on the countries in the East [Japan and Okinawa]), 1471 (modern reprint, Chōsen Sōtokufu 朝鮮總督府, 1933). By Sin Suk-chu 申叔舟.

Haedong yŏksa 海東繹史
(Historical treatises on Korea), 1776–1800 (modern reprint, Chosŏn Kosŏ Kanhaenghoe 朝鮮古書刊行會, 1935). By Han Ch'i-yun 韓致奫.

Han'gyŏng chiryak 漢京識略
(Notes on the capital city [Seoul]), 1800–1834, MS (modern reprint, Sŏul T'ŭkpyŏlsisa P'yŏnch'an Wiwŏnhoe 서울特別市史編纂委員會, 1957).

Honch'ŏn chŏndo 渾天全圖
(Complete map of the celestial sphere), eighteenth century.

Hsi-yang hsin-fa li shu 西洋新法曆書
(Treatises on calendrical science according to new Western methods [revised version of *Ch'ung-chen li shu*]), printed by imperial order, 1645.

Hsin i hsiang fa yao 新儀象法要
(New description of an armillary clock), 1094. By Su Sung 蘇頌.

Hsin T'ang shu 新唐書
(New history of the T'ang Dynasty [618–906]), 1060. Compiled by Ou-yang Hsiu 歐陽修, Sung Ch'i 宋祁, et al.

Hsu Po wu chih 續博物志
(Supplement to the Record of the investigation of phenomena), twelfth century. By Li Shih 李石.

Hsuan-ho feng shih Kao-li t'u ching 宣和奉使高麗圖經
(Illustrated record of an embassy to Korea in the Hsuan-ho Reign Period), 1124. By Hsu Ching 徐兢.

Huang tao tsung hsing t'u 黃道總星圖
(Star maps arranged according to ecliptic coordinates [celestial latitudes and longitudes]), 1746. By Ignatius Kögler (Tai Chin-hsien 戴進賢).

Hun kai t'ung hsien t'u shuo 渾蓋通憲圖說
(On plotting the coordinates of the celestial sphere and vault), 1607. By Li Chih-tsao 李之藻.

"Hwagitogam ŭigwe 火器都監儀軌"
(Record of the Bureau of Firearms), 1615, MS preserved in Seoul National University Library. Officially compiled by the Hwagitogam 火器都監.

Hwangdo chŏnsŏngdo 黃道全星圖
(General map of the stars in ecliptic coordinates), 1742. Preserved in Pŏpchu Temple, Southern Korea. Officially edited by the Kwansanggam 觀象監 (based on the MS by Kim T'ae-sŏ and An Kuk-pin brought from Ch'ing China).

Hwangdo nambuk hangsŏngdo 黃道南北恒星圖
(Star atlas of both southern and northern hemispheres in ecliptic coordinates), 1834. Published by Kim Chŏng-ho 金正浩.

Hwasŏng sŏngyŏk ŭigwe 華城城役儀軌
(Records and machines of the Emergency Capital Construction Service), 1800 (modern photo reprint, Suwŏn Kojŏk Pojŏnhoe, 1967). Compiled by Chŏng Yag-yong 丁若鏞.

Hyangyak chipsŏngbang 鄉藥集成方
(Great collection of native Korean prescriptions), 1433 (several modern reprint editions have been published). Compiled by Yu Hyo-t'ong 俞孝通 and No Chung-ye 盧重禮.

I hsiang k'ao ch'eng 儀象考成
(The imperial astronomical instruments), 1744; enlarged 1757. By Ignatius Kögler (Tai Chin-hsien 戴進賢) et al.

Imwŏn simyukchi 林園十六志
(Sixteen treatises on science and technology), nineteenth-century MS (modern photo reprint, Seoul National University Library, 1965–1968). Edited by Sŏ Yu-gu 徐有榘.

"Kaksŏn tobon 各船圖本"
(Illustration of various ships), Yi dynasty. MS preserved in Seoul National University Library.

K'ao kung chi 考工記
(The artificers' record [a section of the *Chou li* 周禮]), ca. second century B.C. Compiler unknown.

"Kiuje tŭngnok 祈雨祭謄錄"
(Complete records of the prayer service for rain), 1636–1889. MS preserved in Seoul National University Library. Officially compiled by Board of Rites.

Ko chih ching yuan 格致鏡原
(Mirror of scientific and technological origins), 1735. Edited by Ch'en Yuan-lung 陳元龍.

Koryŏ togyŏng (Kao-li t'u ching)
See *Hsuan-ho feng shih Kao-li t'u ching.*

Koryŏsa 高麗史
(History of the Koryŏ Dynasty), 1451 (modern reprints, Kokusho Kankōkai, 1908, and Yonsei University Press, 1955). Officially compiled by Chŏng In-ji 鄭麟趾 et al.

Koryŏsa chŏlyo 高麗史節要
(Abridged history of the Koryŏ Dynasty), 1452 (modern photo reprint, Kojŏn Kanhaenghoe 古典刊行會, 1959). Officially compiled by Kim Chong-sŏ 金宗瑞 et al.

Kosa ch'walyo 攷事撮要
(Concise source book on ancient matters), seventeenth century (modern photo reprint, Kosŏ Kanhaenghoe, 1933). Edited by Ŏ Suk-kwŏn 魚叔權.

Kosa sinsŏ 攷事新書
(New source book on ancient matters), 1771. Edited by Sŏ Myŏng-ung 徐命膺.

Kuang yü t'u 廣輿圖
(Enlarged terrestial atlas), 1320. By Chu Ssu-pen 朱思本. Revised and printed by Lo Hung-hsien 羅洪先, ca. 1555.

Ku-chin t'u-shu chi-ch'eng, Ch'in-ting 欽定古今圖書集成
(Imperial encyclopedia), 1726. Ed. Ch'en Meng-lei 陳夢雷 et al.

Kukcho ore sŏrye 國朝五禮序例
(Introductory remarks on national rituals), 1474. Officially compiled by Board of Rites.

Kukcho pogam 國朝寶鑑
(Precious mirror of the Yi Dynasty), 1730, revised and enlarged in 1908. Officially compiled by the Yi government.

Kukcho yŏksanggo 國朝曆象考
(Compendium of calendrical science and astronomy in the Yi Dynasty), 1795. By Sŏng Chu-dŏk 成周悳.

Kŭmyang chamnok 衿陽雜錄
(Miscellaneous records of farming in the Kŭmch'ŏn district [Kyŏnggi Province]), ca. 1460. By Kang Hi-maeng 姜希孟.

Kunggwŏlchi 宮闕志
(Records on palaces), MS (modern photo reprint, Committee on the History of Seoul, 1959).

K'uo ti chih 括地志
(Comprehensive geography), seventh century. By Wei Wang-t'ai 魏王泰. (Reconstituted from fragments by Sun Hsing-yen in 1797).

Kyŏngguk taejŏn 經國大典
(National code), completed in 1461. Officially compiled.

Kyŏngsang-do chiriji 慶尙道地理志
(Treatise on the geography of Kyŏngsang Province), 1425. MS preserved in Seoul National University Library. (Modern reprint, Chōsen Sōtokufu, 1932). Compiled by Ha Yŏn 河演, Governor of Kyŏngsang Province.

Kyŏngsang-do sokch'an chiriji 慶尙道續撰地理志
(Revised treatise on the geography of Kyŏngsang Province), 1469. MS preserved in Seoul National University Library. (Modern reprint, Chōsen Sōtokufu, 1932). Officially compiled by Kyŏngsang Province.

Kyosik ch'ubopŏp 交食推步法
(Method of calculation of eclipses), 1458. By Yi Sun-ji 李純之 and Kim Sŏk-che 金石悌.

Li hsiang k'ao-ch'eng 曆象考成
(Compendium of calendrical science and astronomy), printed 1724. Ed. Mei Ku-ch'eng 梅瑴成 and Ho Kuo-tsung 何國宗.

Li hsiang k'ao-ch'eng hou-pien 曆象考成後編
(Sequel to the compendium of calendrical science and astronomy), 1742. Ed. Ignatius Kögler et al.

Ling-t'ai i hsiang chih 靈臺儀象志
(On the astronomical instruments in the Imperial Observatory), 1674. By Ferdinand Verbiest and other missionaries.

Man'gi yoram 萬機要覽
(Handbook of economics and military affairs [in the late Yi Dynasty]), 1800–1834 (modern reprint, Chōsen Sōtokufu, 1937). Edited by Sim Sang-kyu 沈象奎.

Mongmin simsŏ 牧民心書
(Maxims for government officials on caring for the people), 1800–1834. By Chŏng Yag-yong 丁若鏞.

Mulmyŏng yugo 物名類考
(A collection of the names of phenomena classified [according to their properties]), MS, 1824 (modern photo reprint, *Chōsen gakuhō*, nos. 16–20). Ed. Yu Hŭi 柳僖.

Munhŏn pigo 文獻備考
Comprehensive study of civilization. See *Chŭngbo munhŏn pigo*.

Myŏngnamnu munjip 明南樓文集
(Collected works of Ch'oe Han-gi), ca. 1860. By Ch'oe Han-gi 崔漢綺.

Nihon shoki 日本書紀
(Chronicle of Japan [to 697]), compiled A.D. 720. (In *Kokushi taikei* 國史大系, vol. 1, Tokyo, 1900).

Nongsa chiksŏl 農事直說
(Theories and practice of farming), 1429. Compiled by Chŏng Ch'o 鄭招 and Pyŏn Hyo-mun 卞孝文.

Nongga chipsŏng 農家集成
(Collected treatises on farming), 1665. By Sin Suk 申洬.

Nulchaejip 訥齋集
(Collected writings of Nulchae = Yang Sŏng-ji), fifteenth century (first published in 1791, reprinted in 1958). By Yang Sŏng-ji 梁誠之.

Nusu t'ongŭi 漏籌通義
(Manual for operation of the clepsydra), 1754. By An Kuk-pin 安國賓.

Oju sŏjong pangmul kobyŏn 五洲書種博物攷辨
(Oju's book on the investigation of phenomena), 1834–1849 (modern photo reprint in the *Oju yŏnmun changjŏn san'go*, vol. 2, Kosŏ Kanhaenghoe, Seoul, 1959). By Yi Kyu-kyŏng 李圭景.

Oju yŏnmun changjŏn san'go 五洲衍文長箋散考
(Collected works of Oju = Yi Kyu-kyŏng), 1834–1849 (modern photo reprint, Kosŏ Kanhaeghoe, Seoul, 1959). By Yi Kyu-kyŏng 李圭景.

Pen-ts'ao kang mu 本草綱目
(The great pharmacopoeia), printed 1596 (supplement published 1871). By Li Shih-chen 李時珍.

Po wu chih 博物志
(Record of the investigation of phenomena), A.D. 290. By Chang Hua 張華.

Pu t'ien ko (Korean *Po ch'ŏn ka*) 步天歌
(Song of the sky pacers), sixth century. By Wang Hsi-ming 王希明

Samguk sagi 三國史記
(History of the Three Kingdoms), 1145 (reprinted in 1394, 1512). By Kim Pu-sik 金富軾.

Samguk yusa 三國遺事
(Memorabilia of the Three Kingdoms), 1274–1308. By Iryŏn 一然 (monk).

San ts'ai t'u hui 三才圖會
(Universal pictorial encyclopedia), 1609. By Wang Ch'i 王圻.

Sasi ch'anyo 四時纂要
(A synopsis of agriculture in the four seasons), 1424–1483. By Kang Himaeng 姜希孟.

"Sech'o yuhwi 細草類彙"
(Manual of calculations of the Shih-hsien calendar), 1710. MS preserved in Seoul National University Library. By Hŏ Wŏn 許遠.

Sejong sillok Chiriji 世宗實錄地理志
(Treatise on geography of the Veritable records of the King Sejong era), 1422, revised 1454 (modern photo reprint, Kuksa P'yŏnch'an Wiwŏnhoe 國史編纂委員會, Seoul, 1957). Compiled by Yun Hoe 尹淮 et al.

Shou-shih li 授時曆
(The Shou-shih calendrical treatise [in *Yuan shih*, history of the Yuan Dynasty]), 1280. By Kuo Shou-ching 郭守敬.

Shu li ching yun 數理精蘊
(Treasury of mathematics), 1723. Ed. Mei Ku-ch'eng 梅瑴成 and Ho Kuotsung 何國宗.

Sihŏn kiyo 時憲紀要
(Principles of cosmography), 1860. By Nam Pyŏng-gil 南秉吉.

Sin'gi pikyŏl 神器秘訣
(Secrets of marvelous weapons), 1603. By Han Hyo-sun 韓孝純.

Sinjŏn chach'obang 新傳煑硝方
(New preparation of saltpeter), 1698 (first printed 1796). By Kim Chi-nam 金指南.

Sinjŏn chach'ui yŏnch'obang 新傳煮取焰硝方
(New preparation of saltpeter), 1635. By Yi Sŏ 李曙.

Sinjŭng Tongguk yŏji sŭngnam 新增東國輿地勝覽
See *Tongguk yŏji sŭngnam*.

Sinpŏp nusu t'ongŭi 新法漏籌通義
(Manual for operation of the clepsydra by a new system), eighteenth century. Edited by Kim Yŏng 金泳.

Sinpŏp po ch'ŏn ka 新法步天歌
(Song of the sky pacers, adapted to the new methods), 1862. Edited by Yi Chun-yang 李俊養.

Sojaejip 疎齊集
(Collected works of Sojae = Yi Yi-myŏng), eighteenth century. By Yi Yi-myŏng 李頤命.

Sŏnggyŏng 星鏡
(Mirror of stars), 1861. By Nam Pyŏng-gil 南秉吉.

Sŏngho sasŏl 星湖僿說
(Sŏngho's detailed discourses), 1724–1776. By Yi Ik 李瀷.

Sŏngho sasŏl yusŏn 星湖僿說類選
(Sŏngho's detailed discourses, classified), 1712–1791. Edited by An Chŏng-bok 安鼎福.

Sŏun'gwanji 書雲觀志
(Treatise on the Bureau of Astronomy), 1818. By Sŏng Chu-dŏk 成周悳.

Suan fa t'ung tsung 算法統宗
(Authoritative treatise on arithmetic), 1593. By Ch'eng Ta-wei 程大位.

Suan hsueh ch'i meng 算學啓蒙
(Introduction to mathematical studies), 1299. By Chu Shih-chieh 朱世傑.

Sŭngjŏngwŏn Ilgi 承政院日記
(Diary of the Royal Secretariat), 1623–1894 (modern photo reprint, Kuksa P'yŏnch'an Wiwŏnhoe, 1961–). Officially compiled by the Royal Secretariat.

Susiryŏk ch'ŏppŏp ipsŏng 授時曆捷法立成
(A ready reckoner for Shou-shih Calendar calculations), 1343 (reprinted in the fifteenth century). By Kang Po 姜保.

Ta Ming i t'ung chih 大明一統志
(Comprehensive geography of the Ming Empire), 1450–1461. Edited by Li Hsien 李賢.

Ta T'ang hsi yü chi 大唐西域記
(Record of Western countries in the time of the T'ang), 646. By Hsuan-chuang (monk) 玄奘.

Ta Yuan i t'ung chih 大元一統志
(Comprehensive geography of the Yuan Empire), presented to the Throne 1303. Ed. Po Lan-shih 孛蘭肹 et al.

Taedong chiji 大東地志
(Treatise on Korean geography), 1864. By Kim Chŏng-ho 金正浩.

Taedong sangwigo 大東象緯考
(History and observational records of astrology and meteorology in the Great Eastern Kingdom [Korea]), 1708. Officially compiled.

Taedong unbu kunok 大東韻府群玉
(A dictionary of arts and sciences of the Yi Dynasty), ca. 1590. Compiled by Kwŏn Mun-hae 權文海.

Taedong yŏjido 大東輿地圖
(Map of the Great Eastern Kingdom, Korea), 1861 (modern photo reprints in 1936 and 1965). By Kim Chŏng-ho 金正浩.

Taejŏn hoet'ong 大典會通
(Collection of national codes), 1865. Officially compiled by the Yi government.

Taengniji 擇里誌
(Notes on choosing a domicile), early eighteenth century (vernacular translation by No To-yang, Seoul, 1968). By Yi Chung-hwan 李重煥.

Tamhŏnsŏ 湛軒書
(Collected writings of Tamhŏn = Hong Tae-yong), eighteenth century (modern reprint, Seoul, 1939). By Hong Tae-yong 洪大容.

T'ang shu 唐書
(History of the T'ang Dynasty), See *Chiu T'ang shu, Hsin T'ang shu.*

T'ien kung k'ai wu 天工開物
(The exploitation of the works of nature), 1637. By Sung Ying-hsing 宋應星.

T'ien wen lüeh 天問略
(Explication of the Celestial Sphere), 1615. By Immanuel Diaz (Yang Mano 陽瑪諾).

T'ien wen ta ch'eng kuan k'uei chi yao 天文大成管窺輯要
(Compendium of astrology: essentials of observation), 1653. By Huang Ting 黃鼎.

Tongguk chiriji 東國地理誌
(Geography of the Eastern Kingdom [Korea]), 1615. By Han Paek-kyŏm 韓百謙.

Tongguk munhŏn pigo 東國文獻備考
(Handbook of the history of Eastern Kingdom civilization), 1770. Officially compiled by Hong Pong-han 洪鳳漢 et al.

Tongguk yŏji pigo 東國輿地備考
(A note on Korean geography), 1865–1883, MS (modern reprint, Sŏul Sisa P'yŏnch'an Wiwŏnhoe, 1963).

Tongguk yŏji sŭngnam 東國輿地勝覽
(Geographical conspectus of the Eastern Kingdom [Korea]), 1530 (completed 1481 by No Sa-sin 盧思愼 et al.; modern photo reprint, Kosŏ Kanhaenghoe, 1958). Officially compiled by Yi Haeng 李荇 et al.

Tonggyŏng chapki 東京雜記
(Miscellaneous records of the Eastern Capital [Kyŏngju]), 1669 (modern reprint ed., Chosŏn Kwangmunhoe, 1913). Edited by Min Chu-myŏn 閔周冕.

Tongmunsŏn 東文選
(Selections from Korean literature), 1478 (modern photo reprint, Kyŏnghŭi Publishing Co., 1967). Compiled by Sŏ Kŏ-chŏng 徐居正.

Tongŭi pogam 東醫寶鑑
(Precious mirror of Eastern medicine), 1613. Edited by Hŏ Chun 許俊.

Torogo 道路考
(Study of Korean roads), 1770. By Sin Kyŏng-jun 申景濬.

Ts'an t'ung ch'i 參同契
(The concordance of the three, an apocryphal tradition of interpretation of the Book of Changes), Later Han, traditional date 142. Attributed to Wei Po-yang 魏伯陽.

T'u shu pien 圖書編
(Bibliographical encyclopedia), 1562, 1577, 1585. Edited by Chang Huang 章潢.

Ŭibang yuch'ui 醫方類聚
(Classified collection of medical prescriptions), 1445 (first printed 1477). Compiled by No Chung-ye 盧重禮 et al.

Ŭigi chipsŏl 儀器輯說
(Collected writings on astronomical instruments), 1860. By Nam Pyŏng-ch'ŏl 南秉哲.

Wu ching tsung yao 武經總要
(Collection of the most important military techniques), 1040–1044. Edited by Tsêng Kung-liang 曾公亮.

Wu li hsiao chih 物理小識
(Notes on the principles of things), 1664. By Fang I-chih 方以智.

Wu pei chih 武備志
(Treatise on armaments), 1628. Edited by Mao Yuan-i 茅元儀.

Wu wei li chih 五緯曆指
(On the principles of the planetary motions), 1634. By James Rho (Lo Ya-ku 羅雅谷).

Yangch'onjip 陽村集
(Collected works of Yangch'on = Kwŏn Kŭn), fourteenth century (first printed 1674). By Kwŏn Kŭn 權近.

Yangdoŭi tosŏl 量度儀圖說
(Illustrated description of the angle measurement instrument), 1885. By Nam Sang-gil 南相吉.

Yi Ch'ungmugong chŏnsŏ 李忠武公全書
(Collected works of Admiral Yi Sun-sin), sixteenth century (first printed 1795). By Yi Sun-sin 李舜臣, officially compiled by the order of King Chŏngjo.

Yijo sillok 李朝實錄
(Veritable records of the Yi Dynasty), 1413–1865 (modern photo reprints, Korea and Japan, 1950s). Compiled by Veritable Records Office, Yi Dynasty.

Yŏamjip 旅菴集
(Complete works of Yŏam = Sin Kyŏng-jun), 1712–1781. By Sin Kyŏng-jun 申景濬.

Yŏjae ch'walyo 輿載撮要
(Essentials of geography), 1894. By O Hong-muk 吳宏默.

"Yŏktae yosŏngnok 歷代妖星錄"
(Record of stars of ill omen in successive ages), seventeenth century, MS preserved in Seoul National University Library. Edited by Kim Ing-nyŏm 金益廉.

Yŏllyŏsil kisul 燃藜室記述
(Narrative of Yŏllyŏsil = Yi Kŭng-ik), eighteenth century (modern reprint, Chosŏn Kwangmunhoe, 1913). By Yi Kŭng-ik 李肯翊.

Yŏnamjip 燕巖集
(Collected writings of Yŏnam = Pak Chi-wŏn), 1770 onward, published 1901. By Pak Chi-wŏn 朴趾源.

Yongjae ch'onghwa 慵齋叢話
(Collected essays of Yongjae = Sŏng Hyŏn), 1439–1504 (modern translation, Korea University Press, 1964). By Sŏng Hyŏn 成俔.

Yŏrha ilgi 熱河日記
(Diary of a mission to Jehol [Korean embassy to congratulate the Ch'ien-lung Emperor on his birthday]), in *Yŏnamjip*.

Yü ti chi sheng 輿地紀勝
(Record of world geography), 1221. By Wang Hsiang-chih 王象之.

Yuan shih 元史
(History of the Yuan Dynasty), ca. 1370.

Yuk hae pŏp 陸海法
(Survey methods on land and sea), ca. 1860. By Ch'oe Han-gi 崔漢綺.

Yungwŏn p'ilbi 戎垣必備
(Manual for a military commander), 1815. Edited by Pak Chong-kyŏng 朴宗慶 et al.

Yuwŏn ch'ongbo 類苑叢寶
(Source book of ancient matters), 1643. Edited by Kim Yuk 金堉.

Korean, Japanese, and Chinese Secondary Sources

Akioka Takejirō 秋岡武次郎
Nihon chizushi 日本地圖史 (A history of Japanese cartography). Tokyo, 1955.

Anak chesamhobun palgul pogo 안악 제 3 호분 발굴 보고 (Report on the excavation of tomb no. 3 in Anak), ed. To Yu-ho 都宥浩 et al. P'yŏngyang, 1958.

Aoyama Sadao 青山定雄
"Gendai no chizu ni tsuite 元代の地圖について" (On maps of the Yuan Dynasty), *Tōhō gakuhō* 東方學報 (Tokyo), 1938, *8*: 103–152.

Aoyama Sadao 青山定雄
"Kochishi chizu nado no chōsa 古地誌地圖等の調査" (In search of old geographical books and maps), *Tōhō gakuhō* (Tokyo), 1935, *5* (suppl. vol.): 123.

Aoyama Sadao 青山定雄
"Mindai no chizu ni tsuite 明代の地圖について" (On maps of the Ming Dynasty), *Rekishigaku kenkyū* 歴史學研究, 1963, *7*.11: 279–294.

Aoyama Sadao 青山定雄
Richō ni okeru nisan no Chōsen zenzu ni tsuite 李朝に於ける二三の朝鮮全圖について" (About some general maps of Korea made during the Yi Dynasty), *Tōhō gakuhō* (Tokyo), 1939, *9*: 143–172.

Aoyama Sadao 青山定雄
"Sōdai no chizu to sono tokushoku 宋代の地圖とその特色" (The maps of the Sung Dynasty and their characteristics), *Tōhō gakuhō* (Tokyo), 1940, *11*: 415–458.

Aoyama Sadao 青山定雄
Tō Sō jidai no kōtsū to chishi chizu no kenkyū 唐宋時代の交通と地誌地圖の研究 (Study of the communication systems of T'ang and Sung China and the development of their gazetteers and maps), Tokyo, 1963.

Arii Tomonori 有井智德
"Richō shoki no yōeki 李朝初期の徭役" (Statutory labor in the early period of the Yi Dynasty, parts 1 and 2), *Chōsen gakuhō* 朝鮮學報, 1964, *30*: 62–106; 1964, *31*: 58–101.

Arima Seiho 有馬成甫
Chōseneki suigunshi 朝鮮役水軍史 (A history of naval forces during the Hideyoshi invasion). Tokyo, 1942.

Arima Seiho 有馬成甫
Kahō no kigen to sono denryū 火砲の起原とその傳流 (The origin of firearms and their early transmission). Tokyo, 1962.

Arimitsu Kyōichi 有光教一
Chōsen kushimemon doki no kenkyū 朝鮮櫛目文土器の研究 (The Kushimemon [comb-pattern] pottery of Korea). Kyoto, 1962.

Arimitsu Kyōichi 有光教一
"Chōsen shoki kinzoku bunka ni kansuru shin shiryō no shōkai to kōsatsu 朝鮮初期金屬文化に關する新資料の紹介と考察" (An introduction to and study of new material on earlier Korean metal culture), *Shirin* 史林, 1965, *48*.2: 120–132.

Asakawa Takumi 淺川巧
Chōsen tojimei kō 朝鮮陶磁名考 (A study of the nomenclature of Korean ceramics). Tokyo, 1931.

Ayuzawa Shintarō 鮎澤信太郎
Chirigakushi no kenkyū 地理學史の研究 (Studies in the history of geography). Tokyo, 1948.

Ayuzawa Shintarō 鮎澤信太郎
"Mateo Ritchi no *Ryōgigenranzu* ni tsuite マテオリッチの兩儀玄覽圖について" (Matteo Ricci's world map published in 1603), *Chirigakushi kenkyū* 地理學史研究 (Researches in the history of geography), 1957, *1* (special number for early cartography): 1–21.

Bukkoku-ji to Sekkutsu-an 佛國寺と石窟庵
(Pulguk Temple and Sŏkkuram Grotto). Seoul, 1938.

Chang Hsiu-min 張秀民
Chung-kuo yin-shua-shu ti fa-ming chi ch'i ying-hsiang 中國印刷術的發明及其影響 (The invention of printing in China and its influence). Peking, 1958.

Chikashige Masumi 近重眞澄
"Tōyō kodōki no kagakuteki kenkyū 東洋古銅器の化學的研究" (Chemical analysis of ancient Oriental bronze wares), *Shirin*, 1919, *3*.2: 24–25.

Cho Chi-hun 趙芝薰
Han'guk munhwasa sŏsŏl 韓國文化史序說 (Introduction to Korean cultural history). Seoul, 1965.

Cho Ki-jun 趙璣濬
Han'guk kyŏngjesa 韓國經濟史 (A history of Korean economics). Seoul, 1962.

Ch'oe Ho-jin 崔虎鎭
Han'guk kyŏngjesa kaeron 韓國經濟史概論 (Outline history of Korean economics). Seoul, 1962.

Ch'oe Nam-sŏn 崔南善
Urinara yŏksa 우리 나라 역사 (A Korean history). Seoul, 1952.

Ch'oe Sang-jun 최상준
"Urinara wŏnsi sidae mit kodaeŭi soebuch'i yumul punsŏk 우리 나라 원시 시대 및 고대의 쇠붙이 유물분석" (Report on the chemical analysis of ancient Korean bronze and iron wares), *Kogo minsok* 고고민속, 1966, *3*: 43–46.

Ch'oe Sang-su 崔常壽
Han'guk chiyŏnŭi yŏn'gu 韓國紙鳶의 研究 (A study of Korean kites), with English summary. Seoul, 1958.

Ch'oe Sŏng-nam 崔碩南
Han'guk sugunsa yŏn'gu 韓國水軍史研究 (Study of the history of Korean naval forces to 1910). Seoul, 1964.

Ch'oe Yŏng-hi 崔永禧
"Kusŏn ko 龜船考" (A study of the turtle ship), *Sach'ong* 史叢 1958, *3*: 3–20.

Ch'ŏn Kwan-u 千寬宇
"Hong Tae-yongŭi chijŏnsŏlŭi chae-gŏmt'o 洪大容의 地轉說의 再檢討" (Reexamination of the rotating-earth theory of Hong Tae-yong), in *Hyosŏng Cho Myŏng-gi paksa hwagap kinyŏm pulgyo sahak nonch'ong* 曉城趙明基博士華甲記念佛敎史學論叢 (Commemorative studies on Buddhism and history for Dr. Cho Myong-gi's sixtieth birthday). Seoul, 1965.

Chŏn Sang-un. See Jeon Sang-woon.

Chŏng Ch'an-yŏng 정찬영
"Urinara kutulŭi yuraewa palchŏn 우리 나라 구들의 유래와 발전" (Origin and development of the Korean heated floor), *Kogo minsok* 1966, 4: 15–24.

Chŏngch'i kyŏngje sa 政治經濟史
(A history of Korean politics and economics), in *Han'guk munhwasa taege* 韓國文化史大系 (Series on Korean cultural history), ed. Minjok Munhwa Yŏn'guso 民族文化研究所 (Korean Classical Research Institute, Koryŏ University). Seoul, 1965.

Ch'ŏngdong yumul torok 靑銅遺物圖錄
(Selected bronze objects of the early metal period in Korea, 1945–1968), National Museum of Korea, Research Materials, vol. 1. Seoul, 1968.

Chŏng In-bo 鄭寅普
Tamwŏn kukhak san'go 薝園國學散藁 (Notes and essays on Korean studies of Tamwŏn = Chŏng In-bo). Seoul, 1955.

Chŏng Paek-un 鄭白雲
"Chōsen ni okeru tekki sivō no kaishi ni tsuite 朝鮮に於ける鐵器使用の開始について" (On the first use of iron ware in ancient Korea), *Chōsen gakuhō*, 1957, *17*: 171–180.

Chŏng Paek-un 鄭白雲 and To Yu-ho 都宥浩
Najin Ch'odo wŏnsi yujŏk palgul pogo 나진 초도 원시유적 발굴보고 (Report on the excavation of primitive ruins in Ch'odo Island, Najin, North Korea). P'yŏngyang, 1956.

Chōsen bunkashi 朝鮮文化史
(Cultural history of Korea), Nihon Chōsen Kenkyūsho 日本朝鮮研究所 (Research Institute for Korea in Japan), from Korean edition published in P'yŏngyang. Tokyo, 1966.

Chōsen jinmei jisho 朝鮮人名辭書
(Cyclopedia of Korean biography), ed. Chōsen Sōtokufu 朝鮮總督府. Seoul, 1937, 1939.

Chōsen koseki chōsa hōkoku 朝鮮古蹟調査報告
(Report of the survey of antiquities of Korea), ed. Chōsen Sōtokufu. Seoul, 1916–1938.

Chōsen koseki zufu 朝鮮古蹟圖譜
(Album of Korean antiquities), ed. Chōsen Sōtokufu, vols. 1–15. Seoul, 1915–1935.

Chōsenshi 朝鮮史
(Chronological history of Korea), ed. Chōsen Sōtokufu, 38 vols. Seoul, 1932–1938.

Chōsenshi taikei 朝鮮史大系
(History of Korea). Chōsenshi Gakkai (Society for Korean History), 5 vols. Seoul, 1927.

Chōsen tosho kaidai 朝鮮圖書解題
(Korean bibliography, annotated), ed. Chōsen Sōtokufu. Seoul, 1915 (revised 1919).

Chosŏn kojŏn haeje (1) 조선 고전 해제
(Remarks on Korean bibliography), vol. 1, ed. Kojŏn Yŏn'guso 고전연구소 (Research Institute of Korean Classics). P'yŏngyang, 1965.

Chosŏn sach'al saryo 朝鮮寺刹史料
(Historical documents of the Korean temples). Seoul, 1911.

Chu Nam-ch'ŏl 朱南哲
"Yijo sidae Sŏulŭi chut'aek 李朝時代 서울의 住宅" (Dwelling houses in Seoul during the Yi Dynasty), *Hyangt'o Sŏul* 鄕土서울, 1965, *25*:85.

Dōno Tsurumatsu 道野鶴松
"Kagakujō yori mitaru kodai Shina no kinzoku to kinzoku bunka ni tsuite 化學上より見たる古代支那の金屬と金屬文化に就いて" (On metallurgy in ancient China seen through chemical analysis), *Tōhō gakuhō* (Tokyo), 1933, *4*: 1–63.

Funakoshi Akio 船越昭生
"Sakoku Nihon ni kita 'kōhizu'—waga kuni kindai chirigaku no zenku— 鎖國日本にきた「康熙圖」—わが國近代地理學の前驅" (The K'ang-hsi Atlas brought to Japan in the period of isolation, a forerunner of modern geography in Japan), *Tōhō gakuhō* (Kyoto), 1967, *38*: 1–132.

Gyosen chōsa hōkoku 漁船調査報告
(A survey report on Korean fishing boats), ed. Chōsen Sōtokufu Marine Laboratory. Seoul, 1924.

Hamaguchi Yoshimitsu 浜口良光
Chōsen no kōgei 朝鮮の工藝 (Korean crafts). Tokyo, 1966.

Han'guk inmyŏng taesajŏn 韓國人名大事典
(Cyclopedia of Korean biography), ed. Sin'gu Munhwasa Co. Seoul, 1967.

Han'guk kunje sa 韓國軍制史
(History of military institutions in the early Yi Dynasty), ed. Military Academy of Korea, Seoul, 1968.

Han'guk munhwasa taege 韓國文化史大系
(Series on Korean Cultural History) vols. 1–7. Minjok Munhwa Yon'guso 民族文化研究所, Seoul, 1968–1972.

Han'guksa 韓國史
(A history of Korea), ed. Chindan Hakhoe 震檀學會 (Kim Che-wŏn, Yi Pyŏng-do, Yi Sang-baek, and Yi Sŏn-gŭn), 7 vols. Seoul, 1959–1965.

Han Sŏng-guk 韓星國
"Kaech'ŏn ko 開川考" (A study on the creeks in Seoul), *Hyangt'o Sŏul*, 1965, *24*: 13.

Hashimoto Manpei 橋本萬平
"Taiyō no shutsunyū ni mirareru Nihon chūsei no jikoku seido 太陽の出入に見られる日本中世の時刻制度" (On the Japanese medieval time system), *Kagakushi kenkyū*, 1955, *56*: 8–12.

Hashimoto Masukichi 橋本増吉
Shina kodai rekihōshi kenkyū 支那古代暦法史研究 (On astronomical chronology in ancient China). Tokyo, 1943.

Hayashi Taisuke 林泰輔
"Chōsen no kappanjutsu 朝鮮の活版術" (On Korean movable-type printing), *Shigaku zasshi* 史學雜誌, 1906, *17*.3: 54–63.

Higuchi Kiyoyuki 樋口清之
Nihon kodai sangyōshi 日本古代産業史 (A history of ancient industry in Japan). Tokyo, 1943.

Hirayama Kiyotsugu 平山清次
"Nihon ni okonawaretaru jikokuhō 日本におこなわれたる時刻法" (The time system of Japan), *Tenmon geppo* 天文月報 1913, *5*: 121–124 and 135–139.

Hŏ Sŏn-do 許善道
"Kach'ŏng ŭlmyomyŏng ch'ŏnja ch'ongt'onge taehayŏ 嘉靖乙卯銘天字銃筒에 對하여" (On the *ch'ŏnja*-type firearm dated 1555), *Misul charyo* 美術資料 (National Museum of Korea art magazine), 1965, *10*: 5–14.

Hŏ Sŏn-do 許善道
"Yijo chunggi hwagiŭi paltal 李朝中期火器의 發達" (The development of firearms in Korea, 1474–1592), *Yŏksa hakpo* 歷史學報, 1966, *30*: 47-107; 1966, *31*: 67-127.

Hŏ Sŏn-do 許善道
"Yŏmal sŏnch'o hwagiŭi chŏllaewa paltal (sang), (chung), (ha) 麗末鮮初火器의 傳來와 發達 上, 中, 下" (The introduction and development of firearms in Korea, 1356–1474, Parts 1, 2, 3), *Yŏksa hakpo*, 1964, *24*: 1–60; 1964, *25*: 39–98; 1965, *26*: 141–165.

Hong I-sŏp 洪以燮
Chŏng Yag-yongŭi chŏngch'i kyŏngje sasang yŏn'gu 丁若鏞의 政治經濟思想研究 (Study of the politico-economic thought of Chŏng Yag-yong). Seoul, 1959.

Hong I-sŏp 洪以燮
Chōsen kagakushi 朝鮮科學史 (A history of Korean science). Tokyo, 1944.

Hong I-sŏp 洪以燮
Chosŏn kwahaksa 朝鮮科學史 (A history of Korean science). Seoul, 1946.

Hong I-sŏp 洪以燮
"Chosŏn kwahaksa sayŏn sugu 朝鮮科學史事緣數齣" (Several phases in the work on *Chosŏn kwahaksa*), in *Ilsan Kim Tu-jong paksa hisu kinyŏm nonmunjip* 一山金斗鐘博士稀壽記念論文集 (Commemoration papers for Dr. Kim Tu-jong's seventieth birthday). Seoul, 1966.

Hong Sa-jun 洪思俊
"Kyŏngju ch'ŏmsŏngdae silch'ŭk chosŏ 慶州瞻星臺實測調書" (Report of the survey of the remains of the observatory in Kyŏngju), *Kogo misul*, 1965, *6*.3–4: 56–60.

Hori Junji 堀淳二
"Kokuri kofun no seishinzu ni tsuite 高句麗古墳の星辰圖について" (On the paintings of the celestial bodies in the ancient tombs of Koguryŏ), *Shiseki* 史蹟, 1954, *42*: 370–391.

Hyŏn Sang-yun 玄相允
"Han'guk sasangsa 韓國思想史" (A history of Korean thought), *Asea yŏn'gu* 亞細亞研究, 1960, *3*.2: 261–312; 1961, *4*.1: 299–355.

Hyŏng Ki-ju 邢基柱
"Han'guk kojidoe kwanhan yŏn'gu charyo 韓國古地圖에 關한 研究資料" (Research materials on ancient Korean maps), *Chirihak* 地理學, 1963, *1*: 97–109.

Iijima Tadao 飯島忠夫
"*Sankoku shiki* no nisshoku kiji ni tsuite 三國史記の日蝕記事について" (On the records of solar eclipses in the *Samguk sagi*), *Tōyō gakuhō* (Tokyo), 1926, *15*.3: 410–424.

Iijima Tadao 飯島忠夫
Shina kodaishi to tenmongaku 支那古代史と天文學 (Ancient Chinese history and astronomy). Tokyo, 1925.

Ikeuchi Hiroshi 池內宏
Chōsen no bunka 朝鮮の文化 (Korean culture), in *Tōyō shichō* 東洋思潮 (Oriental thought) series. Tokyo, 1936.

Ikeuchi Hiroshi 池內宏
T'ung-kou 通溝 (The ancient site of Kao-kou-li [Koguryŏ] in Chi-an district, Northeastern China). Tokyo and Hsin-ching, 1938.

Imamura Tomo 今村鞆
Fune no Chōsen 船の朝鮮 (Maritime Korea). Seoul, 1930.

Imoto Susumu and Hasegawa Ichirō 井本進, 長谷川一郎
"Chūgoku Chōsen oyobi Nihon no ryūseiu kokiroku 中國朝鮮及び日本の流星雨古記錄" (Ancient records of Chinese, Korean, and Japanese meteor showers), *Kagakushi kenkyū*, 1956, *37*: 7–15. English translation, "Historical records of meteor showers in China, Korea and Japan," *Smithsonian Contributions to Astrophysics*, 1958, *2*.6:131–144.

Jeon Sang-woon
"15 segi chŏnban Yijo kwahak kisulsa sŏsŏl 15世紀前半李朝科學技術史序說" (An introduction to the history of science and technology in early fifteenth-century Korea), in *Kim Tu-jong paksa kohi kinyŏm nonmunjip* (Commemoration papers for Dr. Kim Tu-jong's seventieth birthday). Seoul, 1966.

Jeon Sang-woon
"Han'guk ch'ŏnmun kisanghaksa 韓國天文氣象學史" (A history of astronomy and meteorology in Korea), in *Kwahak kisulsa* 科學技術史 (History of science and technology), in the *Han'guk munhwasa taege* series. Seoul, 1968.

Jeon Sang-woon
"Han'guk kwahak kisul chŏngch'aekŭi sachŏk koch'al 韓國科學技術政策의 史的考察" (Historical survey of policy toward science and technology in Korea), unpublished report to the Ministry of Education, Korea. Seoul, 1969.

Jeon Sang-woon
Han'guk kwahak kisulsa 韓國科學技術史 (A history of science and technology in Korea). Seoul, 1966.

Jeon Sang-woon
"*Oju sŏjong pangmul kobyŏn*kwa kŭmsok hwahak 五洲書種博物考辨과 金屬化學" (Oju's book on the investigation of phenomena and metallurgy), *Kwahak segi* 科學世紀, 1965, *9*: 53–55.

Jeon Sang-woon
"Richō jidai ni okeru kōuryō sokuteihō ni tsuite 李朝時代における降雨量測定法について" (On the scientific measurement of precipitation in the Yi Dynasty), *Kagakushi kenkyū*, 1963, *66*: 49–56.

Jeon Sang-woon
"Samguk mit t'ongil Sillaŭi ch'ŏnmun ŭigi 三國 및 統一新羅의 天文儀器" (On astronomical instruments in the Three Kingdoms and Unified Silla periods), *Komunhwa* 古文化, 1964, *3*: 13–23.

Jeon Sang-woon
"Sŏn'gi okhyŏnge taehayŏ 璿璣玉衡에 對하여" (On armillary spheres with clockwork in the Yi Dynasty), *Komunhwa*, 1963, *2*: 2–10. Same article summarized in Japanese, "Senki gyokkō (tenmon tokei) ni tsuite 璇璣玉衡(天文時計)について," in *Kagakushi kenkyū*, 1962, *63*: 137–141.

Jeon Sang-woon
"Sŏun'gwankwa kanŭidae 書雲觀과 簡儀臺" (The Bureau of Astronomy and the observatory in the Yi Dynasty), *Hyangt'o Sŏul*, 1964, 20: 37–51.

Jeon Sang-woon
"Yijo ch'ogiŭi chirihakkwa chido 李朝初期의 地理學과 地圖" (Geography and cartography in the early Yi Dynasty), *Komunhwa*, 1966, *4*: 1–16.

Jeon Sang-woon
"Yissi Chosŏnŭi sige chejak sogo 李氏朝鮮의 時計製作小考" (A study of timekeepers in the Yi Dynasty), *Hyangt'o Sŏul*, 1963, *17*: 49–114.

Kagaku gijutsushi nenpyō 科學技術史年表
(A chronological table of the history of science and technology), ed. Heibonsha 平凡社. Tokyo, 1956.

Kang Man-gil 姜萬吉
"Chosŏn chŏn'gi kongjang ko 朝鮮前期工匠考" (A study of artisans in the first half of the Yi Dynasty), *Sahak yon'gu* 史學研究 (Journal of the study of history), 1961, *12*: 1–72.

Kang Man-gil 姜萬吉
"Yijo chosŏnsa 李朝造船史" (A history of shipbuilding in the Yi Dynasty), in *Kwahak kisulsa*, in the *Han'guk munhwasa taege* series. Seoul, 1968.

Katayama Ryūzō 片山隆三
"Konyō zatsuroku no kenkyū 衿陽雜錄の研究" (A study of the *Kŭmyang chamnok*), *Chōsen gakuhō*, 1958, *13*: 163–178.

Kawasaki Shigetaro 川崎繁太郎
Kobunken ni arawaretaru Chōsen kōsanbutsu 古文獻に顯はれたる朝鮮鑛產物 (Mineral resources of Korea which appear in ancient documents). Seoul, 1935.

Kawase Kazuma 川瀨一馬
Ko katsujibon no kenkyū 古活字本の研究 (Studies on books printed with movable type). Tokyo, 1937.

Kim Hwa-ja 金和子
"Sŏsa chaeryorosŏŭi Han'gukchiŭi palchŏne kwanhan yŏn'gu 書寫材料로서의 韓國紙의 發展에 관한 研究" (A bibliographical study on the development of Korean paper), unpublished master's thesis for Ehwa Women's University. Seoul, 1968.

Kim Sang-gi 金庠基
Koryŏ sidae sa 高麗時代史 (A history of the Koryŏ period). Seoul, 1961.

Kim Tu-jong 金斗鍾
Han'guk ŭihak munhwa taeyŏnp'yo 韓國醫學文化大年表 (Chrononology of Korean medicine). Seoul, 1966.

Kim Tu-jong 金斗鍾
Han'guk ŭihak palchŏne taehan Kumi mit Sŏnambang ŭihakŭi yŏnghyang 韓國醫學發展에 對한 歐美 및 西南方醫學의 影響 (The influence of Western and Central Asian medicine on the development of Korean medicine). Seoul, 1960.

Kim Tu-jong 金斗鍾
Han'guk ŭihaksa 韓國醫學史 (A history of Korean medicine). Seoul, 1966.

Kim Tu-jong 金斗鍾
"Han'gŭl hwalcha ko 한글活字考" (On Han'gŭl printing type), in *Ch'oe Hyŏn-bae sŏnsaeng hwan'gap kinyŏn nonmunjip* 崔鉉培先生還甲記念論文集 (Papers for the sixtieth birthday of Dr. Ch'oe Hyŏn-bae). Seoul, 1954.

Kim Tu-jong 金斗鍾
"Yissi Chosŏn hugi hwalchaŭi kaejuwa Chamgok Kim Yuk sŏnsaeng samdaeŭi konghŏn 李氏朝鮮後期 活字의 改鑄와 潛谷 金堉先生 三代의 貢獻" (Innovations in typecasting during the late Yi Dynasty by Kim Yuk and his immediate descendants), in *Paek Nak-jun paksa hwan'gap kinyŏm kukhak nonch'ong* 白樂濬博士還甲記念論叢 (Commemorative papers for the sixtieth birthday of Dr. Paek Nak-jun). Seoul, 1955.

Kim Wŏn-yong 金元龍
Han'guk kogohak kaeron 韓國考古學概論 (An outline of Korean archeology). Seoul, 1967. This study was based on the article, "Han'guk munhwaŭi kogohakchŏk yŏn'gu 韓國文化의 考古學的研究" (Archeological researches in Korean culture), in *Minjok kukkasa* 民族國家史, in the *Han'guk munhwasa taege* series. Seoul, 1964.

Kim Wŏn-yong 金元龍
Han'guk kohwalcha kaeyo 韓國古活字概要 (An outline of old Korean type), with English summary, Korean National Museum series, no. 1. Seoul, 1959.

Kim Wŏn-yong 金元龍
Han'guk misulsa 韓國美術史 (A history of art in Korea). Seoul, 1968.

Kim Wŏn-yong 金元龍
"Kyemija poyu 癸未字補遺" (Supplementary note on Kemi printing type), *Sŏji*, 1960, *1*: 35–37.

Kim Wŏn-yong 金元龍
"Yijo hugiŭi chuja insoe 李朝後期의 鑄字印刷" (On cast-type printing in the latter period of the Yi Dynasty), *Hyangt'o Sŏul,* 1959, *7*: 7–66.

Kim Wŏn-yong 金元龍
"Yissi Chosŏn chuja insoe sosa—chujasorŭl chungsimŭro 李氏朝鮮鑄字印刷小史—鑄字所를 中心으로" (A short history of cast-type printing in the early Yi Dynasty), *Hyangt'o Sŏul,* 1958, *3*: 123–170.

Kim Yang-sŏn 金良善
"Han'guk kojido yŏn'gu ch'o 韓國古地圖硏究抄" (A study of ancient maps in Korea), *Sungdae* 崇大, 1965, *10*: 62–88.

Kim Yang-sŏn 金良善
"Myŏngmal Ch'ŏngch'o Yasohoe sŏngyosatŭri chejakhan sege chidowa kŭ Han'guk munhwasasange mich'in yŏnghyang 明末淸初 耶蘇會宣敎師들이 製作한 世界地圖와 그 韓國文化史上에 미친 影響" (Jesuit world maps published in the Ming and Ch'ing, and their influence in Korea), *Sungdae,* 1961, *6*: 16–58.

Kim Yong-guk 金龍國
"Imjin waeranhu kusŏnŭi pyŏnch'ŏn kwajŏng 壬辰倭亂後 龜船의 變遷過程" (On the changing process of turtle ship building after the Hideyoshi invasions), *Haksulwŏn nonmunjip* 學術院論文集, 1968, *7*: 137–164.

Kim Yong-sŏp 金容燮
"Chosŏn hugiŭi sudojak kisul—iyŏnggwa surimunje 朝鮮後期水稻作技術—移秧과 水利問題—" (The techniques of paddy rice cultivation in the late Yi Dynasty), *Asea yŏn'gu,* 1965, *8*.2: 281–306.

Ko Chong-kŏn and Ham In-yŏng 高鐘健, 咸仁英
"Pangsasŏn t'ugwapŏpe ŭihan komisulp'umŭi chosa 1, 2 放射線透過法에 의한 古美術品의 調査" (Examination of ancient art objects by gamma radiography, parts 1 and 2), *Misul charyo,* 1963, *8*: 1–5; 1964, *9*: 13–17.

Ko Pyŏng-ik 高柄翊
"Hech'o *Wang o-Ch'ŏnch'ukkuk chŏn* yŏn'gu sosa 慧超往五天竺國傳硏究小史" (Brief history of studies on Hui-ch'ao's *Wang Wu-t'ien-chu-kuo chuan*), in *Paek Sŏng-uk paksa hoeryŏk kinyŏm pulgyo nonch'ong* 白性郁博士回曆記念佛敎論叢 (Commemorative papers on Buddhism for the sixtieth birthday of Dr. Paek Sŏng-uk). Seoul, 1959.

Ko Pyŏng-ik 高柄翊
"Sŏngjong-cho Ch'oe Puŭi p'yoryuwa *P'yohaerok* 成宗朝 崔溥의 漂流와 漂海錄" (Ch'oe Pu and his "Record of drifting," fifteenth century), in *Yi Sang-baek paksa hoegap kinyŏm nonch'ong* 李相栢博士回甲記念論叢 (Commemorative papers for the sixtieth birthday of Dr. Yi Sang-baek). Seoul, 1964.

Ko Sŭng-je 高承濟
Kŭnse Han'guk sanŏpsa yŏn'gu 近世韓國產業史硏究 (A study of the industrial history of the Hermit Kingdom of Korea). Seoul, 1959.

Ko Yu-sŏp 高裕燮
Han'guk misul munhwasa nonch'ong 韓國美術文化史論叢 (Papers on the history of Korean art and culture). Seoul, 1949, 1966.

Ko Yu-sŏp 高裕燮
Han'guk misulsa kŭp mihak non'go 韓國美術史及美學論攷 (Studies on Korean art and aesthetics). Seoul, 1963.

Ko Yu-sŏp 高裕燮
Koryŏ ch'ŏngja 高麗青磁 (Koryŏ celadons). Seoul, 1954. First published in Japanese as *Kōrai seizi*, Tokyo, 1939.

Kobayashi Yukio 小林行雄
Kodai no gijutsu 古代の技術 (Ancient technology in Japan). Tokyo, 1962.

Kobayashi Yukio 小林行雄
Zoku kodai no gijutsu 續古代の技術 (Ancient technology in Japan, sequel). Tokyo, 1965.

Komai Kazuchika 駒井和愛
Chūgoku kokyō no kenkyū 中國古鏡の研究 (A study on Chinese mirrors before the Six Dynasties). Tokyo, 1953.

Komatsu Shigeru and Yamanouchi Yosito 小松茂, 山内淑人
"Kokyō no kagakuteki kenkyū 古鏡の化學的研究" (A chemical and metallographic study of ancient Far Eastern mirrors), *Tōhō gakuhō* (Kyoto), 1937, *8*: 11–31.

Komatsu Shigeru and Yamanouchi Yosito 小松茂, 山内淑人
"Tōyō kodōki no kagakuteki kenkyū 東洋古銅器の化學的研究" (A chemical study of ancient Chinese bronzes, part 1), *Tōhō gakuhō* (Kyoto), 1933, *3*: 295–303.

Kukhak yŏn'gu nonchŏ ch'ongnam 國學研究論著總覽
(Korean studies, a bibliographical guide), ed. Kukhak Yŏn'gu Nonchŏ Ch'ongnam Kanhaenghoe 國學研究論著總覽刊行會. Seoul, 1960.

Kukpo torok 國寶圖錄
(Album of national treasures), ed. Munkyobu 文教部 (Ministry of Education), 6 vols. Seoul, 1959–1962.

Kwahak kisulsa 科學技術史
(History of Korean science and technology). In *Han'guk munhwasa taege* 韓國文化史大系 series. Seoul, 1968.

Kwŏn Sang-no 權相老
Han'guk chimyŏng yŏnhyŏk ko 韓國地名沿革考 (A study of the origin and development of Korean geographical names). Seoul, 1961.

Li Shu-hua 李書華
Chih-nan-chen ti ch'i-yuan 指南針的起源 (The origin of the compass). Taipei, 1954.

Li Shu-hua 李書華
Chih-nan-ch'e yü chih-nan-chen 指南車與指南針 (The south-pointing carriage and the mariner's compass). Taipei, 1959.

Maema Kyōsaku 前間恭作
Kosen satsufu 古鮮册譜 (Bibliography of classical Korean books), 3 vols. Tokyo, 1914–1957.

Marugame Kinsaku 丸龜金作
"Chōsen no katsuji juzōsho ni tsuite 朝鮮の活字鑄造所について" (Type foundries in Korea), *Chōsen gakuhō*, 1953, *4*: 107–116.

Maruyama Kiyoyasu 丸山清康
"Hōken shakai to gijutsu—Wadokei ni shūyaku sareta hōken gijutsu 封建社會と技術—和時計に集約された封建技術" (On Japanese clockmaking and feudal society in Japan), *Kagakushi kenkyū*, 1954, *31*: 16–22.

Maruyama Kiyoyasu 丸山清康
"Nōson suisha no gijutsushi—Kita Kantō ni okeru 農村水車の技術史—北關東における" (On waterwheels in the North Kantō area, Japan), *Kagakushi kenkyū*, 1956, *37*: 1–7.

Meiji zen Nihon butsuri kagaku shi 明治前日本物理化學史 (A history of Japanese physics and chemistry before the Meiji era), ed. Nihon Gakushiin 日本學士院. Tokyo, 1964.

Meiji zen Nihon kōgyō hattatsu shi 明治前日本鑛業發達史 (A history of the development of Japanese mining before the Meiji era). Tokyo, 1958.

Meiji zen Nihon ōyō kagaku shi 明治前日本應用化學史 (A history of Japanese applied chemistry before the Meiji era). Tokyo, 1963.

Meiji zen Nihon tenmongaku shi 明治前日本天文學史 (A history of Japanese astronomy before the Meiji era). Tokyo, 1960.

Mikami Yoshio 三上義夫
Nihon kagaku no tokushitsu: Tenmon 日本科學の特質: 天文 (Characteristics of Japanese science: astronomy), in *Tōyō shichō no tenkai* 東洋思潮の展開 (The development of Oriental thought). Tokyo, 1936.

Miki Sakae 三木榮
Chōsen igakushi oyobi sitsubyōshi 朝鮮醫學史及疾病史 (A history of Korean medicine and disease). Osaka, 1955.

Miki Sakae 三木榮
"Chōsen igaku kyōiku shi 朝鮮醫學教育史" (History of medical education in Korea from the Paekche period to the Yi Dynasty), *Chōsen gakuhō*, 1959, *14*: 73–95.

Miki Sakae 三木榮
Chōsen ishoshi 朝鮮醫書誌 (Bibliography of Korean medicine, ancient and medieval). Osaka, 1956.

Minjok kukka sa 民族國家史
(History of the Korean people and their country), in the *Han'guk munhwasa taege* series. Seoul, 1964.

Miyahara Toichi 宮原兎一
"14.15.16 seiki Chōsen ni okeru kayaku 14.15.16 世紀朝鮮に於ける火藥" (On gunpowder during the fourteenth to sixteenth centuries in Korea), in *Tōyōshigaku ronshū* 東洋史學論集, Tōkyō Kyōiku Daikagu 東京教育大學, 1953.

Miyahara Toichi 宮原兎一
"Chōsen shoki no dōsen ni tsuite 朝鮮初期の銅錢について" (On bronze coins in the early days of the Yi Dynasty), *Chōsen gakuhō,* 1951, *2*: 75–101.

Mok Yŏng-man 목영만
Chido iyagi 지도이야기 (The story of maps). P'yŏngyang, 1965.

Muroga Nobuo and Unno Kazutaka 室賀信夫, 海野一隆
"Nihon ni okonawareta Bukkyōkei sekaizu ni tsuite 日本に行われた佛教系世界圖について" (Buddhist world maps in Japan), *Chirigakushi kenkyū,* 1957, *1*: 67–141; 1962, *2*: 137–229.

Naitō Tadashi 内藤匡
Kotōji no kagaku 古陶磁の科學 (Science in traditional ceramics). Tokyo, 1944.

Nakamura Eikō 中村榮孝
"*Kaitō shokokuki* no senshū to insatsu 海東諸國紀の撰修と印刷" (Compilation and publication of the *Haedong chegukki*), *Shigaku zasshi,* 1928, *38*.8: 770–796;*38*.9: 924–944.

Nakamura Hiroshi 中村拓
"Honbō ni tsutawaru Braū sekaizu ni tsuite 本邦に傳わるブラウー世界圖について" (J. Blaeu's large world maps preserved in Japan), *Chirigakushi kenkyū,* 1957, *1*: 23–65.

Nakamura Kiyoe 中村清兄
"Kokuri jidai no kofun ni tsuite—sono seishō hekiga no kōsatsu o chūshin to shite 高句麗時代の古墳について—その星象壁畫の考察を中心として" (On the paintings of various celestial bodies in ancient tombs of the Koguryŏ period), *Kōkogaku ronsō* 考古學論叢 1937, *4*: 384–401.

Nakashima Satoshi 中島敏
"Shina ni okeru shisshiki shūdō no enkaku 支那における濕式收銅の沿革" (A study on the progress of hydro-metallurgy of copper during the Sung period), *Tōyō gakuhō,* 1940, *27*.3: 393–423.

Nakayama Kushirō 中山久四郎
Sekai insatsu tsūshi 世界印刷通史 (A history of East Asian printing), 2 vols. Tokyo, 1930.

Nakayama Shigeru 中山茂
"Toyō ni kagaku kakumei wa okorietaka 東洋に科學革命は起り得たか" (The possibility of a scientific revolution in the history of the Far East), in *Kagaku kakumei* 科學革命 (Scientific revolution). Tokyo, 1961.

Nihon kagaku gijutsushi 日本科學技術史
(A history of science and technology in Japan), ed. Yajima Suketoshi. Tokyo, 1962.

Nihon kagaku koten zensho 日本科學古典全書
(Comprehensive collection of classical works on Japanese science), ed. Saigusa Hiroto 三枝博音, 13 vols. Tokyo, 1942–1948.

No Chŏng-u 盧正祐
"Han'guk ŭihaksa 韓國醫學史" (A history of Korean medicine), in *Kwahak kisulsa,* in the *Han'guk munhwasa taege* series. Seoul, 1968.

Nōda Chūryō 能田忠亮
Shūhi sankei no kenkyū 周髀算經の研究 (An inquiry concerning the *Chou pi suan ching*), Academy of Oriental Culture, Kyoto, Institute Monograph Series, no. 3. Kyoto, 1933.

Nōda Chūryō 能田忠亮
Tōyō tenmongakushi ronsō 東洋天文學史論叢 (Collected papers on the history of astronomy in the Far East). Tokyo, 1944.

Oda Shōgo 小田省吾
Chōsen tōjishi bunken kō 朝鮮陶磁史文獻考 (Historical documents on Korean ceramics). Tokyo, 1936.

Ogawa Takuji 小川琢治
"Shina chizugaku no hattatsu 支那地圖學の發達" (Development of Chinese cartography), in *Shina rekishi chiri kenkyū*. Tokyo, 1928.

Ogawa Takuji 小川琢治
Shina rekishi chiri kenkyū 支那歷史地理研究 (Studies in Chinese historical geography). Kyoto, 1928.

Ōtsuka Noboru 大塚鐙
"Keimi katsuji ni tsuite 癸未活字について" (On books printed with the Kemi-ja [the movable type cast in 1403]), *Chōsen gakuhō,* 1961, *21/22*: 393–402.

Paek Nam-un 白南雲
Chōsen shakai keizai shi 朝鮮社會經濟史 (History of Korean society and its economy). Tokyo, 1933.

Pak Mun-guk and Arimitsu Kyōichi 朴文國, 有光教一
"Rashin Sōtō genshi iseki hakkutsu hōkoku 羅津草島原始遺跡發掘報告" (Report on the excavation of primitive ruins, abstract), *Chōsen kenkyū nenpō,* Kyoto, 1959, *1*:76–90.

Pak Si-hyŏng 박시형
"Chosŏnesŏ kŭmsok hwalchaŭi palmyŏnggwa kŭ sayong 조선에서 금속활자의 발명과 그 사용" (On the invention of metallic type and its employment in Korea), *Yŏksa kwahak,* 1959, *5*: 33–41.

Pak .Tong-hyŏn 朴同玄
"*Koryŏsa Ch'ŏnmunji*e kiroktoen hesŏnge kwanhaesŏ 高麗史天文志에 記錄된 彗星에 關해서" (Records of comets in the Treatise on astronomy of the *Koryŏsa*), *Han'guk munhwa yon'guwŏn nonch'ong* 韓國文化研究院論叢, 1960, 2.1: 179–196.

Pak Yŏng-su 朴泳洙
"Koryŏ *Taejanggyŏng*-p'anŭi yŏn'gu 高麗大藏經板의 研究" (A study of the printing blocks of the Koryŏ Tripitaka), in *Paek Sŏng-uk paksa hoeryŏk kinyŏm pulgyo nonch'ong* 白性郁博士回曆記念佛教論叢 (Commemorative papers on Buddhism for the sixtieth birthday of Dr. Paek Sŏng-uk). Seoul, 1959.

Sasaki Chūgi 佐佐木忠義
"Chōsen no shakudo 朝鮮の尺度" (Korean measures), *Kagakushi kenkyū,* 1949, *9*: 2–11.

Satake Bunji 佐竹文治
"Kankoku no kanshōdai oyobi Daikan Kōbu jūichinen Meishireki 韓國の觀象臺及大韓光武十一年明時曆" (The Korean Meteorological Observatory and the Korean calendar of 1907), *Chigaku zasshi* 地學雜誌, 1907, *19*: 746.

Satō Seiji 佐藤政治
Nihon rekigakushi 日本曆學史 (A history of Japanese calendrical science). Tokyo, 1968.

Satō Taketoshi 佐藤武敏
Chūgoku kodai kōgyōshi no kenkyū 中國古代工業史の研究 (Studies on the history of ancient industries in China). Tokyo, 1962.

Seiyō kohan Nihon chizushū 西洋古版日本地圖集
(Early printed maps and atlases of Japan made in western countries, 1552–1840), Tenri University Central Library, Photo Series, no. 4. Tenri, 1954.

Seki Yoshikuni 關義城
Kokon kamisuki zushū 古今紙漉圖集 (Collection of pictures of ancient and modern hand papermaking with one hundred specimens of actual handmade papers). Tokyo, 1965.

Sekino Tadashi 關野貞
Chōsen no kenchiku to geijutsu 朝鮮の建築と藝術 (Studies in Korean architecture and art). Tokyo, 1941

Sekino Tadashi 關野貞 et al. (eds.)
Rakurō-gun jidai no iseki 樂浪郡時代の遺蹟 (Archeological researches on the ancient Lolang district [Korea]), 2 vols. Seoul, 1925, 1927.

Shinjō Shinzō 新城新藏
Koyomi to tenmon こよみと天文 (The calendar and astronomy). Kyoto, 1930.

Shinjō Shinzō 新城新藏
Tōyō tenmongaku-shi kenkyū 東洋天文學史研究 (Researches on the history of astronomy in East Asia). Tokyo, 1929.

Sin T'ae-hyŏn 辛兌鉉
Samguk sagi chirijiŭi yŏn'gu 三國史記地理志의 硏究 (A study of the Treatise on Geography of the *Samguk sagi*). Seoul, 1958.

Sŏ Su-in 徐樹仁
"*Taengniji* yŏn'gu sŏsŏl 擇里誌硏究序說" (Preliminary study of the *T'aengniji*), *Chirihak*, 1963, *1*: 83–90.

Son Po-gi 孫寶基
"Han'guk insoe kisulsa 韓國印刷技術史" (A history of Korean printing technology), in *Kwahak kisulsa,* in the *Han'guk munhwasa taege* series. Seoul, 1968.

Sŏul t'ŭkpyŏlsi sa kojŏk-p'yŏn 서울特別市史古蹟篇
(A history and historic places of Seoul, the capital city of the Yi Dynasty), ed. Committee on the History of Seoul. Seoul, 1963.

Tabohashi Kiyoshi 田保橋潔
"Chōsen sokuchishi-jō no ichi gyōseki 朝鮮測地史上の一業蹟" (A contribution to the history of Korean surveying), *Rekishi chiri* (Kyoto), 1932, *60*.6: 531.

Tagawa Kōzō 田川孝三
"Richō kōbutsu kō 李朝貢物考" (Taxation and tribute in the Yi Dynasty), *Chōsen gakuhō*, 1956, *9*: 103–163.

Takabayashi Hyōgo 高林兵衛
Tokei hattatsu shi 時計發達史 (A history of the development of timekeepers). Tokyo, 1924.

Takabayashi Hyōgo 高林兵衛
Tokei no hanashi 時計の話 (Story of timekeepers). Tokyo, 1925.

Takagi Kikusaburō 高木菊三郎
Nihon ni okeru chizu sokuryō no hattatsu ni kansuru kenkyū 日本に於ける地圖測量の發達に關する研究 (A study on the development of cartographic surveying). Tokyo, 1966.

Takahashi Tadashi 高橋正
"Konichi kyōri rekidai kokuto no zu saikō 混一疆理歴代國都之圖再考" (Addendum on the Korean World Map of 1402), *Ryūkoku shidan*, 1966, *56/57*: 204–215.

Takahashi Tadashi 高橋正
"Tōzen seru chūsei Isuramu sekaizu 東漸せる中世イスラーム世界圖" (The introduction of Islamic world maps to China and Korea; on the Korean world map of 1402), *Ryūkoku University Theses* 龍谷大學論集, 1966, no. 374, pp. 77–95.

Tamura Sennosuke 田村專之助
"Chōsen Richō gakusha no chikyū kaitensetsu ni tsuite 朝鮮李朝學者の地球回轉說について" (On the rotating-earth theory of Yi Dynasty scholars), *Kagakushi kenkyū*, 1954, *30*: 23–24.

Tamura Sennosuke 田村專之助
"Richō kishōgaku seiritsu no kyakkanteki jōken 李朝氣象學成立の客觀的條件" (Objective conditions for the formation of meteorology in the Yi Dynasty), *Kagakushi kenkyū*, 1958, *48*: 7–10.

Tamura Sennosuke 田村專之助
Tōyōjin no kagaku to gijutsu 東洋人の科學と技術 (Science and technology of the East Asian peoples). Tokyo, 1958.

Tasaka Okimichi 田坂興道
"Seiyō rekihō no tōzen to *fui-fui* rekihō no unmei 西洋曆法の東漸と回回曆法の運命" (The introduction of European astronomy to China, and the fate of Islamic astronomy), *Tōyō gakuhō*, 1947, *31*.2:141.

Tasaka Okimichi 田坂興道
"Tōzen seru Isruamu bunka no issokumen ni tsuite 東漸せるイスラム文化の一側面に就いて" (About an aspect of the eastward movement of Islamic culture), *Shigaku zasshi*, 1942, *53*.4–5: 17–49.

Tokyo Kagaku Hakubutsukan (ed.) 東京科學博物館
Edo jidai no kagaku 江戸時代の科學 (Science during the Tokugawa period). Tokyo, 1934.

Tsuboi Kumazō 坪井九馬三
"*Keishōdō chirishi* ni tsuite 慶尙道地理志について" (On the treatise on geography of Kyŏngsang province [*Kyŏngsang-do chiriji*]), *Shigaku zasshi*, 1919, *30*.8: 867–871.

Tsuboi Kumazō 坪井九馬三
"*Keishōdō chirishi* hoi 慶尙道地理志補遺" (Supplement on the treatise on geography of Kyŏngsang province), *Shigaku zasshi*, 1919, *30*.9: 963–966.

Tsukada Taisaburō 塚田泰三郎
Wadokei 和時計 (Japanese clocks). Tokyo, 1960.

Tung Tso-pin, Liu Tun-chen, and Kao P'ing-tzu 董作賓, 劉敦楨, 高平子
Chou kung ts'e-ying-t'ai tiao ch'a pao kao 周公測景臺調査報告 (Report of the investigation of the Duke of Chou's tower for measurement of the shadow of the sun), Ch'angsha, 1938.

Umehara Sueji 梅原末治
"Dōtaku no kagaku seibun ni tsuite 銅鐸の化學成分に就いて" (On the chemical composition of bronze *to* bells), in *Shiratori hakase kanreki kinen Tōyōshi ronsō* 白鳥博士還曆記念東洋史論叢 (Commemorative essays on Oriental history for Dr. Shiratori's sixtieth birthday). Tokyo, 1925.

Umehara Sueji 梅原末治
"Kokyō no kagaku seibun ni kansuru kōkogaku-teki kōsatsu 古鏡の化學成分に關する考古學的考察" (Archeological notes on the chemical analysis of ancient Far Eastern mirrors), *Tōhō gakuhō* (Kyoto), 1937, *8*: 32–55.

Umehara Sueji 梅原末治
"Shina kodōki no kagakuteki kenkyū ni tsuite 支那古銅器の化學的研究に就いて" (An appreciation of chemical study of ancient Chinese bronzes by Prof. Komatsu and Mr. Yamanouchi), *Tōhō gakuhō* (Kyoto), 1933, *3*: 305–323.

Umehara Sueji and Fujida Ryōsaku 藤田亮策
Chōsen kobunka sōkan 朝鮮古文化綜鑑 (Album of ancient Korean cultural remains), 4 vols. Tokyo, 1945–1966.

Wada Yūji 和田雄治
Chōsen kodai kansoku kiroku chōsa hōkoku 朝鮮古代觀測記錄調査報告 (Report on the survey of the ancient records of observation in Korea). Seoul, 1917.

Wada Yūji 和田雄治
"Chōsen Kōka shima no iseki 朝鮮江華島の遺蹟" (Sites ln Kanghwa Island, Korea), *Rekishi chiri*, 1911, *18*.1: 53–57.

Wada Yūji 和田雄治
"Keijō no Suihyōkyō 京城の水標橋" (Sup'yo Bridge in Seoul), *Rekishi chiri,* 1912, *19*.2: 63–66.

Wada Yūji 和田雄治
"Kōkatō no senjōdan 江華島の塹城壇" (Ch'amsŏngdan at Kanghwa Island), *Kōkogaku zasshi,* 1911, *1*.8: 525–532.

Wada Yūji 和田雄治
"Seisō Eiso ryōchō no sokuuki 世宗英祖兩朝ノ測雨器" (Korean rain gauges from the reigns of Sejong and Yŏngjo), *Chōsen kodai kansoku kiroku chōsa hōkoku,* Seoul, 1917, 1–7.

Wada Yūji 和田雄治 (ed.)
Kankoku kansokusho gakujutsu hōbun 韓國觀測所學術報文 (Reports of the Korean Meteorological Observatory to 1910). Tokyo and Seoul, 1910.

Wang Chen-to 王振鐸
"Chih-nan-ch'e chi-li-ku-ch'e chih k'ao-cheng chi mo-chih 指南車記里鼓車之考證及模製" (Critical study and models of the south-pointing chariot and hodometer carriage), *Shih-hsueh chi-k'an* 史學集刊, 1937, *3*: 1.

Wang Chen-to 王振鐸
"Ssu-nan chih-nan-chen yü lo-ching-p'an 司南指南針與羅經盤 (上, 中, 下)" (Discovery and application of magnetic phenomena in China, parts 1, 2, and 3), *Chung-kuo k'ao-ku hsueh pao* 中國考古學報, 1948, *3*: 119; 1950, *4*: 185; 1951, *5*: 101.

Wang Chin 王璡
"Wu-shu-ch'ien hua-hsueh ch'eng-fen chi ku-tai ying-yung ch'ien hsi hsin la k'ao 五銖錢化學成分及古代應用鉛錫鋅鑞考" (The chemical composition of five-*shu* coins and the use of lead, tin, zinc, and pewter in ancient times), *K'o-hsueh* 科學, 1933, *8*.8: 839–854.

Wang Yung 王庸
Chung-kuo ti-li hsueh shih 中國地理學史 (History of geography in China). Ch'angsha, 1938.

Watanabe Masao 渡邊正雄
"Meijiki ni okeru makyō no kenkyū 明治期における魔鏡の研究" (A study of magic mirrors during the Meiji era), *Kagakushi kenkyū,* 1961, *61*: 25–29.

Watanabe Toshio 渡邊敏夫
Koyomi 暦 (The calendar). Tokyo, 1937.

Watanabe Yosuke 渡邊世祐
"Chōseneki to waga zōsen no hattatsu 朝鮮役と我が造船の發達" (The Hideyoshi invasion and the development of shipbuilding in Japan), *Shigaku zasshi,* 1935, *46*.5: 574–597.

Wu Ch'eng-lo 吳承洛
Chung-kuo tu liang heng shih 中國度量衡史 (History of Chinese metrology). Shanghai, 1937.

Yabuuchi Kiyoshi 藪內清
"Asuka Nara jidai no shizen kagaku 飛鳥奈良時代の自然科學" (Natural

science during the Asuka and Nara periods), in *Asuka Nara jidai no bunka* 飛鳥奈良時代の文化, ed. Haneda Tōru 羽田通. Osaka, 1955.

Yabuuchi Kiyoshi 藪內清
"Chūgoku ni okeru Isuramu tenmongaku 中國に於けるイスラム天文學" (The introduction of Islamic astronomy to China), *Tōyō gakuhō* (Kyoto), 1950, *19*: 65.

Yabuuchi Kiyoshi 藪內清
Chūgoku no tenmongaku 中國の天文學 (Chinese astronomy). Tokyo, 1949.

Yabuuchi Kiyoshi 藪內清
"Chūgoku no tokei 中國の時計" (Timekeepers in ancient China), *Kagaku-shi kenkyū,* 1951, *19*: 19–22.

Yabuuchi Kiyoshi 藪內清
"Chūsei kagaku gijutsushi josetsu 中世科學技術史序說" (Introduction to the history of medieval sciences and technology), *Tōhō gakuhō* (Kyoto), 1962, *32*: 313–332.

Yabuuchi Kiyoshi 藪內清
"Gen Min rekihō-shi 元明曆法史" (Calendrical science in the Yuan and Ming dynasties), *Tōhō gakuhō* (Kyoto), 1944, *14*: 264.

Yabuuchi Kiyoshi 藪內清
"Kinsei Chūgoku ni tsutaerareta seiyō tenmongaku 近世中國に傳えられた西洋天文學" (The introduction of Western astronomy into China), *Kagaku-shi kenkyū,* 1955, *32*: 15–18.

Yabuuchi Kiyoshi 藪內清
"Richō gakusha no chikyū kaitensetsu 李朝學者の地球回轉說" (On the rotating-earth theory of scholars in the Yi Dynasty), *Chōsen gakuhō,* 1968, *49*: 427–434.

Yabuuchi Kiyoshi 藪內清
"Sōdai no seishuku 宋代の星宿" (Description of the constellations of the Sung Dynasty), *Tōhō gakuhō* (Kyoto), 1936, *1*: 42–89.

Yabuuchi Kiyoshi 藪內清
"Sōgen jidai ni okeru kagaku gijutsu no tenkai 宋元時代における科學技術の展開" (Development of sciences and techniques in the Sung and Yuan periods), *Tōhō gakuhō* (Kyoto), 1966, *37*: 1–40.

Yabuuchi Kiyoshi 藪內清
"Tō Sō rekihō-shi 唐宋曆法史" (Calendrical science in the T'ang and Sung dynasties), *Tōhō gakuhō* (Kyoto), 1943, *13*: 491. This and other articles are reprinted in *Chūgoku no tenmon rekihō* (Chinese astronomy), Tokyo, 1969.

Yabuuchi Kiyoshi 藪內清
Zui Tō rekihōshi no kenkyū 隋唐曆法史の研究 (Studies on the calendrical science of the Sui and T'ang periods). Tokyo, 1944.

Yabuuchi Kiyoshi 藪內清 (ed.)
Chūgoku chūsei kagaku gijutsushi no kenkyū 中國中世科學技術史の研究 (Studies the history of science and technology in medieval China). Tokyo, 1963.

Yabuuchi Kiyoshi 藪內清 (ed.)
Sōgen jidai no kagaku gijutsu shi 宋元時代の科學技術史 (History of science and technology during the Sung and Yuan periods). Kyoto, 1967.

Yabuuchi Kiyoshi 藪內清 (ed.)
Tenkō kaibutsu no kenkyū 天工開物の研究 (Studies in the *T'ien kung k'ai wu*). Tokyo, 1953.

Yabuuchi Kiyoshi 藪內清 (tr.)
Tenkō kaibutsu 天工開物 (A Japanese translation of the *T'ien kung k'ai wu*). Tokyo, 1969.

Yamaguchi Ryūji 山口隆二
Nihon no tokei 日本の時計 (Japanese clocks). Tokyo, 1942.

Yamaguchi Seiji 山口政之
"Nan Kaijin no *Konyōzenzu* ni tsuite 南懷仁の坤輿全圖について" (On Verbiest's world map), *Chōsen gakkai kaihō*, 1952, *14*: 10.

Yamaguchi Seiji 山口政之
"Shinchō ni okeru zai-Shi Ōjin to Chōsen shisin 清朝に於ける在支歐人と朝鮮使臣" (Jesuits and Korean envoys in Ch'ing China), *Shigaku zasshi*, 1933, *44*.7: 13–15.

Yamaguchi Seiji 山口政之
"Shōken Seshi to Tō Jakubō 昭顯世子と湯若望" (On Prince Sohōn and Adam Schall von Bell), *Seikyū gakusō* 青丘學叢, 1931, *5*: 112–117.

Ye Yong-hae 芮庸海
In'gan munhwajae 人間文化財 (Human cultural assets). Seoul, 1963.

Yi Ch'an 李燦
"Han'guk chirihaksa 韓國地理學史" (A history of Korean geography), in *Kwahak kisulsa*, in the *Han'guk munhwasa taege* series. Seoul, 1968.

Yi Chin-hi 李進熙
"Kaihōgo Chōsen kōkogaku no hatten—Kokuri hekiga kofun no kenkyū 解放後朝鮮考古學の發展—高句麗壁畫古墳の研究" (The postwar development of Korean archeology—Tombs with wall paintings of the Koguryŏ Dynasty), *Kōkogaku zasshi*, 1959, *45*.3: 43–64.

Yi Ch'un-yŏng 李春寧
Chosŏn nongŏp kisul sosa 朝鮮農業技術小史 (A short history of Korean agricultural technology). Seoul, 1950.

Yi Ch'un-yŏng 李春寧
Yijo nongŏp kisulsa 李朝農業技術史 (A history of agricultural technology in the Yi Dynasty). Seoul, 1964.

Yi Hong-jik (ed.) 李弘稙
Kuksa taesajŏn 國史大事典 (Encyclopedia of Korean history). Seoul, 1962.

Yi Ka-wŏn 李家源
Yŏnam sosŏl yŏn'gu 燕巖小說研究 (Studies in the novels of Yŏnam = Pak Chi-wŏn). Seoul, 1965.

Yi Ka-wŏn 李家源
"Taejanggyŏng kakp'angwa kŭ chŏnsŏl 大藏經刻板과 그 傳說" (The woodblocks of the Korean Tripitaka and legends concerning them), *Tongguk sasang*, 1958, *1*: 40–44.

Yi Ki-baek 李基白
Han'guksa sillon 韓國史新論 (New interpretations of Korean history). Seoul, 1967.

Yi Kwang-nin 李光麟
Yijo surisa yŏn'gu 李朝水利史研究 (A study of irrigation under the Yi Dynasty). Seoul, 1961.

Yi Kwang-nin 李光麟
"Yijo ch'ogiŭi chejiŏp 李朝初期의 製紙業" (Paper manufacturing in the early Yi Dynasty), *Yŏksa hakpo*, 1958, *10*: 1–48.

Yi Kwang-nin 李光麟
"Yijo hubangiŭi sach'al chejiŏp 李朝後半期의 寺刹製紙業" (Paper manufacture in Buddhist temples during the second half of the Yi Dynasty), *Yŏksa hakpo*, *17/18*: 201–219.

Yi Nŭng-hwa 李能和
Chosŏn kitokkyo kŭp oegyosa 朝鮮基督教及外交史" (History of the Christian missions and foreign relations in Korea). Seoul, 1928, reprinted in 1968.

Yi Nung-hwa 李能和
Han'guk togyosa 韓國道教史 (A history of Taoism in Korea). Seoul, 1956.

Yi Pyŏng-do 李丙燾
"Chŏng Sang-giwa *Tongguk chido* 鄭尙驥와 東國地圖" (Chŏng Sang-gi and his map of Korea, *Tongguk chido*), *Sŏji*, 1960, *1*.1: 5–16.

Yi Pyŏng-do 李丙燾
Han'guksa taegwan 韓國史大觀 (An outline of Korean history). Seoul, 1964.

Yi Pyŏng-do 李丙燾
"Kangdo chido 江都地圖" (On the Kangdo chido [Map of Kanghwa Island]), *Sŏji*, 1960, *1*.2:1.

Yi Pyŏng-do 李丙燾
"Heijō no zaijō oyobi rajō 平壤の在城及び羅城" (On the castles in P'yŏngyang), *Seikyū gakusō*, 1931, *1*.3: 113–125.

Yi Pyŏng-do 李丙燾
"Toch'ame taehan il-iŭi koch'al 圖讖에 對한 一. 二의 考察" (A study of prognostication), *Chindan hakpo*, 1939, *10*: 1–18.

Yi Pyŏng-do 李丙燾
"Chosŏn T'aejoŭi kaegukgwa tangsiŭi toch'amsŏl 朝鮮太祖의 開國과 당시의 圖讖說" (King T'aejo, founder of the Yi Dynasty, and prognostication theory), in *Koryŏ sidaeŭi yŏn'gu* 高麗時代의 研究 (Koryŏ era studies). Seoul, 1948.

Yi Sang-baek 李相佰
Han'guk munhwasa yŏn'gu non'go 韓國文化研究論攷 (Studies on Korean cultural history). Seoul, 1947.

Yi Sung-yŏng 李崇寧
"Sejongŭi ŏnŏ chŏngch'aeke kwanhan yŏn'gu 世宗의 言語政策에 關한 研究" (A study of Sejong's policy toward the Korean language), *Asea yŏn'gu*, 1958, *1*.2: 29–84.

Yi Tŏk-pong 李德鳳
"Han'guk saengmulhaksa 韓國生物學史" (A history of Korean biology), in *Kwahak kisulsa*, in the *Han'guk munhwasa taege* series. Seoul, 1968.

Yi Tŏk-pong 李德鳳
"Han'guk saengmulhakŭi sachŏk koch'al 韓國生物學의 史的考察" (The dawn of biological knowledge in Korea), *Asea yŏn'gu*, 1959, *2*.1: 101–142.

Yi Yong-pŏm 李龍範
"Pŏpchusa sojangŭi sinpŏp ch'ŏnmuntosŏle taehayŏ 法住寺所藏의 新法天文圖說에 대하여" (On the astronomical map of 1743 preserved in Pŏpchu temple, Korea), *Yŏksa hakpo*, 1966, *31*.1–66; *32*: 59–119.

Yoneda Miyoji 米田美代治
Chōsen jōdai kenchiku no kenkyū 朝鮮上代建築の研究 (Studies in ancient Korean architecture). Tokyo, 1944.

Yoshida Mitsukuni 吉田光邦
"Bōseki kangai—Chūgoku Nihon gijutsushi no shihyō 紡績灌漑—中國日本技術史の指標" (Spinning and irrigation—on indices for the history of technology), *Kagakushi kenkyū*, 1956, *56*: 33–37.

Yoshida Mitsukuni 吉田光邦
"Chūgoku kodai no kinzoku gijutsu 中國古代の金屬技術" (Metallurgy in ancient China), *Tōhō gakuhō* (Kyoto), 1959, *29*: 51–110.

Yoshida Mitsukuni 吉田光邦
"Chūsei no kagaku (rentan-jutsu) to senjutsu 中世の化學 (煉丹術) と仙術" (Medieval chemistry [alchemy] and the arts of immortality), in *Chūgoku chūsei kagaku gijutsushi no kenkyū*. Tokyo, 1963.

Yoshida Mitsukuni 吉田光邦
"Kongi to konshō 渾儀と渾象" (Celestial globes and armillary spheres in China), *Silver Jubilee Volume of the Zinbun Kagaku Kenkyusyo, Kyoto University* 京都大學人文科學研究所創立二十年記念論文集. Kyoto, 1954.

Yoshida Mitsukuni 吉田光邦
"Mindai no heiki 明代の兵器" (Weapons in the Ming Dynasty), in *Tenkō kaibutsu no kenkyū*. Tokyo, 1953.

Yoshida Mitsukuni 吉田光邦
Nihon kagakushi 日本科學史 (A history of Japanese science). Tokyo, 1955.

Yoshida Mitsukuni 吉田光邦
Renkin-jutsu 錬金術 (Alchemy). Tokyo, 1963.

Yoshida Mitsukuni 吉田光邦
"Ryūkotsusha to ryūbisha 龍骨車と龍尾車" (Chain pumps and Archimedean screws), *Kagakushi kenkyū*, 1959, *51*: 28–30.

Yoshida Mitsukuni 吉田光邦
"Tenkō kaibutsu ni tsuite 天工開物について" (On the *T'ien kung k'ai wu*), *Kagakushi kenkyū*, 1951, *18*: 12–16.

Yoshida Mitsukuni 吉田光邦
"Wachūsan to *Kikizusetsu* 和中散と奇器圖說" (*Wachū* powder and the *Ch'i ch'i t'u shuo*), *Kagakushi kenkyū*, 1958, *45*: 35–37.

Yoshida Mitsukuni 吉田光邦
Yakimono やきもの (Japanese ceramics). Tokyo, 1963.

Yoshida Mitsukuni 吉田光邦
"Yōhan (Igata) ni tsuite 熔范(鑄型)について" (On the casting mold in Japan), *Kagakushi kenkyū*, 1954, *32*: 5–8.

Yu Hong-yŏl 柳洪烈
Han'guk Ch'ŏnju-kyohoe sa 韓國天主教會史 (A history of the Roman Catholic Church in Korea). Seoul, 1962.

Yu Kyo-sŏng 劉教聖
"Han'guk sanggongŏpsa 韓國商工業史" (A history of Korean commerce and industry), in *Chŏngch'i kyŏngjesa*, in the *Han'guk munhwasa taege* series, vol. 2. Seoul, 1965.

Yu Yŏng-bak 유영박
"Sejongŭi sahoe chŏngch'aek 世宗의 社會政策" (Social policies of King Sejong), *Chindan hakpo*, 1966, *29/30*: 129–144. Part of *Tuge paksa kohi kinyŏm nonmunjip* 斗溪博士古稀記念論文集 (Commemorative papers for Dr. Yi Pyŏngdo's seventieth birthday).

Zusetsu sekai bunkashi taikei 圖說世界文化史大系, 朝鮮. 東北アジア (Illustrated history of world culture), vol. 19, Korea and Northeast Asia, ed. Kadokawa Shoten. Tokyo, 1959.

Western-Language Sources

Adams, Frank Dawson
The Birth and Development of the Geological Sciences. Baltimore, 1938.

Aitken, M. J.
Physics and Archaeology. London, 1961.

Baillie, C. H.
Clocks and Watches: An Historical Bibliography. London, 1951.

Barnard, Noel
Bronze Casting and Bronze Alloys in Ancient China. Canberra, 1961.

Bedini, Silvio A.
"Scent of Time," *Transactions of the American Philosophical Society*, 1963, *53.5*: 1–51.

Bernal, J. D.
Science in History. London, 1954.

Bibliography of Korean Studies—A Bibliographical Guide to Korean Publications on Korean Studies Appearing from 1945 to 1958, ed. Asiatic Research Center. Seoul, 1961.

Boots, T. L.
"Korean Weapons and Armor," *Transactions of the Royal Asiatic Society, Korea Branch*, 1934, *23.2*: 1.

Bowman, N. H.
"The History of Korean Medicine," *Transactions of the Royal Asiatic Society, Korea Branch,* 1915, *6*.1: 1–34.

Britten, F. J.
Old Clocks and Watches, and Their Makers. London, 1932.

Butterfield, H.
The Origins of Modern Science. London, 1949.

Cammann, S.
"The TLV Pattern on the Cosmic Mirrors of the Han Dynasty," *Journal of the American Oriental Society,* 1948, *68*: 159–167.

Carter, T. F., and Goodrich, L. C.
The Invention of Printing in China and Its Spread Westward, 2d ed. New York, 1955.

Chatley, H.
"Ancient Chinese Astronomy," *Asiatic Review,* 1938, *34*: 140–146.

Ch'en, Kuan-sheng
"Matteo Ricci's Contribution to, and Influence on, Geographical Knowledge in China," *Journal of the American Oriental Society,* 1939, *59*: 325, 509.

Chikashige, Masumi
Oriental Alchemy—Alchemy and other Chemical Achievements of the Ancient Orient. Tokyo, 1936.

Chu, K'o-chen
"Some Chinese Contributions to Meteorology," *Geographical Review,* 1918, *5*: 136–139.

Courant, Maurice
Bibliographie coréenne. Paris, 1895.

Cranmer, G. E.
"Denver's Chinese Sundial," *Popular Astronomy,* 1950, *58*: 119.

Cressey, G. B.
"The Evolution of Chinese Cartography," *Geographical Review,* 1934, *24*: 497.

Daland, Judson
"The Evolution of Modern Printing and the Discovery of Movable Metal Type by the Chinese and the Koreans in the Fourteenth Century," *Journal of the Franklin Institute,* 1931, *212*: 208–234.

Danneman, Friedrich
Die Naturwissenschaften in ihrer Entwicklung und in ihrem Zusammenhänge, 2d ed., 4 vols. Leipzig, 1920–1928.

Davis, T. L., and Chao Yun-tsung
"An Alchemical Poem by Kao Hsiang-hsien," *Isis,* 1939, *30*.2: 236–240.

Davis, T. L., and Feng Chia-sheng
"The Early Development of Firearms in China," *Isis,* 1946, *36*: 114–123, 250.

Diels, H.
Antike Technik. Leipzig and Berlin, 1914; 2d ed., 1920.

Divers, Edward
"The Manufacture of Calomel in Japan," *Journal of the Society of Chemical Industry*, 1894, *13*: 108–111.

Emoto, Yoshimichi
"Characteristics of Antiques and Art Objects by X-Ray Fluorescent Spectrometry," in *Archeological Chemistry, A Symposium*, ed. Martin Levey, pp. 75–85. Philadelphia, 1967.

Feng, Han-yi and Shryock, J. K.
"The Black Magic in China Known as Ku," *Journal of the American Oriental Society*, 1935, *55*: 1–30 .

Gale, J. S.
"The Korean Alphabet," *Transactions of the Royal Asiatic Society, Korea Branch*, 1912, *4*.1:13.

Gunther, R. T.
The Astrolabes of the World, 2 vols. Oxford, 1932.

Hartner, W.
"The Astronomical Instruments of Cha-Ma-Lu-Ting, Their Identification, and Their Relations to the Instruments of the Observatory of Maragha," *Isis*, 1950, *41*:184.

Haskins, C. H.
Studies in the History of Medieval Science. Cambridge, Mass., 1927.

Hatada, Takashi
A History of Korea, tr. W. W. Smith, Jr. and B. H. Hazard, Santa Barbara, 1969.

Henderson, Gregory
"Chŏng Tasan—A Study in Korea's Intellectual History," *Journal of Asiatic Studies*, 1957, *16*.3: 377.

Higgins, K.
"The Classification of Sundials," *Annals of Science*, 1953, *9*: 342.

Ho, Peng Yoke. See Ho Ping-yü.

Ho, Ping-yü
"Ancient and Mediaeval Observations of Comets and Novae in Chinese Sources," *Vistas in Astronomy*, 1962, *5*: 127–225.

Ho, Ping-yü
"Elixir Poisoning in Medieval China," *Janus*, 1959, *48*: 221–251.

Ho, Ping-yü and Needham, Joseph
"The Laboratory Equipment of the Early Medieval Chinese Alchemists," *Ambix*, 1957, *7*: 57–115.

Ho, Ping-yü and Needham, Joseph
"Theories of Categories in Early Medieval Chinese Alchemy," *Journal of the Warburg and Courtauld Institutes*, 1959, *22*: 173–210.

Ho, Ping-yü and Ts'ao, T'ien-ch'in
"An Early Medieval Chinese Alchemical Text on Aqueous Solutions," *Ambix*, 1959, *7*: 122–158.

Hoang, Peter
A Notice of the Chinese Calendar. Nan King, 1885.

Holmyard, E. J.
Alchemy. Baltimore, 1957.

Hong I-sŏp
"Korean Studies of Natural Science," in *UNESCO Annual*. Seoul, 1960.

Huard, P., and Wong, M.
Chinese Medicine. New York, 1968.

Hulbert, H. B.
"An Ancient Gazetteer of Korea," *Korean Repository*, 1897, *4*: 407–416.

Hunter, Dard
Papermaking, The History and Technique of an Ancient Craft, 2d ed. New York, 1947.

Hunter, Dard
A Papermaking Pilgrimage to Japan, Korea, and China. New York, 1936.

Iba, Yasuaki
"Fragmentary Notes on Astronomy in Japan," *Popular Astronomy*, 1934, *42*: 243; 1937, *45*: 301; 1938, *46*: 89, 141, 263.

Ichihara, M.
"Coinage of Old Korea," *Transactions of the Royal Asiatic Society, Korea Branch*, 1913, *4*.2; 45–74.

Jeon, Sang-woon
"Meteorology in the Yi Dynasty, Korea," *Theses Collection of Sungshin Women's Teachers College*, 1968, *1*: 61–75.

Jeon, Sang-woon
"Understanding of Science in History of Korea, with Emphasis on Scientists in the Early 15th Century," *Japanese Studies in the History of Science*, 1967, *6*: 124–137.

Johnson, M. C.
"Greek, Muslim, and Chinese Instrument Design in the Surviving Mongol Equatorials of 1279," *Isis*, 1947, *32*: 27.

Johnson, Obed S.
A Study of Chinese Alchemy. Shanghai, 1928.

Jugaku, Bunsho
Paper-making by Hand in Japan. Tokyo, 1959.

Karlbeck, O.
Catalogue of the Collection of Chinese and Korean Bronzes at Hollwyl House, Stockholm. Stockholm, 1938.

Karlgren, B.
Analytical Dictionary of Chinese and Sino-Japanese. Paris, 1923.

Kim, Che-won and Kim, Won-yong
Treasures of Korean Art: 2000 Years of Ceramics, Sculpture and Jeweled Arts. New York, 1966. Also published as *The Arts of Korea: Ceramics, Sculpture, Gold, Bronze and Lacquer.* London, 1966.

Kim, Won-yong
Early Moveable Type in Korea. Seoul, 1954.

Koop, A. J.
Early Chinese Bronzes. London, 1924.

Laufer, Berthold
Chinese Pottery of the Han Dynasty, 2d ed. Rutland, Vermont, and Tokyo, 1962 (1st ed., Leiden, 1909).

Laufer, Berthold
"Optical Lenses," *T'oung Pao,* 1915, *16*: 169–228.

Laufer, Berthold
Sino-Iranica: Chinese Contributions to the History of Civilization in Ancient Iran, with Special Reference to the History of Cultivated Plants and Products. Chicago, 1919.

Lee, T'ao
"Achievements of Chinese Medicine in the Sui (589–617 A.D.) and T'ang (618–907 A.D.) Dynasties," *Chinese Medical Journal,* 1953, *71*: 301–320.

Levey, Martin
Chemistry and Chemical Technology in Ancient Mesopotamia. Amsterdam, 1959.

Levey, Martin
"Medical Ethics of Medieval Islam with Special Reference to Al-Ruhwi's Practical Ethics of the Physician," *Transactions of the American Philosophical Society,* 1957, *57*.2 (N.S.):1.

Levey, Martin
"Medieval Arabic Bookmaking and its Relation to Early Chemistry and Pharmacology," *Transactions of the American Philosophical Society,* 1962, *52*.4:1.

Levey, Martin (ed.)
Archeological Chemistry, A Symposium. Philadelphia, 1967.

Li Ch'iao-p'ing
The Chemical Arts of Old China. Easton, Pa., 1948.

Lyons, H. G.
"An Early Korean Rain-Gauge," *Quarterly Journal of the Royal Meteorological Society of London,* 1924, *50*: 26.

McCune, Evelyn
The Arts of Korea: An Illustrated History. Rutland, Vermont, and Tokyo, 1962.

McCune, Evelyn
"Old Korean World Maps," *Korean Review,* 1949, *2*.1: 14–17.

McGovern, Melvin P.
Specimen Pages of Korean Movable Types. Los Angeles, 1966.

Mills, Edwin. W.
"Gold Mining in Korea," *Transactions of the Royal Asiatic Society, Korea Branch*, 1916, 7.1: 1–39.

Mody, N. H. N.
Japanese Clocks, 2nd ed. Rutland, Vermont, and Tokyo, 1967.

Nakamura, H.
"Old Chinese World-Maps Preserved by the Koreans," *Imago Mundi: Yearbook of Early Cartography*, 1947, *4*: 3.

Nakamura, H.
"Old Chinese World Map Preserved by the Koreans," *Chōsen gakuhō* 1966, 39–40: 1–73. Correction to previous article.

Nakayama, Shigeru
A History of Japanese Astronomy, Chinese Background and Western Impact. Cambridge, Mass., 1969.

Needham, Joseph
The Development of Iron and Steel Technology in China. London, 1958.

Needham, Joseph
Science and Civilisation in China, 7 vols. projected. Cambridge, England, 1954–.

Needham, Joseph and Lu Gwei-djen
"A Korean Astronomical Screen of the Mid-Eighteenth Century from the Royal Palace of the Yi Dynasty (Chosŏn Kingdom, 1392 to 1910)," *Physis*, 1966, *8*.2: 137–162.

Needham, Joseph and Lu Gwei-djen
"A Contribution to the History of Chinese Dietetics," *Isis*, 1951, *42*: 13–20.

Needham, Joseph, Wang Ling, and Price, D. J.
Heavenly Clockwork: the Great Astronomical Clocks of Medieval China. Cambridge, 1960.

Neugebauer, O.
"The Early History of the Astrolabe," *Isis*, 1949, *40*: 240–256.

Neugebauer, O.
The Exact Sciences in Antiquity. Princeton, 1952.

Neugebauer, O.
"The Water-Clock in Babylonian Astronomy," *Isis*, 1947, *37*: 37.

Osgood, Cornelius
The Koreans and their Culture. New York, 1951.

Partington, J. R.
Origins and Development of Applied Chemistry. London, 1935.

Price, D. J.
"Clockwork before the Clock," *Horological Journal*, 1955, *97*: 810; 1956, *98*: 31.

Price, D. J.
"A Collection of Armillary Spheres and other Antique Scientific Instruments," *Annals of Science*, 1954, *10*: 172.

Price, D. J.
"Medieval Land Surveying and Topographical Maps," *Geographical Journal*, 1955, *121*: 1.

Price, D. J.
"The Prehistory of the Clock," *Discovery*, 1956, *17*: 153.

Ramming, M.
"The Evolution of Cartography in Japan," *Imago Mundi*, 1937, *2*: 17.

Read, Bernard E., and Pak, C.
A Compendium of Minerals and Stones Used in Chinese Medicine, 2nd ed. Peking 1936.

Reischauer, Edwin O., and Fairbank, John F.
East Asia, The Great Tradition: A History of East Asian Civilization, vol. 1. Boston, 1958, 1960.

Rigge, W. F.
"A Chinese Star-Map Two Centuries Old," *Popular Astronomy*, 1915, *23*: 29.

Rudolph, R. C.
"Chinese Movable Type Printing in the Eighteenth Century," in *Silver Jubilee Volume of The Zinbun Kagaku Kenkyusho*, Kyoto University, pp. 317–335. Kyoto, 1954.

Rudolph, R. C.
"Early Chinese References to Fossil Fish," *Isis*, 1946, *36*: 155.

Rudolph, R. C.
"The Jumar in China," *Isis*, 1949, *40*: 35–37.

Rufus, W. C.
"Astronomy in Korea," *Transactions of the Royal Asiatic Society, Korea Branch*, 1936, *26*: 1–52.

Rufus, W. C.
"The Celestial Planisphere of King Yi Tai-Jo," *Journal of the Royal Asiatic Society, Korea Branch*, 1913, *4*.3:23–72; *Popular Astronomy*, 1915, *23*: 6.

Rufus, W. C.
"A Political Star Chart of the Twelfth Century," *Journal of the Royal Astronomical Society of Canada*, 1945, *39*: 33.

Rufus, W. C., and Chao, Celia
"A Korean Star-Map," *Isis*, 1944, *35*: 316–326.

Rufus, W. C., and Lee, Won-chul
"Marking Time in Korea," *Popular Astronomy*, 1936, *44*: 252.

Rufus, W. C., and T'ien Hsing-Chih
The Soochow Astronomical Chart. Ann Arbor, 1945.

Sarton, George
"Early Observations of Sun-Spots," *Isis*, 1947, *37*: 69.

Sarton, George
A History of Science, 2 vols. Cambridge, Mass., 1959.

Sarton, George
Introduction to the History of Science, 3 vols. Baltimore, 1927–1947.

Schafer, Edward H.
"Early History of Lead Pigments and Cosmetics in China," *T'oung Pao*, 1956, *44*: 413–438.

Schove, D. J.
"The Sun-Spot Cycle, −649 to +2000," *Journal of Geophysical Research*, 1955, *60*: 127.

Schove, D. J.
"Sun-Spots, Aurorae and Blood Rain: The Spectrum of Time," *Isis*, 1951, *42*: 133–138.

Singer, Charles
The Earliest Chemical Industry: An Essay in the Historical Relations of Economics and Technology, Illustrated from the Alum Trade. London, 1948.

Singer, Charles
A Short History of Science, to the Nineteenth Century. Oxford, 1941.

Singer, Charles (ed.)
A History of Technology, 5 vols. Oxford, 1954–1958.

Sivin, Nathan
Chinese Alchemy: Preliminary Studies. Cambridge, Mass., 1968.

Sivin, Nathan
"Copernicus in China," In Union Internationale d'Histoire et de Philosophie des Sciences, Comité Nicolas Copernic (ed.), *Colloquia Copernicana II. Études sur l'audience de la théorie héliocentrique*. Conférences du Symposium de l'UIHPS, Toruń 1973. Warsaw et al., 1973.

Sivin, Nathan
"Cosmos and Computation in Early Chinese Mathematical Astronomy," *T'oung Pao*, 1969, *55*: 1–73.

Sivin, Nathan
"On 'China's Opposition to Western Science During Late Ming and Early Ch'ing,' " *Isis*, 1965, *56*: 201–205.

Smith, Cyril Stanley
"Note on a Japanese Magic Mirror," *Archives of the Chinese Art Society of America*, 1963, *17*: 29–31.

Smith, E. W.
"The Making of Brass Ware," *The Korea Review*, 1905, *5.9*: 322–323.

Sohn, Pow-key (Son Po-gi)
"Early Korean Printing," *Journal of the American Oriental Society*, 1959, *79* (2): 99–103.

Soymié, M., and Litsch, F.
"Bibliographie du Taoisme. Études dans les langues occidentales," *Dokkyo kenkyū* 道教研究, 1968, *3*: 249–318; *4*: 225–290.

Sudzuki, Osamu
"A Concave Mirror of Koryŏ Dynasty and its Earlier Phases," *Chōsen gakuhō*, 1959, *14*: 625–664.

Sung Ying-hsing
T'ien-kung K'ai-wu: Chinese Technology in the Seventeenth Century (tr. E-tu Zen Sun and Shiou-chuan Sun). University Park and London, 1966.

Szczesniak, B.
"Matteo Ricci's Maps of China," *Imago Mundi*, 1955, *11*: 127–136.

Taylor, R. Sherwood
The Alchemists. London, 1951.

Taylor, R. Sherwood
A History of Industrial Chemistry. London, 1957.

Thomson, J. O.
History of Ancient Geography. Cambridge, 1948.

Thorndike, L.
A History of Magic and Experimental Science, 8 vols. New York, 1923–1958.

Thorndike, L.
"A Weather Record for 1399–1406 A.D.," *Isis*, 1940, *32*:304.

Tooley, R. V.
Maps and Map-Makers. London, 1949.

Trollope, M. N.
"Book Production and Printing in Korea," *Transactions of the Royal Asiatic Society, Korea Branch*, 1936, *25*: 103–108.

Underwood, H. H.
"Korean Boats and Ships," *Transactions of the Royal Asiatic Society, Korea Branch*, 1934, *23*: 1–89.

Usher, A. P.
A History of Mechanical Invention, revised edition. Cambridge, Mass., 1954.

Viessman, W.
"Ondol—Radiant Heat in Korea," *Transactions of the Royal Asiatic Society, Korea Branch*, 1948, *31*: 9–23.

Wada, Yuji
"A Korean Rain-Gauge of the 15th Century," *Quarterly Journal of the Royal Meteorological Society of London*, 1911, *37*: 83.

Wang, Ling
"On the Invention and Use of Gunpowder and Firearms in China," *Isis*, 1947, *37*: 160–178.

Ware, J. R
Alchemy, Medicine, Religion in the China of A.D. 320, the Nei P'ien of Ko Hung. Cambridge, Mass., 1966.

Watson, William
"The Earliest Buddhist Images of Korea," *Transactions of the Oriental Ceramic Society*, 1957.

Weems, G.N. (ed.)
Hulbert's History of Korea, 2 vols. New York, 1962.

Weinberger, W. M.
"An Early Chinese Bronze Foot Measure," *Oriental Art,* 1949, *2*: 35.

Welch, Holmes H.
"The Bellagio Conference on Taoist Studies," *History of Religions,* 1970, *9*: 107–136.

Wen, Chion-tsu
"Observations of Halley's Comet in Chinese History," *Popular Astronomy,* 1934, *42*: 191.

Wen, Chion-tsu
"A Statistical Survey of Eclipses in Chinese History," *Popular Astronomy,* 1934, *42*: 136.

Werner, E. T. C.
Chinese Weapons. Shanghai, 1932.

White, W. C., and Millman, P. M.
"An Ancient Chinese Sun-Dial," *Journal of the Royal Astronomical Society of Canada,* 1938, *32*: 417.

Wiborg, Frank B.
Printing Ink. New York, 1926.

Wilson, William J.
"Alchemy in China," *Ciba Symposia,* 1940, *2*: 594–624.

Wolf, A.
A History of Science, Technology and Philosophy in the 16th and 17th Centuries. London, 1935.

Wolf, A.
A History of Science, Technology and Philosophy in the 18th Century. London, 1938.

Wong, George H. C.
"China's Opposition to Western Science during Late Ming and Early Ch'ing," *Isis,* 1963, *54*: 29–49.

Wong, K. Chimin and Wu Lieh-teh
History of Chinese Medicine, Being a Chronicle of Medical Happenings in China from Ancient Times to the Present Period, 2d ed. Shanghai, 1936.

Wu, Lu-ch'iang and Davis, T. L.
"An Ancient Chinese Treatise on Alchemy Entitled Ts'an T'ung Ch'i," *Isis,* 1932, *18*:210–289.

Yabuuchi, Kiyoshi
"Indian and Arabian Astronomy in China," in *Silver Jubilee Volume of the Zinbun Kagaku Kenkyusyo,* Kyoto University, 583–602. Kyoto, 1954.

Yamasaki, Kazuo
"Pigments Employed in Old Paintings of Japan," in *Archeological Chemistry,* ed. Martin Levey, pp. 347–365. Philadelphia, 1967.

Yi Ik-sop
"A Map of the World," *Korean Repository,* 1892, *1*: 336.

Index

www.ingramcontent.com/pod-product-compliance
Lightning Source LLC
Chambersburg PA
CBHW021026210326
41598CB00016B/921

* 9 7 8 0 2 6 2 6 0 0 5 2 1 *